LIQUID SCINTILLATION COUNTING

Volume 3

LIQUID SCINTILLATION COUNTING

Volume 3

Proceedings of a Symposium on
Liquid Scintillation Counting
organised by
The Society for Analytical Chemistry
Brighton, England
September 3-6 1973

Editors:
M. A. Crook (Polytechnic of the South Bank)
P. Johnson (Hoechst Pharmaceuticals Research Limited)

HEYDEN
London · New York · Rheine

Heyden & Son Ltd., Spectrum House, Alderton Crescent, London NW4 3XX
Heyden & Son Inc., 225 Park Avenue, New York, N.Y. 10017, U.S.A.
Heyden & Son GmbH, 4440 Rheine/Westf., Münsterstrasse 22, Germany

Library of Congress Catalog Card No. 70-156826

ISBN 0 85501 084 3

Designed by C. J. Newsome & Associates, Ringwood, Hants BH24 1ND.
Printed in Great Britain by Galliard Ltd., Gt. Yarmouth, Norfolk, England

Contents

V. SPECIAL TOPICS

Preface

This volume, the third in the series on Liquid Scintillation Counting, records the proceedings of the International Symposium on this topic, held in Brighton, England, for the second time, on September 3rd-6th, 1973. The Symposium was again organised by the Radiochemical Methods Group of the Society for Analytical Chemistry (Analytical Division, The Chemical Society). The Committee of the Radiochemical Methods Group, the Plenary Lecturers and the Chairmen of the five sessions (Mr. M.A. Crook and Drs. P. Johnson, B. Scales, B.W. Fox and E. Rapkin) all contributed to the undoubted success of a symposium supported by 250 delegates from 20 countries. A particularly successful feature of the symposium was the excellent introductory 'teach-in' on liquid scintillation counting conducted by Drs. A. Dyer and E. Rapkin (under the chairmanship of Mr. McMillan).

As in Volume 2, we are pleased to record the mounting of an excellent exhibition by many manufacturers:

W. and R. Balston (Whatman)
Beckman R.I.I.C.
Camlab (Berthold Frieseke)
E.M.I. Electronics
E.S.I. Nuclear
Fisons Scientific Apparatus
I.C.N. (Tracerlab)
Intertechnique
Koch-Light
LKB Instruments
NEN Chemicals, GmbH
G.D. Searle and Co.
Nuclear Enterprises
Packard Instruments
The Radiochemical Centre

and the support of their representatives on the Organising Committee, Messrs. D.A. Ginger, D. Moore and A.B. Raven.

The support for the meeting, of whose 250 delegates only 50 attended the 1971 Symposium, suggests that there is still scope for international, but possibly less frequent, meetings on this topic. We would suggest that the rapid progress in recent years, which stemmed both from the advances in instrumentation by the manufacturers and from the innovations in sample preparation and application by the users, has levelled off to an extent which suggests that progress in the immediate future may be limited to more diverse applications rather than fundamental advances in technique. Nevertheless there are many samples in many fields which are still not easy to handle routinely by liquid scintillation counting, there are many potential users who have yet to apply the technique and there are many existing users for whom computer processing is a goal still to be achieved. In the light of these needs further meetings on this invaluable technique seem assured.

April 1974

M.A. Crook
P. Johnson

Some of the material presented at the Symposium and reported in certain chapters in this publication is contributed by authors from commercial organisations which have a financial interest in the technique of liquid scintillation counting. Note should be taken of this when assessing such contributions.

SECTION I

MODERN CONCEPTS IN LIQUID SCINTILLATION COUNTING

Chapter 1

Applied Liquid Scintillation Counting

D.L. Horrocks

Beckman Instruments, Inc., Fullerton, Ca 92634, U.S.A.

INTRODUCTION

The scintillation process in solutions is comprised of several different steps. The ultimate result of the scintillation process is the conversion of part of the energy from some nuclear transformation into a number of photons. The number of photons is related to the energy deposited in the solution and the type of particle which imparts that energy. The excited scintillator molecules which produce the observed fluorescence determine the energy (wavelength) distribution of the emitted photons. The number of photons is also dependent upon such factors as the amount of scintillator solute present in the solution, the presence of certain impurities (quenching), the optical transmission properties (colour quenching) and scintillation efficiency of the total system.

Composition of a typical scintillator solution

Most liquid scintillator solutions are composed of three main parts; the solvent(s), the solute(s) and the sample. In special applications there may be more than one solvent for such purposes as increasing the solubility of the sample or increasing the energy transfer probability. Also it is often necessary to use more than one scintillator solute in order to shift the wavelength spectrum of the photons to more efficiently match the spectral response characteristics of the photon measuring device (multiplier phototube) or to reduce the effects of a high extinction coefficient for a given wavelength distribution by one of the components in the solution.

To understand the role of the solvent(s) and the solute(s) in the scintillation processes it is necessary to study the solutions without foreign substances present, such as the sample. These types of studies can be done using radioactive sources external to the solution for the production of excited molecules. Another type of external source is an ultra-violet radiation source.

The solvent plays at least two very important roles in the scintillation process. The solvent molecules accept a part of the energy from the exciting source and, in so doing, excited solvent molecules and ionised solvent molecules are produced. Upon recombination of solvent ions with electrons some of the solvent molecules are left in an excited state. The solvent molecules also act as an efficient means of transferring excitation energy through the solution to the fluor molecules. This solvent-solvent transfer process is very efficient. The solvent molecules themselves are very poor scintillators for several reasons, some of which are as follows:

(a) Solvent molecules have low probabilities for the emission of photons, i.e. a low quantum efficiency.
(b) The energy (wavelength) distribution for the photons emitted by the solvent molecules is of a range (2000-3000 Å) where most multiplier phototubes have reduced sensitivity.
(c) The probability of re-absorption of emitted photons by the solvent itself (self-absorption) is high due in part to the high solvent concentration.
(d) The life of the excited solvent molecule in the fluorescing state is relatively long (about 30 ns) which renders it very vulnerable to competitive processes which do not lead to fluorescence, i.e. quenching or energy transfer.

The role of the solute(s) in this process is to trap the excitation energy of the solvent molecules and efficiently release part of that energy in the form of photon energy. Whereas solvent molecules have low fluorescence efficiencies (0.07 to 0.40), the commonly used solute molecules have fluorescence efficiencies near 0.90. The probability of solute self-absorption is greatly reduced by the low solute concentration in the scintillation solution. The probability of competitive processes, such as quenching, are greatly reduced by the short decay time of these solute excited molecules, usually between 1 and 2 ns.

At the relatively low concentration of 3 to 10 g/litre, the solute molecules are capable of trapping 100% of the solvent excitation energy, i.e. the energy transfer is 100% efficient. Before the excited solute molecule emits a photon there is an intramolecular de-excitation by a non-radiative process by which a small part of the excess energy is lost in the form of vibrational energy of the molecule. All fluorescence occurs between the zero vibrational level of the first excited singlet energy level of the solute molecule and any of several vibrational energy levels of the ground state of the molecule. Figure 1 shows some of the many processes which can occur during the scintillation process.

Excitation processes

As an ionising particle (electron, alpha particle, proton, etc.) passes through a scintillator solution, slows down and is finally stopped, many different events occur. Figure 2 depicts a highly simplified representation of some of the more important events. A more detailed description of these events is given by Laustriat.[1] The largest part of the particle energy is expended in the form of kinetic (thermal) energy of the solution molecules and does not lead to the production of excited solvent molecules. Along the particle track ions, excited molecules, free radicals and secondary electrons are among the different products formed. The secondary electrons will produce their own track of events, which have been called 'spurs'. Many of the excited molecules that are produced along the track of the primary particle interact with other species such as free radicals, ions and non-excited (ground state) molecules. These types of interactions lead to the non-radiative de-excitation of the excited molecules. This is referred to as 'track quenching'.

Both ions and excited molecules can lead to fluorescence. An appreciable part of the fluorescence (about 60%) is the result of ion recombination.[2,3] Most

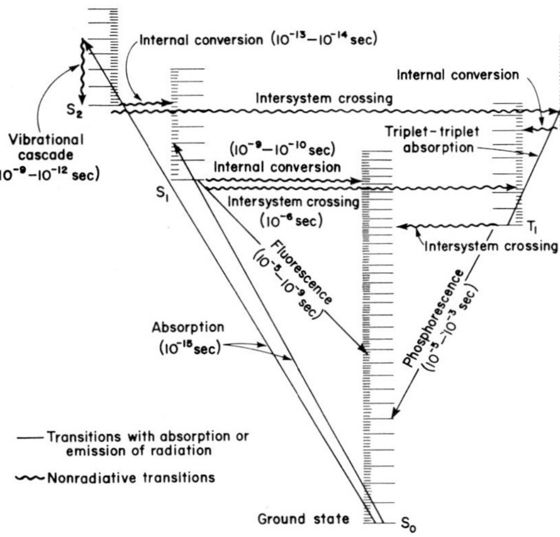

Fig. 1. A modified Jablonski diagram showing the various energy levels and the processes of energy absorption and release of energy (radiatively and non-radiatively) by an organic molecule.

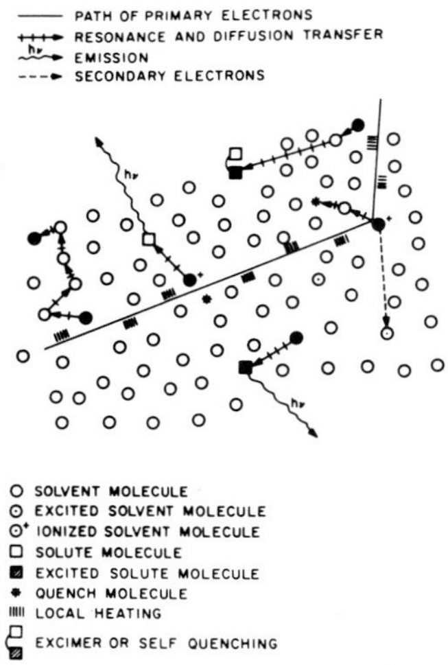

Fig. 2. A simplified diagram of some of the processes which occur in a liquid scintillator solution (solvent and solute) when an energetic electron interacts with the molecules of the solution.

of the primary excitations lead to excited solvent molecules in their upper excited singlet energy states. The upper excited singlet states rapidly undergo an internal conversion by non-radiative processes to give finally the first excited singlet energy level. It is this state which is responsible for the energy transfer and migration. However, recent studies have shown that some energy transfer can occur from upper excited states of the solvent molecules.[3,4]

Energy transfer between solvent molecules involves neighbouring solvent molecules. Two theories are presently in debate as to the mode of the solvent-solvent energy transfer. Birks et al.[5] explain the transfer as being due to the formation and dissociation of solvent excimer molecules with the energy transfer to the previously unexcited solvent molecule when the excimer breaks apart. Voltz et al.[6] theorise that the energy actually jumps from one solvent molecule to its neighbour by a non-radiative process.

The solvent-solute energy transfer is quite different. First, it is not necessary that the solvent (donor) molecule and the solute (acceptor) molecule be adjacent. Secondly, the intra-molecular partial de-excitation (vibrational relaxation) of the excited solute molecule traps the energy on the solute molecule, thus preventing the reverse transfer process (solute to solvent). Finally, the solvent-solute transfer occurs by a non-radiative process at moderate solute concentrations.[7] (At very low solute concentrations there is evidence of a radiative transfer process.[8,9]) The solvent-solute energy transfer efficiency is dependent on the solute concentration as is shown in Fig. 3. The energy transfer is essentially 100% at the optimum solute concentration.

If a second solute is present in the scintillator solution there can be an energy transfer between the solute molecules. The primary solute (higher concentration, higher energy singlet excited level) will transfer its energy by a non-radiative process to the secondary solute (lower concentration, lower energy singlet excited level). The transfer is quantitative at the optimum concentrations of the two solutes.

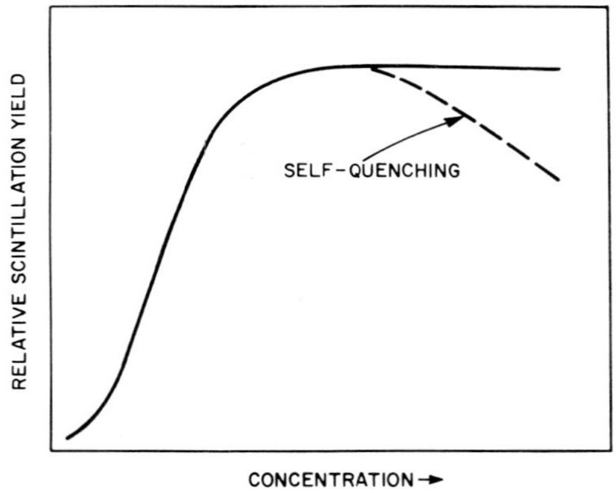

Fig.3. The effect of solute concentration on the relative scintillation yield. This shows the relationship for a solute which self-quenches at high concentrations (i.e. PPO) and a solute which shows no self-quenching at any concentration (i.e. butyl-PBD).

Figure 4 shows a schematic representation of the scintillation processes in a liquid scintillator solution with a primary and a secondary solute.

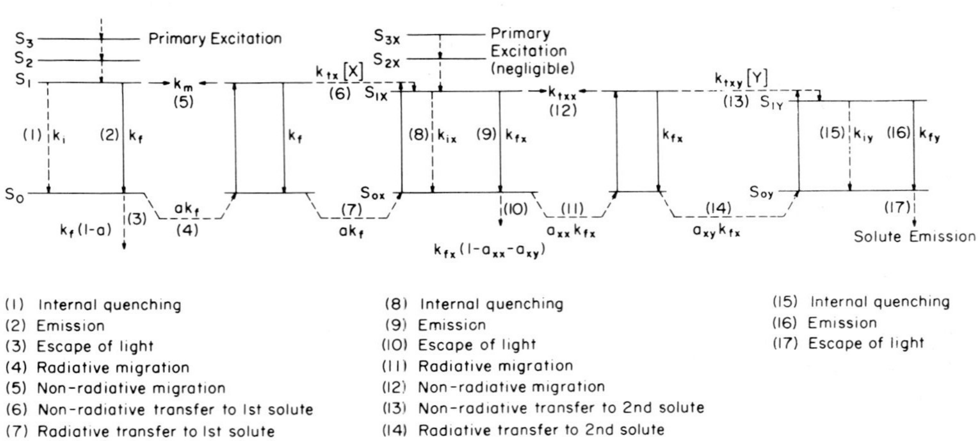

Fig.4. A diagram showing the many steps involved between energy absorption by the solvent molecules and the emission of photons by the solute molecules.

Scintillation efficiency

The scintillation efficiency is defined as the fraction of the energy of the exciting particle which is converted into photon energy, nhν, where n is the number of photons and $h\nu$ is the energy of the photon. However, there is a whole spectrum of photon energies and the total photon energy is given by the equation

$$E_{ph} = h\int_{\nu_1}^{\nu_2} n(\nu)d\nu$$

where $n(\nu)$ is the number of photons of frequency ν and the integration limits ν_1 and ν_2 are the minimum and maximum frequencies of the fluorescence spectrum of the scintillator solute. The scintillation efficiency, S_x, is the ratio of total photon energy, E_{ph}, to excitation energy, E_{ex}, given by the equation

$$S_x = E_{ph}/E_{ex} = \left(h\int_{\nu_1}^{\nu_2} n(\nu)d\nu\right)/E_{ex}$$

Certain approximations can be made by assuming that the average number of photons produced is $\overline{N}_{ph}$ and that the energy of a photon with the average frequency of the fluorescence spectrum is $\overline{E}_{ph}$. Using these approximations the equation for S_x becomes

$$S_x = \frac{(\overline{N}_{ph})\ (\overline{E}_{ph})}{E_{ex}}$$

The solvent used in the scintillation solution determines the value of S_x. Lipsky and co-workers[10] determined the value of S_x for Carbon-14 beta particle excitation of a benzene solution. Hastings and Weber[11] determined the value of S_x for Carbon-14 and Hydrogen-3 beta particle excitations of toluene scintillator solutions. Table 1 summarises their data.

Table 1. Experimentally determined values of the scintillation efficiency, S_x.

Solvent	S_x for excitation with the radionuclide Hydrogen-3	Carbon-14
Benzene	—	0.042 [a]
Toluene	0.044 [b]	0.052 [b]

[a] Reference 10.
[b] Reference 11.

The solute (or solutes) determine the energy (wavelength) distribution of the emitted photons. Using the fluorescence spectra reported by Berlman[12] it is possible to calculate the average photon energy. For the scintillator solute PPO the average wavelength of the fluorescence spectrum is about 380 nm. The energy of a photon of this wavelength is calculated:

$$E = h\nu = hc/\lambda$$

$$= (4.143 \times 10^{-15}\ \text{eV-s})\ (2.99 \times 10^{10}\ \text{cm/s})/(3.8 \times 10^{-5}\ \text{cm})$$

$$= 3.2\ \text{eV}$$

For a toluene solution the average number of photons, using the data of Hastings and Weber[11] in Table 1, can be equated to

$$\overline{N}_{ph} = 13.7\ E_{ex}\ \text{(in keV)} \qquad \text{For Hydrogen-3}$$

$$\overline{N}_{ph} = 16.3\ E_{ex}\ \text{(in keV)} \qquad \text{For Carbon-14}$$

The scintillation efficiency is not constant with the energy of the electron as can be seen in Fig. 5 taken from Horrocks.[13] Above 300 keV electron energy the scintillation efficiency is constant and is about 1.2 times greater than the scintillation efficiency for an electron energy of 50 keV, the average energy of a Carbon-14 beta spectrum. The value of S_x above 300 keV is 0.062 which gives the relationship to the average number of photons as

$$\overline{N}_{ph} = 19.4\ E_{ex}\ \text{(in keV) for } E_{ex} \geq 300 \text{ keV}$$

The scintillation efficiencies, average number of photons, average number of photoelectrons at 28% photocathode efficiency and a theoretical coincidence counting efficiency as a function of electron energy are listed in Table 2.

Table 2. Scintillation data and counting predictions for a toluene scintillator solution with PPO as the scintillator solute for different energy electrons.

Electron energy (keV)	S_x	$\overline{N}_{ph}$	Number of photoelectrons (at 28% eff.)	Theoretical coincidence counting eff.%
1000	0.062	20,600	5,768	100
500	0.062	10,300	2,884	100
158	0.058	3,060	857	100
50	0.052	802	225	100
5	0.042	66	18	100
1	0.031	10	2.8 (0.94)[a]	63[b]
0.5	0.024	4	1.1 (0.67)[a]	0[b]

[a] Poisson probability of measuring a pulse from the designated number of photoelectrons.

[b] Probability of a coincidence pulse calculated from the relationship. Probability $= 1 - 2^{1-Y}$ where Y is the expected number of photoelectrons. Probability is zero for any value of Y less than two, i.e. one photoelectron in each multiplier phototube.

Responses for different types of particles

Different types of ionising particles, i.e. beta, alpha, protons, etc., induce different responses per unit energy in typical liquid scintillator solutions. As a general rule the higher the specific ionisation of the exciting particle the lower the photon yield per unit energy (scintillation efficiency). Figure 6 shows the response of a typical scintillator solution to alpha and beta particles as a function of particle energy. Figure 7 shows the response for electrons of energy less than 100 keV. The responses for several particles are shown together in Fig. 8.

Gamma rays and X-rays can be measured with liquid scintillator solutions only through their interaction by way of scattering which leads to the production of electrons (Compton scattered electrons) or, if the energy is very low, by the total absorption (photoelectric effect). Neutrons are measured through the transfer of all or part of the neutron energy to hydrogen atoms in the solution by a scattering process which gives energetic protons.

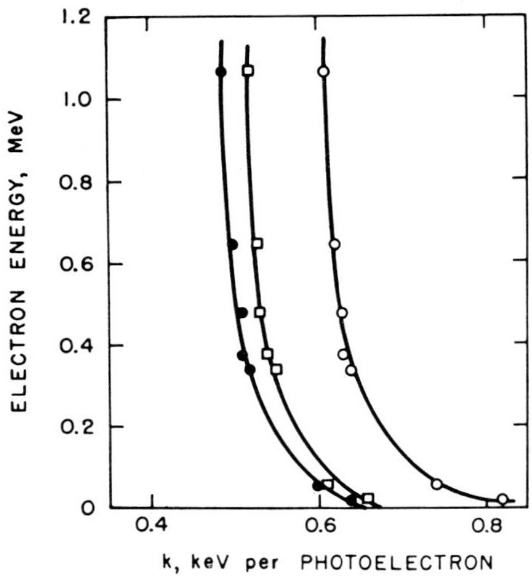

Fig. 5. The average energy required to produce a photoelectron as a function of the energy of the electron which excites the liquid scintillator solution.[13]

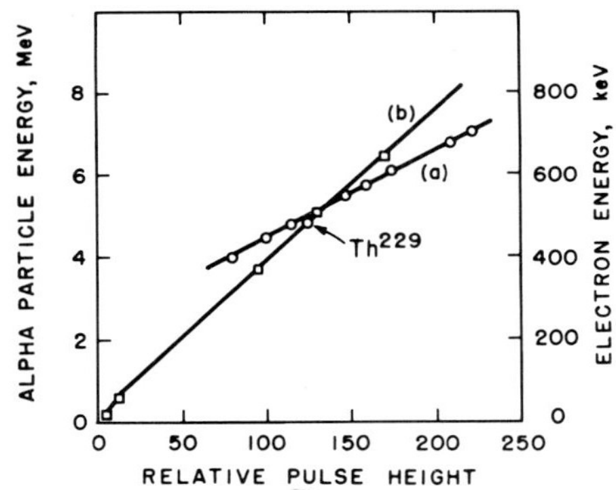

Fig. 6. Pulse height-energy relationship of a liquid scintillator solution for excitations by (a) alpha particles and (b) electrons. Note the factor of ten in the two energy scales.

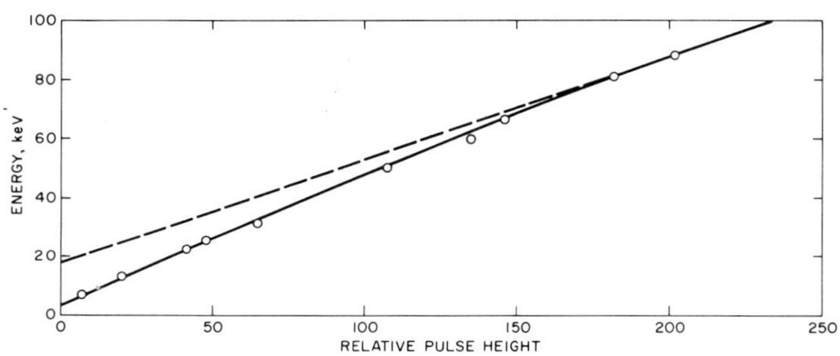

Fig. 7. Pulse height-energy relationship of a liquid scintillator solution for excitation by electrons with energies less than 100 keV.

SPECIAL APPLICATIONS

Determination of absolute disintegration rates

(a) Alpha particles Alpha particles can be counted with 100% efficiency in a liquid scintillation solution. The alpha emitting radionuclide has to be intimately in contact with or dissolved in the solution to count with 100% efficiency. If the radionuclide is in the form of a solid or on the surface of a solid support which is suspended in the solution the counting efficiency will possibly be less than 100%.

Because most naturally occurring alpha emitting radionuclides are present with anywhere from one to several daughter radionuclides which themselves emit beta and alpha particles, the activity of the parent radionuclide can not be determined by merely counting the sample in a liquid scintillator system. Under conditions where the parent radionuclide is in total equilibrium with its daughter

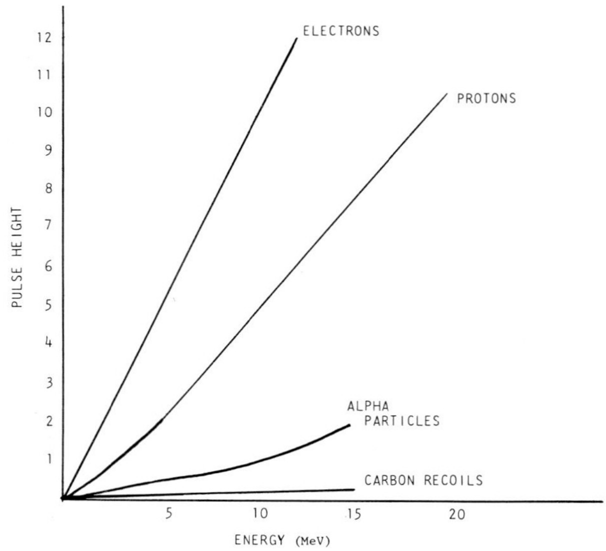

Fig.8. Relative pulse heights for excitations of equal energy electrons, protons, alpha particles and recoil carbon atoms. (From R. Batchelor, W.B. Gilboy, J.B. Parker and J.H. Towle, Nuc. Instr. and Meth. **13**, 70 (1961).

radionuclides and the decay scheme of all the daughter radionuclides is known, a simple total count can be used to calculate the decay rate of the parent radionuclide. Usually there is no total equilibrium and the pulses due to the parent and daughter radionuclides are separated by measurement of the pulse height distribution with a multichannel analyser. Figure 9 shows a pulse height distribution for a sample of thorium dissolved in a liquid scintillator solution.[14] The radionuclide Thorium-232 produced only a part of the total pulses measured.

The determination of small amounts of normal uranium in water and urine[15] illustrates some of the problems encountered when daughter radionuclides grow into secular equilibrium very slowly. The growth of the radionuclides Thorium-234 and Protactinium-234m, daughters of Uranium-238, occurs with a 25 day half life. This is too long a period to wait until the daughters reach equilibrium (in about 170 days the growth will be 99.2% of the true secular equilibrium). Some of the beta particles from Protactinium-234m produce the same pulse height response as the 4.2 and 4.8 MeV alpha particles and a simple counting of pulses of that amplitude would lead to a too high evaluation of the uranium content of the sample. The beta emitters produce a continuum of pulse heights while the single energy alpha particles produce a peaked pulse height response. Figure 10 shows the measured pulse height distribution from a sample of uranium dispersed in an emulsion type liquid scintillator solution.[15] The beta pulse distribution can be obtained by interpolation of the beta continuum and this subtracted from the total pulses to obtain the alpha decay rate.

(b) Beta emitters Radionuclides which decay the emission of beta particles only can be counted very efficiently in a liquid scintillator solution. By dissolving the radionuclide in the scintillator solution it is possible to determine the maximum sample count rate by the integral count method[16] and the use of wide open counting channels. For most radionuclides which have an E_{max} of the beta spectrum

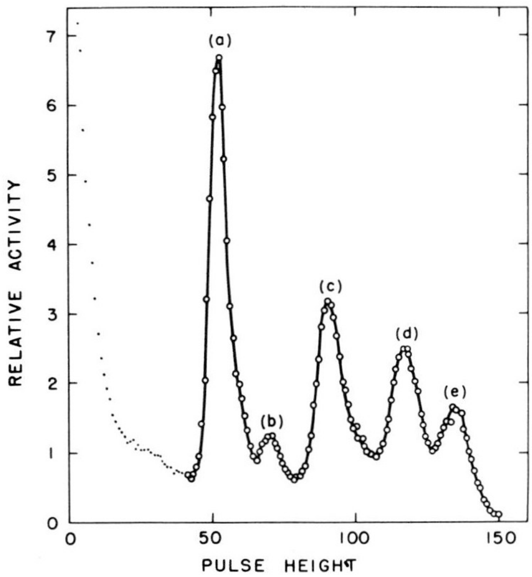

Fig. 9. Differential pulse height spectrum for a sample of thorium dissolved in a liquid scintillator solution. The various peaks are due to (a) 4.0 MeV alpha particles of Thorium-232, (b) 4.7 MeV alpha particles of Thorium-230, (c) 5.3 MeV alpha particles of Thorium-228 plus 5.7 MeV alpha particles of Radium-224 (unresolved), (d) 6.3 MeV alpha particles of Radon-220 and (e) 6.8 MeV alpha particles of Polonium-216.

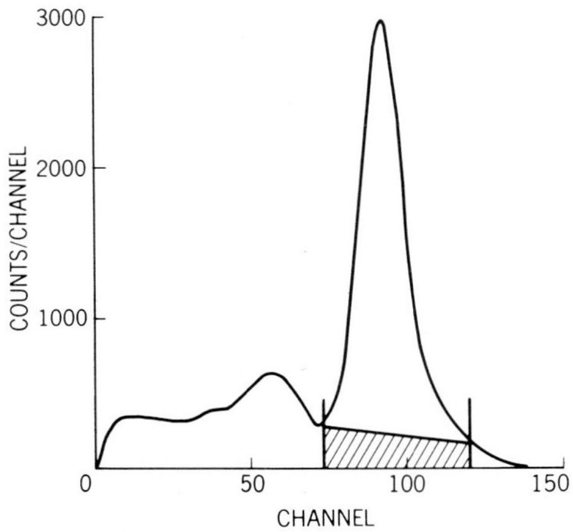

Fig. 10. Differential pulse height distribution for a sample of 1.0 mg of normal uranium in 1.0 ml of water which is dissolved in an emulsifier containing liquid scintillator solution.

greater than 150 keV, the extrapolated integral count rate will be within $\pm 2\%$ of the true sample disintegration rate. Since all beta emitters produce a continuum of beta particle energies from zero to the E_{max}, there are always some electrons which have insufficient energy to produce enough photons to be measured. The fraction of beta particles which have energies below this threshold is not only dependent upon E_{max} but also upon the shape of the beta continuum. Beta spectra weighted towards the low energy particles (called forbidden transitions) have a greater fraction of their beta particles below the threshold. These radionuclides will be measured with a lower counting efficiency.

When the shape of the beta spectrum of the radionuclide is known,[17] it is possible to predict theoretical maximum counting efficiency. Several investigators have developed theoretical methods for the calculation of the disintegration rate of beta emitters from the measured integral count rate and the known beta spectral shape when the energy required to produce a photoelectron in the counting system is known.[18-22]

(c) Conversion electron sources Certain radionuclides which decay by gamma ray emission will for part of the time release their decay energy as kinetic energy of an extranuclear electron in the process called internal conversion. The emitted electrons have a given energy equal to the difference between the energy of the gamma transition and the binding energy of the extranuclear electron:

$$E_e = E\gamma - BE_e$$

The radionuclide Indium-113m decays from an energy level of the nucleus of 393 keV. A gamma ray is emitted in 65% of the decay events while an internal conversion electron is emitted in the other 35% of the events. These approximately 369 keV electrons are counted with 100% efficiency. The decay rate of an Indium-113m sample is equal to the conversion electron count rate divided by 0.35. Figure 11 shows a pulse height distribution obtained for a sample of Tin-113—Indium-113m dissolved in a liquid scintillator solution. The 369 keV electrons are easily detected.

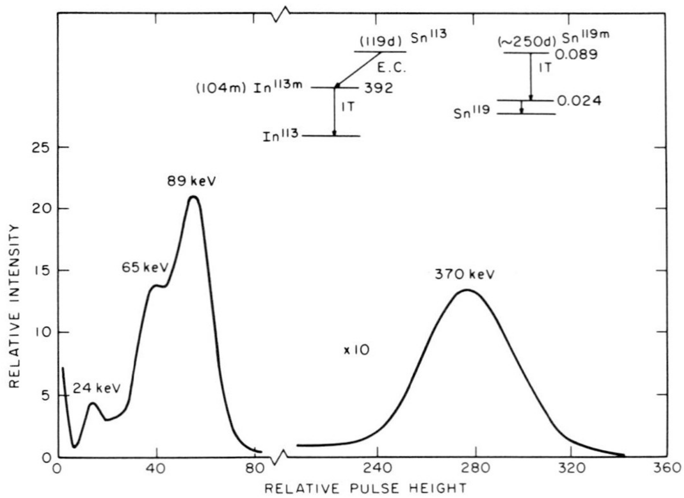

Fig. 11. Differential pulse height distribution for a sample of Tin-113 - Indium-113m dissolved in a liquid scintillator solution showing the 369 keV conversion electrons from the decay of Indium-113m.

(d) **Other sources** Gamma ray sources can be counted by their production of electrons in the liquid scintillator solution by the Compton scattering and photoelectric effects. The counting efficiencies for gamma ray sources are normally very low for at least two main reasons; (1) the probability of gamma ray interaction with the materials of the solution is low because of the low atomic numbers of the atoms in the material which makes up the solution, and (2) when the source is outside the solution the geometry for the gamma rays passing through the solution is low. The counting efficiency can be increased, especially for low energy gamma rays and X-rays, by the addition of a compound which contains an atom with a high atomic number to the solution.[23,24] In this way the electron density of the solution is increased and the probability of scattering or stopping a gamma ray is increased. Another technique increases the geometry by placing the gamma ray source, usually contained in a second smaller tube, inside the scintillator solution.

Neutrons are detected by the neutron-proton scattering process or by thermalising the neutrons and absorbing them in atoms with a high cross section which have been added to the scintillator solution.[25-28]

Even fission fragments can be counted in liquid scintillator solutions. By dissolving a radionuclide in the solution which will undergo fission, either spontaneously or by external stimulation, the fission fragments will excite the solution causing it to scintillate. In one report[29] a source of Californium-252, which decays by both alpha particle emission and spontaneous fission, was dissolved in a scintillator solution and it was shown that the fission fragments were counted with 100% efficiency.

Determination of beta spectra E_{max}

From the pulse height-energy relationship for electrons in a liquid scintillator[30-32] it is possible to convert a normal pulse height distribution into an energy distribution for beta particles from a given radionuclide. These energy distributions are used to obtain Kurie plots from which it is possible to obtain the E_{max} value of the beta spectrum. In some cases it is necessary to use a second order derivative correction to correct for the energy resolution effect of the liquid scintillator solution.[33] Table 3 lists several values of E_{max} for give radionuclides as

Table 3. The E_{max} values of the beta spectra of several radionuclides as determined by liquid scintillator methods.

Radionuclide	Mode of decay	Accepted[a] E_{max} (keV)	Measured E_{max} (keV)	Ref.
Rubidium-87	β^-	275	272 3	46 ±
Sodium-22	β^+	541	533 7	47 ±
Calcium-45	β^-	254	253 3	46 ±
Nickel-63	β^-	67	67 2	48 ±
Silicon-32	β^-	210	213 7	49 ±
Carbon-14	β^-	156	154 5	49 ±
Phosphorus-32	β^-	1707	1690 30	50 ±

[a] From Reference 45.

determined from liquid scintillation counting. Figures 12, 13 and 14 show Kurie plots and the extrapolated values of E_{max} for Calcium-45, Nickel-63 and Sodium-22, respectively. It is worth noting that Sodium-22 is a positron emitter for which a positron spectrum was measured.

Pulse shape discrimination

The intensity of photons from a de-aerated scintillator solution as a function of time can be divided into at least two components; a fast (prompt) and a slow component. Figure 15 shows a typical intensity-time distribution. The fast component is the result of normal fluorescence from the excited molecules. The slow component has all the fluorescence properties of the fast component except it is produced at a time delayed from the initial excitation process. While the prompt component has a decay time equal to the fluorescing molecule (usually 1-2 ns), the slow component has a decay time independent of the fluorescing species and at least two orders of magnitude longer; 200-300 ns.

The slow component is due to the diffusion of two triplet excited molecules until they contact each other and undergo an annihilation reaction leading to the production of a singlet excited molecule which then proceeds to fluoresce:

$$T + T \rightleftharpoons S_1^* + S_0$$

In aerated solutions this delayed fluorescence is not observed because the oxygen

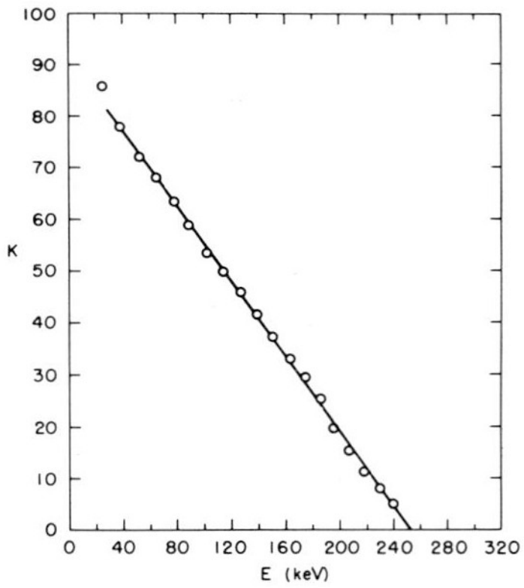

Fig. 12. A Kurie plot for a Calcium-45 sample in a liquid scintillator solution; the extrapolated E_{max} value is 253 keV.

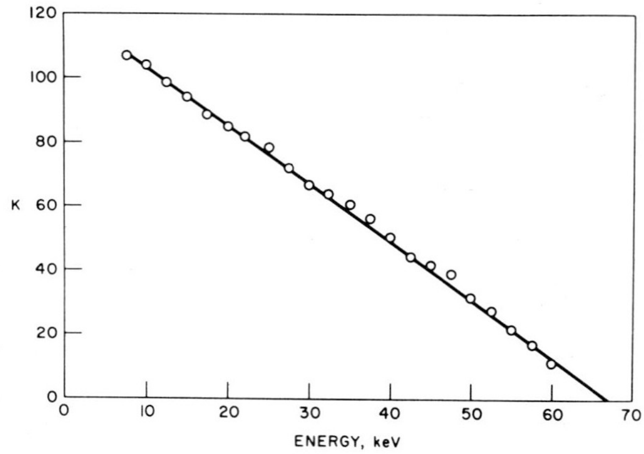

Fig. 13. A Kurie plot for a Nickel-63 sample in a liquid scintillator solution; the extrapolated E_{max} value is 67 keV.

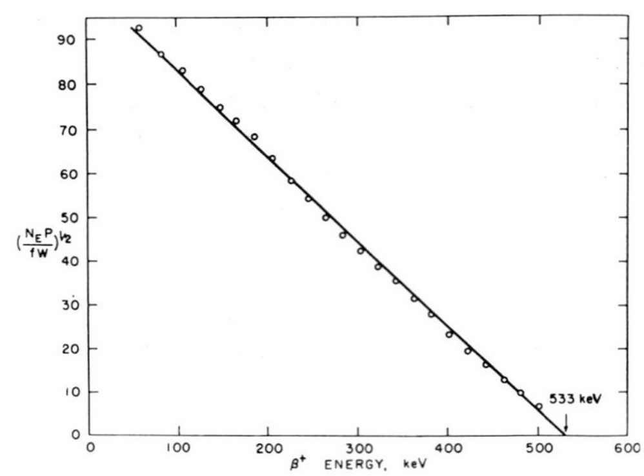

Fig. 14. A Kurie plot for a Sodium-22 sample in a liquid scintillator solution; the extrapolated E_{max} value is 533 keV. Sodium-22 is a positron emitter.

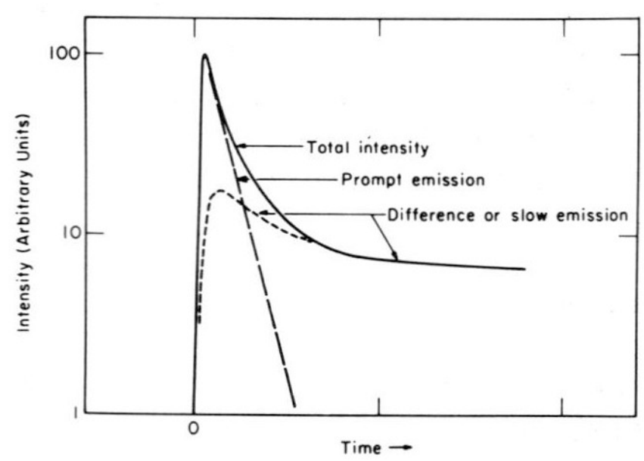

Fig. 15. A representation of the scintillation intensity from an organic liquid scintillator solution as a function of the time after the start of the scintillation event showing the division of the total intensity into a prompt and a slow component.

will quench the triplet excited molecules before they can undergo the annihilation reaction.

The amount of triplet excited molecules and hence the amount of delayed fluorescence is related to the specific ionisation of the exciting particle. Electrons which have a low specific ionisation produce much less triplet excitation than the greater specific ionisation particles like protons and alpha particles. For pulses which give the same intensity in the fast component, the intensity of the slow component is much greater for the proton excitation compared to the electron excitation, as shown in Fig. 16.

The integrated intensity as a function of time shows that the electron generated pulses reach their maximum intensity much faster than the proton generated pulses, as shown in Fig. 17. In the first 50 ns the electron produced pulse has

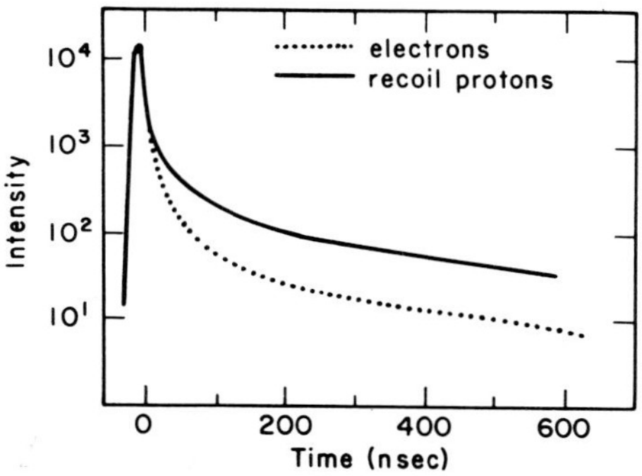

Fig. 16. The intensity of the scintillation as a function of time for a liquid scintillator solution which is excited by neutrons which produced recoil protons in the solution and gamma rays which produced Compton scattered electrons in the solution. The intensities are normalised at the peak intensity of the prompt emission.

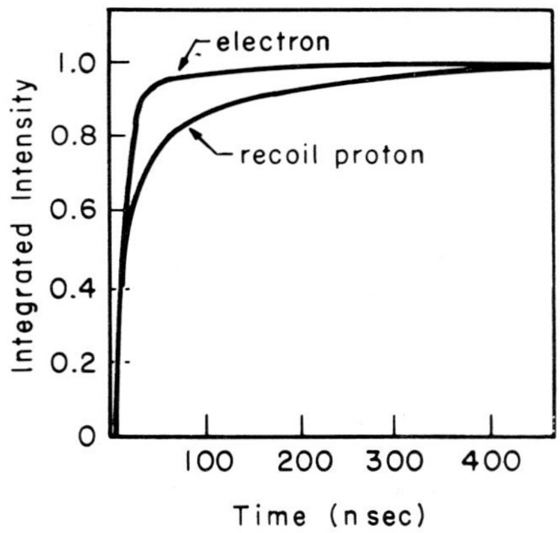

Fig. 17. Integrated intensity curves showing time dependency upon the exciting particle: electrons or recoil protons. The intensities are normalised at the maximum value at a time long after the start of the scintillation.

reached 90% of its total integrated intensity while the proton produced pulse has reached only about 75% of its total integrated intensity. The proton produced pulse does not reach 90% of its total integrated intensity until 180 ns after the initiation of the pulse.

This difference in pulses produced by different types of particles is the basis for the method of pulse shape discrimination. There are at least three different types of electronic circuits that are used for distinguishing between different types of particles which are being counted at the same time in the same scintillator solution. Three of these methods are based on:

(a) Measurement of the amplitude of the slow component relative to the amplitude of the fast component of a single scintillation pulse.[34]

(b) Measurement of the time required for a single pulse to reach a specific fraction of its total integrated intensity.[35,36]

(c) Measurement of the zero amplitude cross-over time for a single pulse using a double differentiated dynode pulse of the multiplier phototube.[37-40]

Pulse shape discrimination techniques are used to detect and measure recoil protons, alpha particles, fission fragments and other particles with high specific ionisation in the presence of electrons and vice versa. Figure 18 shows the pulse height distribution (total integrated intensity) from a liquid scintillator solution which is being excited by alpha particles and Compton scattered electrons from a gamma ray source. The alpha particles come from a radionuclide dissolved in the scintillator solution. The gamma rays come from a radionuclide which is external to the scintillator solution vial. When the pulse generated by the pulse shape discriminator circuit is used in a coincidence mode, only the alpha particle produced pulses are allowed to be counted, as shown by the distribution obtained in Fig. 18(b). On the other hand, when that pulse is used in an anti-coincidence mode, only the Compton scattered electron generated pulses are counted, as shown by the distribution obtained in Fig. 18(c). The pulse shape discriminator circuit was adjusted so that it generated a pulse only for an alpha particle which was determined by the relative amplitude of the slow component of the scintillation pulse.

Pulse shape discrimination techniques have been used extensively to measure the relative contributions of gamma rays and neutrons to the background around an operating nuclear reactor. The gamma rays are counted through the Compton scattered electrons produced in the liquid scintillator solution. The neutrons are often thermalised (by passing through a hydrogenous material like paraffin) and absorbed by an isotope of boron (Boron-10) which undergoes a fission-like event with the emission of an energetic alpha particle. These alpha particles are counted and are proportional to the neutron flux in the area. Figure 19 shows the pulse height spectra obtained with no pulse shape discrimination and with pulse

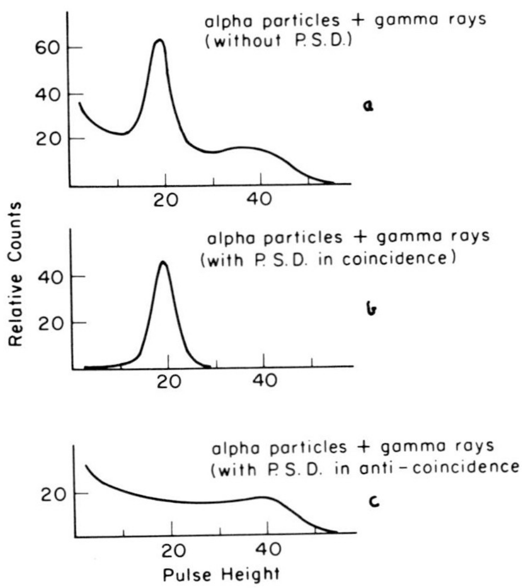

Fig. 18. Relative pulse height spectra from a liquid scintillator solution which was excited by electrons and alpha particles showing the use of the pulse shape discrimination output in coincidence to eliminate electron pulses (curve b) and in anti-coincidence to reject alpha particle pulse (curve c).

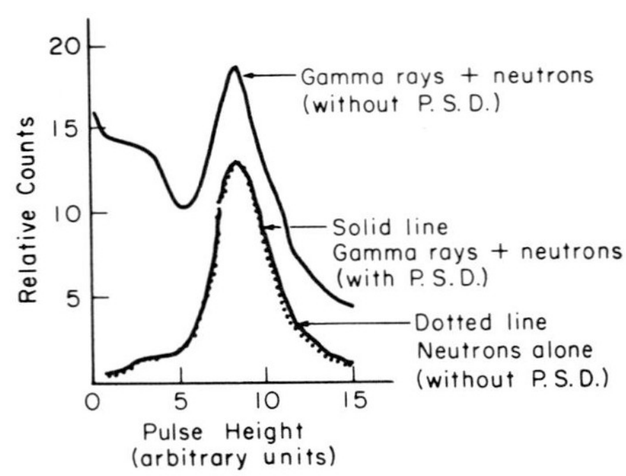

Fig. 19. Pulse height spectra showing the use of pulse shape discrimination to accept or reject pulses produced in a liquid scintillator solution containing boron when a neutron is absorbed.

shape discrimination circuit set to reject all electron generated pulses. The spectrum obtained is identical with that obtained when only a neutron source was present.

A more detailed discussion of the pulse shape discrimination technique is given in a review article by Horrocks.[41]

Counting conversion electrons

The conversion electron process gives rise to mono-energetic electrons. The internal conversion process is competitive with gamma ray emission. Indeed, there are many radionuclides which decay by emission of both gamma rays and conversion electrons. These mono-energetic electrons will give a constant scintillation yield per electron. A pulse height spectrum of a conversion electron decay process will have a peak or peaks for each mono-energetic electron. If the electrons are sufficiently different in energy to be resolved, two peaks in the pulse height spectrum will be measured.

The radionuclide Xenon-131m decays by gamma ray emission 3% of the time and by the internal conversion process 97% of the time. The internal conversion will give rise to 134 keV conversion electrons part of the time and 164 keV conversion electrons for the remaining time. Figure 20 shows a pulse height distribution obtained for a sample of Xenon-131m dissolved in a liquid scintillation solution.[42] This detector system (a single multiplier phototube) had sufficiently good resolution that the two energy groups could be distinguished.

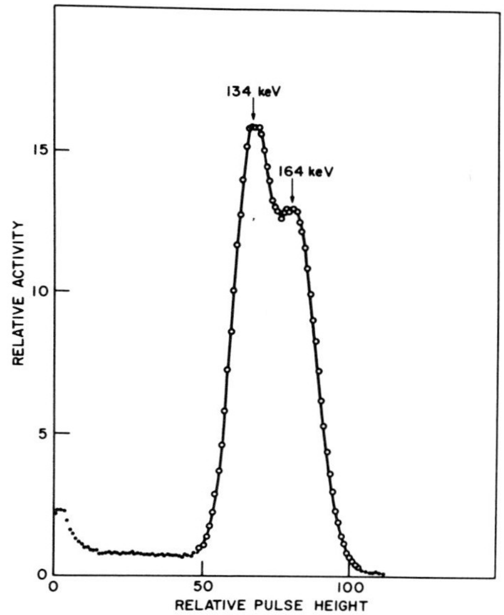

Fig. 20. Differential pulse height spectrum of a Xenon-131m sample dissolved in a liquid scintillator solution.

Another radionuclide which decays partly by the internal conversion process is Indium-113m. Figure 21 shows the pulse height distribution obtained for a liquid scintillator solution containing Indium-113m and its parent radionuclide Tin-113.[43] The 369 keV conversion electrons can easily be measured. The lower energy peaks are due to Auger electrons as a result of the electron capture processes of the radionuclide Tin-113. In one case, the 369 keV conversion electrons were used as a monitor of the amount of quench present in a liquid scintillator solution. Figure 22 shows the shift in the pulse height distribution peak as the amount of quench in the solution changes.[44]

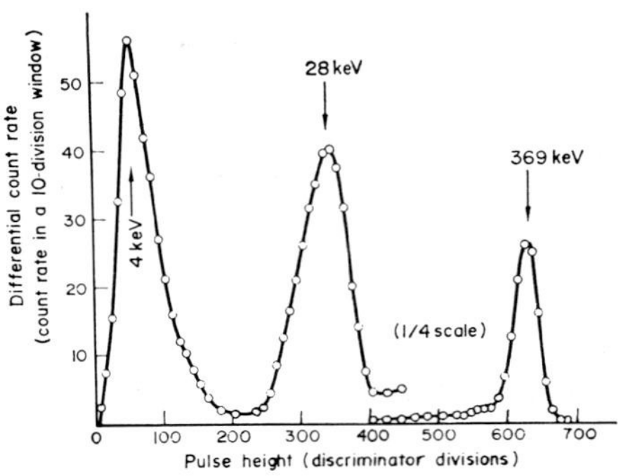

Fig. 21. Differential pulse height spectrum of a sample of Tin-113 – Indium-113m dissolved in a liquid scintillator solution. Pulse height scale is proportional to the logarithm of the energy.

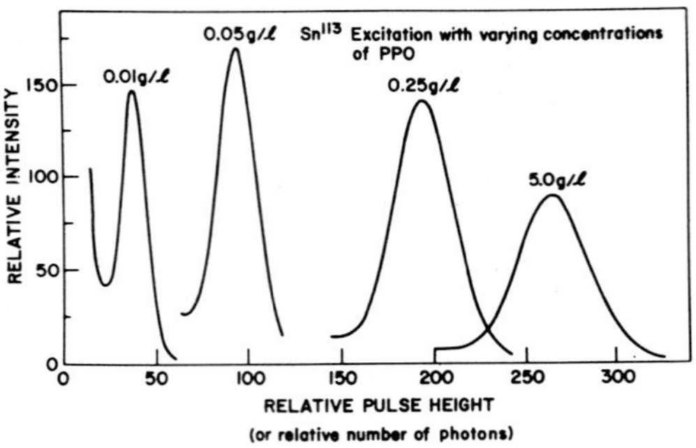

Fig. 22. Differential pulse height spectra for the 369 keV conversion electron peak as a function of the scintillation yield of a liquid scintillator solution.

REFERENCES

1 G. Laustriat, in Organic Scintillators (Ed. D.L. Horrocks), Gordon & Breach, New York, 1968, p.127.

2 R. Cooper and J.K. Thomas, J. Chem. Phys. **48**, 5097 (1968).

3 D.L. Horrocks, J. Chem. Phys. **52**, 1566 (1970).

4 G. Laustriat, R. Voltz and J. Klein, in The Current Status of Liquid Scintillation Counting (Ed. E.D. Bransome, Jr.), Grune & Stratton, New York, 1970, p.13.

5 J.B. Birks, J.C. Conte and G. Walker, IEEE Trans. Nucl. Sci. **NS-13**, No. 3, 148 (1966).

6 R. Voltz, G. Laustriat and A. Coche, C.R. Acad. Sci. (Paris) **257**, 1473 (1963).

7 I.B. Berlman, J. Chem. Phys. **33**, 1124 (1960).

8 J.B. Birks and K.N. Kuchela, Proc. Phys. Soc. **77**, 1083 (1961).

9 D.L. Horrocks, in Organic Scintillators (Ed. D.L. Horrocks), Gordon & Breach, New York, 1968, p.45.

10 P. Skarstad, R. Ma and S. Lipsky in Organic Scintillators (Ed. D.L. Horrocks), Gordon & Breach, New York, 1968, p.3.

11 J.W. Hastings and G. Weber, J. Opt. Soc. Amer. **53**, 1410 (1963).

12 I.B. Berlman, Handbook of Fluorescence Spectra of Aromatic Molecules, Academic Press, New York, 1972, 2nd edition.

13 D.L. Horrocks, Nuc. Instr. and Meth. **27**, 253 (1964).

14 D.L. Horrocks, Rev. Sci. Instr. **35**, 334 (1964).

15 D.L. Horrocks, submitted for publication (1973).

16 J. Steyn, Proc. Phys. Soc. **A69**, 865 (1956).

17 O.H. Hogan, P.E. Zigman and J.L. Mackin, Beta Spectra II. Spectra of Individual Negatron Emitters, U.S. Naval Radiological Defence Laboratory, San Francisco, 1964, Report No. USNRDL-TR-802.

18 D.L. Horrocks and M.H. Studier, Analyt. Chem. **33**, 615 (1961).
19 H.R. Lukens, Jr., Intern. J. Applied Radiation and Isotopes **12**, 134 (1961).
20 J.A.B. Gibson and H.J. Gale, J. Sci. Instr. (Physics E) Ser. **2**, **1**, 99 (1968).
21 N. Kaczmarczyk, in Organic Scintillators and Liquid Scintillation Counting (Eds. D.L. Horrocks and C.T. Peng), Academic Press, New York, 1971, p.977.
22 K.F. Flynn, L.E. Glendenin and V. Prodi, in Organic Scintillators and Liquid Scintillation Counting (Eds. D.L. Horrocks and C.T. Peng), Academic Press, New York, 1971, p.687.
23 A.R. Ronzio, Intern. J. Applied Radiation and Isotopes **4**, 196 (1959).
24 J. Ashcroft, Anal. Biochem. **37**, 268 (1970).
25 C.O. Muehlhaus and G.E. Thomas, Phy. Rev. 85, 926 (1952).
26 C.O. Muehlhaus and G.E. Thomas, Nucleonics **11**, No. 1, 44 (1953).
27 L.M. Bollinger and G.E. Thomas, Rev. Sci. Instr. 28, 489 (1957).
28 H.H. Ross and R.E. Yerick, Nucl. Sci. Eng. 20, 23 (1964).
29 D.L. Horrocks, Rev. Sci. Instr. **34**, 1035 (1963).
30 K.F. Flynn, L.E. Glendenin, E.P. Steinberg and P.M. Wright, Nuc. Instr. and Meth. 27, 13 (1964).
31 D.L. Horrocks, Nuc. Instr. and Meth. **30**, 157 (1964).
32 T. Cless-Bernert, K.E.Duftschmid and H. Vonach, Nukleonik **4**, 162 (1962).
33 E.D. Steinberg and L.E. Glendenin, Argonne National Laboratory, Argonne, Ill., unpublished results (1960).
34 W. Daehnick and R. Sherr, Rev. Sci. Instr. **32**, 666 (1961).
35 L.M. Bollinger and G.E. Thomas, Rev. Sci. Instr. **32**, 1044 (1961).
36 F.T. Kuchnir and F.J. Lynch, IEEE Trans. Nucl. Sci. **NS–15**, No. 3, 107 (1968).
37 T.K. Alexander and F.S. Goulding, Nuc. Instr. and Meth. **13**, 244 (1961).
38 C.M. Ciaella and J.A. Devanney, Nuc. Instr. and Meth. **60**, 269 (1968).
39 T.G. Miller, Nuc. Instr. and Meth. **63**, 121 (1968).
40 G.W. McBeth, R.A. Winyard and J.E. Lutleim, Pulse Shape Discrimination with Organic Scintillators, Koch-Light Laboratories Ltd., Colnbrook, Bucks, (1971).
41 D.L. Horrocks, Appl. Spectrosc. **24**, 397 (1970).
42 D.L. Horrocks and M.H. Studier, Analyt. Chem. **36**, 2077 (1964).
43 D.L. Horrocks, Intern. J. Applied Radiation and Isotopes **24**, 49 (1973).
44 D.L. Horrocks, in Organic Scintillators, (Ed. D.L. Horrocks), Gordon & Breach, New York, 1968, p.45.
45 C.M. Lederer, J.M. Hollander and I.Perlman, Table of Isotopes, John Wiley & Sons, New York, 1967, 6th Edition.
46 K.F. Flynn and L.E. Glendenin, Phys. Rev. **116**, 744 (1960).
47 D.W. Engelkemier, K.F. Flynn and L.E. Glendenin, Phys. Rev. **126**, 1818 (1962).
48 D.L. Horrocks and A.L. Harkness, Phys. Rev. **125**, 1619 (1962).
49 R.L. Brodzinski, J.R. Finkel and D.C. Conway, J. Inorg. Nucl. Chem. **26**, 677 (1964).
50 T. Cless-Bernert, K.E. Duftschmid and H. Vonach, Nukleonik **4**, 162 (1962).

DISCUSSION

R. Berthold: What is the present state regarding the energy resolution when measuring α-particles? In the proceedings of the San Francisco conference, the resolution reported with commercial instruments is only around 25%, while you reported around 6% in an earlier paper with a special single tube instrument.

D.L. Horrocks: There has not been much progress in improving resolution with commercial instruments, and there cannot really be expected to be much due to the light being divided between two photomultipliers. This dual system of light collection leads to a very much broadened pulse height spectrum.

A.R. Ware: You implied in your lecture that there was a difference in the energy transfer scheme between samples containing oxygen and those from which oxygen had been removed. For practical purposes, when great numbers of samples are being counted, how important is de-aeration when one bears in mind that the quantity of oxygen present will vary with time as a result and in any case quench corrections will probably be applied?

D.L. Horrocks: For routine work, with only screw cap type vials, oxygen may diffuse back into samples and de-aeration should not be used as the counting efficiency would depend upon the amount of oxygen that had diffused into the sample.

P.E. Stanley: The overlap of the fluorescence and absorption spectra for PPO is much worse than has been indicated and this is because of the relatively high concentrations of PPO used in scintillation fluors. Thus the O.D. is less than around 1.0 for wavelengths above 350nm. These photons are absorbed and re-emitted at wavelength within the fluorescence wavelength of PPO. I was also interested to learn of the change in scintillation efficiencies with beta-energy. This bears out what we have found recently from a mathematical model we have been building of the liquid scintillation process (for colour and chemical quenching as well as vial and detector response). Thus to obtain the observed efficiencies we need to use about 11 photons per keV for Carbon-14 and around 13 photons per keV for Hydrogen-3. This is approximately in the ratio 0.044 : 0.052 that you mentioned.

D.L. Horrocks: The change in scintillation efficiency is shown in more detail in Table 2 of this paper. Actually the scintillation yield varies over a factor of two between the E_{max} of tritium (18 keV) and 1 keV.

Chapter 2

Recent Developments in Photomultipliers for Liquid Scintillation Counting

D.E. Persyk and T.T. Lewis

RCA Corporation, Lancaster, Pennsylvania, U.S.A.

INTRODUCTION

Considerable new interest is being expressed in the field of liquid scintillation counting. This interest exists due to increased efforts in the clinical, research and environmental monitoring areas. The latter is of principal concern because of the increase in tritium-producing nuclear power facilities. Attendant with this interest there has been an introduction of new photomultipliers for use in liquid scintillation counters.

This paper discusses a new photomultiplier for liquid scintillation counting, the 4501V4. It is contrasted with the generic type 4501V3 in terms of E^2/B in the tritium and carbon windows, crosstalk, background and accidental count rates. A discussion of the effect of tube materials upon photomultiplier performance is given. Future trends in photomultipliers are also discussed.

PHOTOMULTIPLIER DISCUSSION

The 4501 photomultiplier types employ the Matheson-type photocathode-to-first dynode ('front end') geometry[1] coupled to a 12-stage BeO, focussed electron-multiplier structure. The anode output structure consists of a grid-type anode and three-element, parallel plane transmission line. The main features are shown diagrammatically in Fig. 1.

The 2-in diameter spherical-section front end affords two advantages over alternative planar faceplate designs:

1. The photoelectron trajectories are nearly isochronous, and the electric field strength near the photocathode is higher, affording faster time response, and thus a narrower time coincidence window for improved noise rejection.
2. The spherical section permits a reduced mass of faceplate material which in turn provides a lower background from the photomultiplier.

A bialkali (K—Cs—Sb) photocathode is used which has a quantum efficiency of approximately 31% at 385 nm. The 12 BeO dynodes provide typical gains of 1.4×10^7 at 2 kV applied voltage. Anode pulse risetime for delta-function excitation of the photocathode is approximately 2.5 ns, affording time response that exceeds the requirements of most present-day liquid scintillation counting systems.

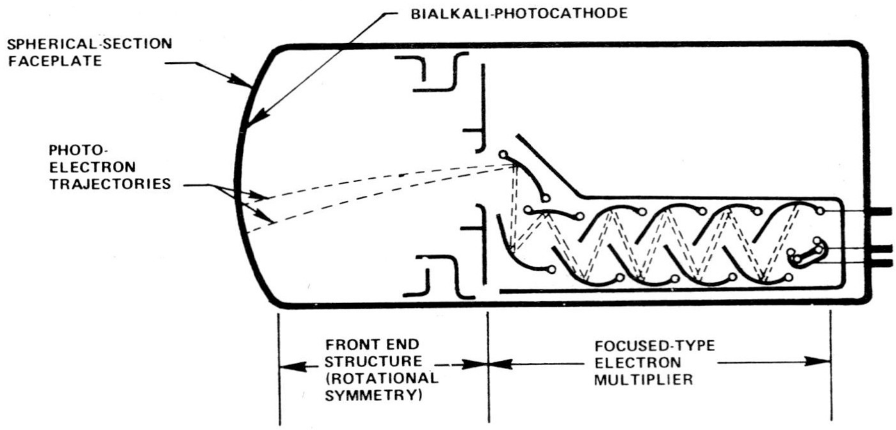

Fig. 1. Diagrammatic representation of 4501 photomultiplier

The focussed electron-multiplier structure has the following advantages over venetian blind-type multiplier structures which are also used in photomultiplier detectors:

1. The time response of the focussed multiplier is typically an order of magnitude faster than that of venetian blind multipliers.
2. The focussed multiplier has a higher total collection efficiency than the venetian blind multiplier. Coates reported a collection efficiency of ~95% for a focussed structure multiplier (8850, which is generically similar to a 4501V4) against 66% for a venetian blind multiplier (9558QB).[2]

MATERIALS CONSIDERATIONS

The envelope and internal tube parts materials chosen for photomultipliers used in liquid scintillation counting applications are very critical. One obvious requirement is that all tube materials shall exhibit negligible radioactivity. Of prime concern in this regard is the glass used in the vacuum envelope housing the photomultiplier.

The glasses most commonly used in photomultiplier construction contain some ^{40}K and thorium daughter-products. These radioactive contaminants are found in the raw materials comprising the glass, such as sand, and are also introduced into the glass when it is melted in thoria refractories.[3] The former problem can be overcome by judicious monitoring of the basic constituents prior to melting, and the latter problem can be solved by proper choice of melting vessel, platinum being an ideal choice.

Low background counting experiments performed on glasses from manufacturers throughout the world indicate a considerable spread in radioactive content. Some data on activity of glasses used in photomultipliers are presented in Table 1.

Less well understood, but of equal importance, is the fluorescent activity of the faceplate glass used in the photomultiplier. All glasses scintillate to some extent when excited by X-rays or γ-rays, and such scintillations lead to undesired background counts in the case of liquid scintillation counters. A pronounced scintillation-like effect due to cosmic ray bombardment is commonly seen in photomultipliers.[5,6] Unpublished measurements at this facility indicate that commonly used photomultiplier window materials vary by several orders of magnitude in their scintillation efficiencies. The problem of scintillating windows can be approached from at least two avenues:

1. Employ special window materials with very few luminescent centres ($<1:10^9$), or
2. Add dopants to common window materials to quench the luminescence (or shift it to a sufficiently long wavelength such that it is of no consequence).

Both approaches have been tried in photomultipliers with good success.

Table 1. Radioactive content of some glasses commonly used in photomultiplier construction (activity is stated in counts/min-kg for 2π counting geometry and a 70 keV to 2.6 MeV energy window).

Glass type	Manufacturers	Activity (counts/min-kg)
7740	Corning Glass Works, Corning, N.Y.	83
KG-33	Owens-Illinois, Vineland, N.J.	53
7070	Corning Glass Works	290
7050	Corning Glass Works	171
EN-1	Owens-Illinois	567
7052	Corning Glass Works	465
7720	Corning Glass Works	208
0080	Corning Glass Works	69

Data from A.W. Consylman and D.D. Crawshaw.

It is clear that considerable improvements can be made to photomultipliers by incorporating the best possible choice of materials in their construction. This approach has been followed in the 4501V3 and 4501V4 photomultipliers, as will be described in the next section.

TUBE EXPERIMENTS

The experiments were performed in a commercially available liquid scintillation spectrometer.[7] Data are reported on a large (>200) tube sample, ensuring statistically valid average values. In all experiments the tubes were 'paired' in the sense that tubes with comparable gains were used together. This requirement stemmed from the limited gain—adjust capability of the instrument (a characteristic of nearly all commercially available spectrometers), and was therefore an instrument-imposed constraint.

Unless otherwise stated, all measurements were made in the tritium energy window and with a tritium counting efficiency of 60%. The carbon counting efficiencies for carbon-above-tritium measurements averaged approximately 70%. The figure-of-merit E^2/B (efficiency, in percent, squared, divided by background, in counts per minute) is stated, in keeping with common usage.

The crosstalk measurements were made with the spectrometer elevator down and with an empty counting chamber. The background measurements were made with a Packard background standard. The accidental count rate was measured with the elevator shaft blocking the tubes.

To prove the magnitude of the cosmic ray/Cerenkov effect, certain measurements were repeated with the entire spectrometer rotated 90° so that the photomultiplier axis was vertical instead of horizontal.

Results and interpretation of experiments

Table 2. Results of experiments comparing 4501V3 with 4501V4.

Parameter	4501V3	4501V4	Remarks
Tritium E^2/B (60% eff.)	160	210	
Background in tritium window (c.p.m.)	22.5	17.1	Measured with Packard standard
Crosstalk in tritium window (c.p.m.)	12	7	
Accidental rate (c.p.m.)	0.8	0.8	Measured with elevator shaft blocking tubes, calculated value ~0.07 c.p.m.
Carbon-above-tritium E^2/B (70% eff.)	563	546	A difference was noted between the pulse height distribution of the background between the V3 and V4 tubes in the carbon energy window

The results of the experiments are tabulated in Table 2. In brief, the E^2/B has been increased from 160 to 210 in the 4501V4 as contrasted to the 4501V3. It should be noted that this increase was achieved through a reduction in background (22.5 c.p.m. to 17.1 c.p.m.) and crosstalk (12 c.p.m. to 7 c.p.m.). From these data one may conclude that the material-related improvement in the 4501V4 was due largely to a reduction in scintillation efficiency of the window, as opposed to a reduction in radioactivity of the envelope, as the background standard is an efficient scintillator.

To assess the magnitude of cosmic ray interactions, the crosstalk and accidental rates were measured with the spectrometer in the horizontal (normal operating) position and in the vertical position. This experiment was performed on only one pair of tubes, owing to the mechanical difficulties of rotating the spectrometer. The data are summarised in Table 3.

The crosstalk in the tritium window increased from 5.7 to 8.6 c.p.m. in going from horizontal to vertical, a ratio of 1.5. The accidental rate increased from 0.34 to 0.95 c.p.m. horizontal to vertical, a ratio of 2.8. The higher ratio for the accidental rate is ascribed to the increased probability of an energetic particle interacting with both tubes while the system is vertical.

The accidental rate may be calculated from the relationship $A = 2\ N_1N_2\lambda$, where N_1 and N_2 are the dark count rates of the two individual photomultipliers and λ is the resolving time of the coincidence system. Typical values in our experiments were $N_1 = N_2 \sim 10 \times 10^3$ c.p.m. and $\lambda = 20$ ns, giving $A = 0.07$ c.p.m. The order-of-magnitude increase in measured accidental rate over calculated rate was attributed to cosmic ray events, as demonstrated by the greatly increased accidental rate for vertical operation of the system.

Table 3. Effect of cosmic rays upon spectrometer orientation — 4501V4

	Spectrometer horizontal (normal position)	Spectrometer vertical	Remarks
Crosstalk in tritium window (c.p.m.)	5.7	8.6	Measured on one pair of tubes
Accidental rate (c.p.m.)	0.34	0.95	Made on one pair of tubes

FUTURE IMPROVEMENTS

Further improvements can be made to photomultipliers to reduce their background and fluorescent properties, and research is being carried out in this area. Additional refinements can also be made to systems to improve performance with existing photomultipliers.[8] A future possibility would be a single tube liquid scintillation counter utilising pulse shape and pulse height discrimination to eliminate the need for a two-tube coincidence system. Such a system with a suitable photomultiplier could distinguish between tube dark pulses (single-electron in origin), tube Cerenkov pulses (fast pulses of multiple-electron origin) and scintillator signal pulses (relatively slower pulses of multiple-electron origin) on the basis of pulse height and pulse shape (time) information. Photomultipliers such as the developmental type C31024 may have the basic time response and electron resolution capabilities to render such a system feasible in today's state-of-the-art.

SUMMARY

Materials used in photomultiplier construction have been investigated for radioactive and fluorescent properties. The results of this investigation have led to the introduction of the 4501V4 photomultiplier which offers a tritium E^2/B of 210 as opposed to 160 for the present 4501V3 type. Further improvements in these photomultipliers are being researched, and possibilities of a single tube liquid scintillation spectrometer utilising pulse shape and pulse height discrimination are being studied.

ACKNOWLEDGEMENTS

The authors are grateful to R.M. Shaffer and J.D. Doering for providing data, and to R.C. Pontz for encouragement in writing this paper.

REFERENCES

1 R.W. Engstrom and R.M. Matheson, IRE Trans. Nucl. Sci. NS-7, No. 2–3, 52 (1960).

2 P.B. Coates, J. Phys. D: Appl. Phys. 6, 153 (1973).

3 Private communication with D.C. Smith, Vineland, N.J.

4 Data provided by A.W. Consylman and D.D. Crawshaw, RCA Corp.

5 Andrew T. Young, Rev. Sci. Instrum. 37, 1472 (1966).

6 S. Roodbergen, R. Kroondijk and H. Verheul, Nuc. Instr. and Meth. **105**, 551 (1972).
7 Packard Tri Carb Spectrometer, Model 3300, Packard Instrument Corp., Chicago, Ill.
8 John E. Noakes, Michael P. Nearly and James D. Spaulding, Nuc. Instr. and Meth. **109**, 177 (1973).

DISCUSSION

P.E. Stanley: Do you have any data on the amount of light reflected at the quartz/photocathode interface and does this vary with the wavelength of the incident light and the thickness of the photocathode?

D.E. Persyk: We do not have the data to answer that question.

P.E. Stanley: What order of variation in quantum efficiency (at a given wavelength, say 380 and 425 nm) do you observe? What I'm really after is the variation in response across the photocathode. Would it be of the order ±20% of the observed quantum efficiency?

D.E. Persyk: Yes, it's of that order.

T.A. Wilkins: With reference to Table 2 (Comparison of Performance of P.M. Tubes 4501V3 and 4501V4), were the results indicated subjected to analysis of variance and statistical tests? These yield (1) further information of validity of results, (2) reproducibility of manufacture of tubes.

D.E. Persyk: Yes, the results were subjected to routine analysis of variance and statistical relevance. Photocathode processing of the tubes is done by a process-control computer which also performs an on-line test of each tube's important performance parameters. The computer processing affords excellent reproducibility, which is important as the tubes are matched in liquid scintillation counting applications.

T.A. Wilkins: Some of the differences between measurements on different tubes were very small, whilst errors on these could be quite large. No errors were presented nor were any confidence levels presented with respect to significance of differences.

D.E. Persyk: The values are cited with an uncertainty in the least significant digit to a confidence level of 68%. For example, in Table 2 background in the tritium window for the 4501V4 is shown as 17.1 c.p.m. This was obtained with a 1000 minute count so that the measurement results could be expressed more precisely as 17.1 c.p.m. ±0.1 c.p.m. The authors stress that the key point regarding the statistical relevance of these measurements is not the small uncertainty associated with each measurement, but rather the large (greater than 200 tube) sample size that was employed.

R.P. Parker: What is the speed of response of your photomultipliers? Whilst a single photomultiplier system is excellent in terms of the physics of the system, can chemiluminescent events be discriminated against using pulse-shape discrimination?

D.E. Persyk: Rise time is 1 to 2 nanoseconds, 800 picoseconds at best. Chemiluminescence will be an important problem and we have not investigated this. Chemiluminescence (and phosphorescence) is also a problem in a conventional two-tube coincidence system. The lower level discriminator is set to ignore single-electron events, which are associated with chemiluminescence, owing to the single photon nature of the process. However, at high count rates pulse pile-up gives rise to false counts. This condition can be recognised by unbalancing the system: A delay line having delay greater than the coincidence window time width is inserted in one leg of the coincidence circuit — then, if the count rate does not change appreciably, the counts are attributed to pulse pile-up from chemiluminescence. In a single-tube system the lower level discriminator is likewise set at the one-electron level, and therefore discriminates against chemiluminescent signals up to some maximum count rate. The presence of false counts due to chemiluminscence could be recognised by a count rate circuit which monitors the single-electron rate and issues a warning when an unacceptably high signal rate is detected.

Chapter 3

Liquid Scintillation Counting of Radionuclides Used in Radioimmunoassay

D. L. Horrocks

Beckman Instruments, Inc., Fullerton, Ca 92634, U.S.A.

INTRODUCTION

Many of the radionuclides used in radioimmunoassay- and radioassay-type in vitro diagnostic tests emit gamma rays or X-rays which can be measured in inorganic crystal (NaI) scintillation counters. At the same time there are some radioimmunoassay tests which use tritium as the radioactive label. Tests which use tritium as the label have to use liquid scintillation counting techniques as the means to measure the amount of tritium in the antibody-bound and/or free antigen-separated fractions. It is often not realised that many of the so-called 'gamma and X-ray emitters' upon decay also produce electrons. In some cases the electrons are the result of beta decay while in others they are the results of conversion or Auger electron productions. It is actually possible to measure these electrons by liquid scintillation methods.

Some radionuclides decay by beta emission followed by gamma ray emission. These radionuclides can be measured with high efficiency in a liquid scintillator system depending upon the energy of the beta transition. The radionuclides Iodine-131, Cobalt-60 and Iron-59 decay by the emission of beta particles in 100% of their transitions followed by the emission of one or more gamma ray groups. The maximum energy of the beta emission is such that these radionuclides are measured with high efficiencies (nearly 100%) by liquid scintillation methods. The radionuclides Iodine-125, Cobalt-57 and Chromium-51 all decay 100% by electron capture. As a result they produce emissions of characteristic X-rays and a certain fraction of these events results in the production of Auger electrons. These Auger electrons, because they are mono-energetic, can be counted with high efficiency, depending upon their energy, by liquid scintillation counting methods.

When trying to decide upon measurement of these dual emission radionuclides by either solid inorganic scintillation crystals; gamma and X-rays; or liquid scintillation solutions; beta, conversion electrons and Auger electrons; it should be remembered that any time liquid scintillation methods are used quenching can be a serious problem which will require correction terms applied to the measured results to normalise all data to a common counting efficiency. In measurement of gamma and X-rays with solid inorganic scintillation crystals there are no quenching corrections required.

The advantage of antigen determination by radioimmunoassay and radioassay techniques is primarily the result of the specificity of the antigen-antibody reactions. This property, along with the sensitivity afforded by the use of radionuclides, allows for the measurement of picogramme (10^{-12} g) amounts of a specific antigen without separation of that antigen from all the other antigens present in the serum sample. A single antigen molecule can be labelled with at least one, and often more than one, radioactive atom. The specific activity of the radionuclide will determine the amount of labelled antigen necessary for minimum detection. The short half life of Iodine-125 (60 days) enables one to measure 10,000 c.p.m. from as little as 2×10^{-15} mole of an antigen labelled with one Iodine-125 atom per molecule.

COUNTING OF BETA-GAMMA RADIONUCLIDES

Several radionuclides used as labels for radioimmunoassay and radioassay measurements emit one beta particle per disintegration. Examples of these types of radionuclides are Iodine-131, Iron-59, Cobalt-60 and Sodium-22. Table 1 lists the beta energy groups for each of these radionuclides and their relative abundances. The beta energy groups are all sufficiently energetic to be counted with high efficiency in a liquid scintillator solution.

Table 1. Beta energies and abundance of Iodine-131, Cobalt-60, Iron-59 and Sodium-22 decay modes and their counting efficiencies in a liquid scintillator solution.

Radionuclide	Mode of decay	E_{max} and abundance (%)	Counting efficiency by LSC (%)
Iodine-131	β^-	0.251 (2.8)	95
		0.333 (9.1)	
		0.606 (87.4)	
		0.810 (0.7)	
Cobalt-60	β^-	0.314 (100)	90
Iron-59	β^-	0.130 (0.9)	85
		0.268 (44.9)	
		0.459 (54.9)	
		1.560 (0.3)	
Sodium-22	β^+	0.540 (100)	89

Also included in Table 1 is a typical counting efficiency of each of the radionuclides in a small amount of water (0.05 ml) dissolved in an emulsion-type scintillator solution. As can be seen from this data the counting efficiencies are indeed high and the measurement by liquid scintillation methods appears to be very reasonable. However, the high efficiency still has to be corrected for any quench variation between samples. In all radioimmunoassay and radioassay techniques it is not necessary to know the absolute counting efficiency, but all samples have to be counted with the same efficiency.

Interpretation of the pulse height distributions for samples of these radionuclides in liquid scintillator solutions is fairly simple. Figures 1 (a-d) show the differential pulse height distributions obtained with a liquid scintillation counter which has a logarithmic amplification system (Beckman LS-250 Model). The output of the summing amplifier of the liquid scintillation counter is input directly into the amplifier of a multichannel analyser (Nuclear Data Model 1100) and the coincidence signal from the liquid scintillation counter is used as a gating pulse for the multi-

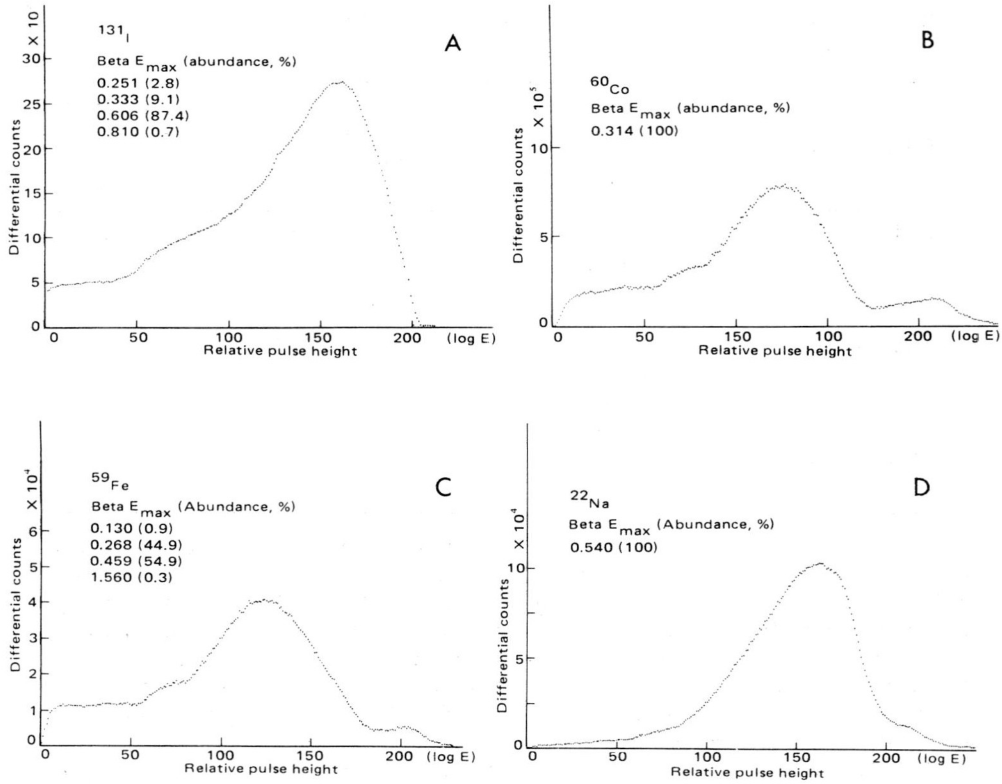

Fig. 1. Differential pulse height distribution for (a) Iodine-131 sample (b) Cobalt-60 sample (c) Iron-59 sample (d) Sodium-22 sample (all aqueous solution) in an emulsifier-containing liquid scintillator solution. Relative pulse height is proportional to the logarithm of the beta particle energy.

channel analyser (after appropriate shaping of the pulse to satisfy the input requirements of the multichannel analyser).

In at least three of the pulse height distributions an additional number of pulses are observed which have pulse amplitudes greater than the E_{max} of the beta spectrum for that radionuclide. These pulses are due to a coincidence between some of the beta particles and Compton scattered electrons produced by the gamma rays which are also emitted by the radionuclide. These can be seen in the pulse height distributions for Iron-59, Sodium-22 and Cobalt-60 but for unknown reasons are not evident in the Iodine-131 pulse height distribution.

COUNTING ELECTRON CAPTURE — AUGER AND CONVERSION ELECTRON RADIONUCLIDES

Table 2 lists the nuclear properties of three other radionuclides used in radioimmunoassay and radioassay tests. These radionuclides all decay by electron capture. The electron capture process will produce either a characteristic X-ray or Auger electron(s). Figures 2 (a-c) show the differential pulse height spectra for Iodine-125, Cobalt-57 and Chromium-51 samples (0.050 ml of aqueous solution in an emulsifier liquid scintillator solution) in a liquid scintillation counter.

Table 2. Liquid scintillation counting of Iodine-125, Cobalt-57 and Chromium-51.

Radionuclide	Mode of decay	Major energy groups (keV)	Counting efficiency by LSC (%)
Iodine-125	Electron capture, conversion electron	12, 40	76
Cobalt-57	Electron capture, conversion electron	14, 136	65
Chromium-51	Electron capture	5.5	30

The Iodine-125 spectrum (Fig. 2a) is composed of two peaks which correspond to the excitations produced by electrons or groups of electrons (Auger electrons and conversion electrons) of energy about 12 keV and 40 keV. These two energy groups are calculated from the decay modes and the probabilities of different processes of energy release. The groups are not due to the stoppage of X-rays in the liquid scintillator solution. Theoretical calculations and experimental measurements both indicate that only about 8% of 27.5 keV X-rays are absorbed in 20 ml of a toluene solution.

The Cobalt-57 spectrum (Fig. 2b) is composed of two peaks which correspond to 14 keV conversion electrons (coincident with low energy Auger electrons) and a somewhat lower intensity 136 keV conversion electron peak. The electron capture of Cobalt-57 produces an excited atom of Iron-57 which decays by the emission of gamma rays, X-rays and conversion electrons. In the liquid scintillator solution only the Auger and conversion electrons are detected.

Figure 2c shows the spectrum for Chromium-51 which is the result of excitations from the 5.5 keV Auger electrons produced by the Vanadium-51 atom which is produced by the electron capture process of Chromium-51. The Chromium-51 process of electron capture occurs 100% of the time. A small contribution to the measured counts comes from the Compton scattering of the 320 keV gamma rays which are produced in 9% of the total decay events.

Table 2 also lists the counting efficiency obtained for samples of these radionuclides dissolved in an aqueous solution and incorporated into an emulsifier containing liquid scintillator solution. The counting efficiency of Iodine-125 is high because nearly every decay event produces a conversion electron or some Auger electrons. Actually the counting efficiency in a 3 in x 3 in NaI(T1) crystal with a well is also about 75% because nearly every decay event produces either one or two 27.5 keV Te K X-rays.

The counting efficiency of Cobalt-57 is also high because of the high probability of the production of conversion electrons (14 and 136 keV). The counting efficiency for Cobalt-57 in a 3 in x 3 in NaI(T1) crystal with a well is also high because of the nearly high probability of emission of at least one gamma ray per decay event. These intermediate energy gamma rays (136 and 122 keV) are detected with a high efficiency by the crystal. The counting efficiency in the crystal is about 85%.

The counting efficiency for Chromium-51 in the liquid scintillator solution is low even though 80% of all decay events produce Auger electrons. The low

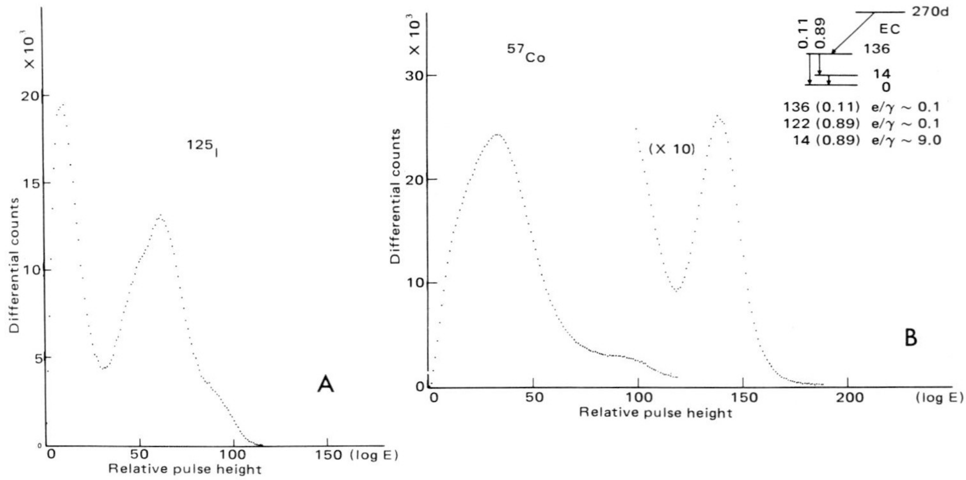

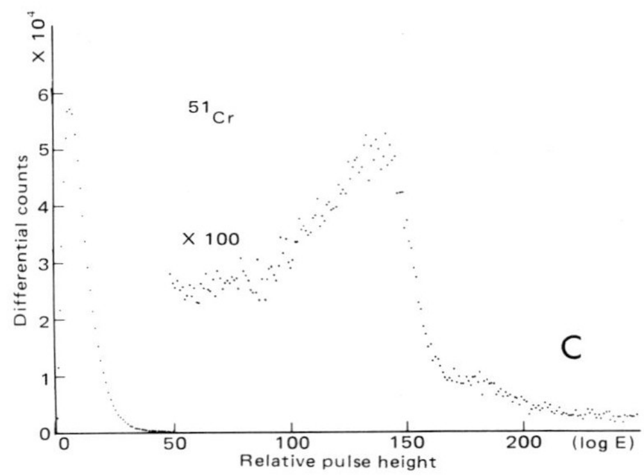

Fig. 2. Differential pulse height distribution for (a) Iodine-125 (b) Cobalt-57 (c) Chromium-51 (all aqueous solutions) in an emulsifier-containing liquid scintillator solution. Relative pulse height is proportional to the logarithm of the electron energy.

efficiency is primarily due to the low energy of these Auger electrons. The most energetic Auger electrons have only 5.5 keV energy. Many of the Auger electrons are less energetic. Even so nearly 40% of the Auger electrons are counted which is reflected in the 30% counting efficiency in the scintillator solution. The counting efficiency is very low in a 3 in x 3 in NaI(Tl) crystal; only about 5.9%. This is primarily due to the fact that only 9% of the total decays produce the 320 keV gamma ray which can be measured in the crystal. Actually this gamma ray is counted with better than 60% efficiency when the abundance is considered. The low energy X-rays from Vanadium-51 are only 5 keV and will not be detected by the crystal. (The low energy X-rays will not penetrate the Aluminium housing of the detector.)

CONCLUSIONS

All these radionuclides can be counted with high efficiencies in a liquid scintillation system. However, there are two main concerns which have to be considered. First, it is necessary to monitor all samples for quench differences between samples with subsequent computation of a quench correction. The second concern involves the possibility of contamination of the liquid scintillation counter with gamma-emitting radionuclides. The high penetration of the electromagnetic radiation would cause an increase in background which would require extensive decontamination before low count rate beta-emitting samples could be measured with any degree of accuracy. Equivalent contamination by beta-emitting radionuclides would be less serious because the beta particles have a low probability of penetration of the bottle walls which is necessary to cause scintillations in the liquid scintillator.

DISCUSSION

J.F.Stoutjesdijk: A difficulty in counting radionuclides emitting γ-rays with a liquid scintillation counter is the influence on the background when samples with greatly different activities are present in the sample changer. A highly active sample adjacent to a low activity sample which is being counted may cause an appreciably higher background. What are the volumes of samples which have been compared in γ-scintillation and liquid scintillation counting? With well type NAI-crystals mostly larger amounts of liquid can be measured, resulting in a higher sensitivity.

D.L.Horrocks: With emulsion-type liquid scintillator solutions it is practical to count up to 3 ml of aqueous samples. Quench correction is possible with external standards and/or sample channels ratio methods.

B.E.Gordon: Why do you prefer emulsion to direct solution in dioxane? Is there any evidence of loss of count rate versus time in dioxane due to adsorption on the walls?

D.L.Horrocks: Emulsion-type scintillation counting solutions are considered less subject to chemiluminescence problems than the dioxane base solutions. Wall adsorptions will be more a function of the samples. However, with emulsions the sample remains in an aqueous environment, and is thus less likely to be salted-out of solution.

Chapter 4

An Intercomparison of Commonly Used Liquid Scintillation Counting Vials

K. Painter[a] and M.J. Gezing

Amersham/Searle Corporation, Arlington Hts., Illinois, U.S.A.

INTRODUCTION

The science of liquid scintillation spectrometry has probably advanced more rapidly than any other instrumental technique. From the beginnings of the early 1950's to the present we have witnessed tremendous progress in instrumentation and sample chemistry. The liquid scintillation analyst is now confronted with push button instrumentation and premixed cocktails which seemingly make the technique as simple to perform as operating the ordinary television set.

Unfortunately, these technological advances have removed the necessity for many users to become familiar with the technique. All too often we have seen users place their samples in vials, add premixed cocktail, push a button on the counter and read a number off a computer tape. Due to the pitfalls and idiosyncrasies of the technique, however, the novice can frequently find himself in considerable difficulty. The liquid scintillation literature is replete with warnings related to instrumentation and sample chemistry. We can only add our voices to those who have cautioned newcomers in this field to review the literature and be aware of the limitations of the liquid scintillation technique.

ADVANTAGES AND DISADVANTAGES OF COMMON COUNTING CONTAINERS

The choice of liquid scintillation vial is of utmost importance in critical counting applications. A recent paper[1] has reviewed in detail the advantages and disadvantages of various counting containers. To summarise the findings the disadvantages of counting vials are pointed out in Table 1.

Several types of counting bottles found in routine use are shown in Fig. 1. The disadvantages of each counting vial are keyed to Tables 2 and 3 and are provided as a warning of the most frequent disadvantages to which a particular vial may be prone.

---

a Present address: Micromedic Diagnostics, Fort Collins, Colorado, U.S.A.

Table 1. Disadvantages of common counting vials.

A.	High background count rate
B.	Low counting efficiency
C.	Limited volume capacity
D.	Breakage
E.	Swelling
F.	Reactive to solvents/solubilisers
G.	Inability to inspect samples visually
H.	High cost
I.	Incompatibility with external standard
J.	Adsorption of sample
K.	Permeability to solvent/sample
L.	Photoluminescence
M.	Static electricity
N.	Acute volume:efficiency dependency

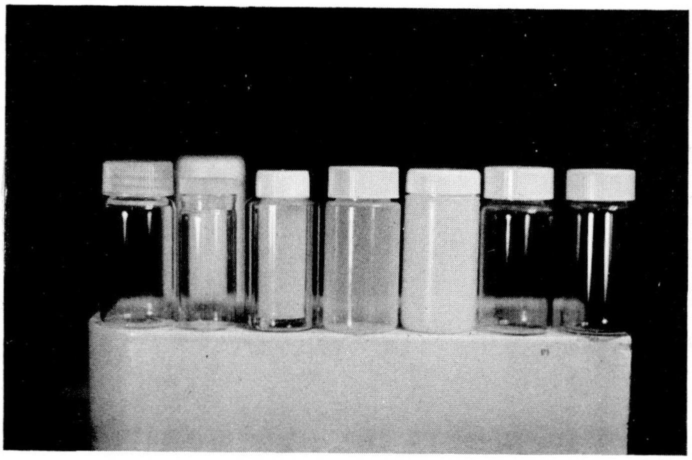

Fig. 1. Various liquid scintillation vials, Left to right: borosilicate glass, general purpose glass, polyethylene, nylon, Mini-vial™, 3 dram mini-vial and quartz. Not shown, Sterilin mini-vial.

INTERCOMPARISON OF COMMON COUNTING VIALS

Tables 2 and 3 provide intercomparison data for the counting vials shown in Fig. 1 with respect to two figures of merit for Carbon-14 and tritium. E^2/B (where E is efficiency squared divided by background count rate) figure of merit gives a good measure of the optimum performance of a liquid scintillation spectrometer for unquenched or very slightly quenched samples. The E^2/B figure of merit was determined using an air quenched toluene solution containing 6% Spectrafluor[a] PPO-POPOP.

Although EV (where E is efficiency and V is volume) has frequently been used as a sample figure of merit,[2-4] a more practical figure of merit is $EV/\sqrt{B}$.[2,5] The latter takes into account efficiency, volume and background. The $EV/\sqrt{B}$ figure of merit is calculated on a 40% (v/v) water sample in PCS[a] liquid scintillation cocktail. Relative

[a] Trademark Amersham/Searle.

Table 2. Intercomparison of various liquid scintillation vials — tritium

	Relative E^2/B	Relative $EV/\sqrt{B}$	Disadvantages
Borosilicate glass	35.9	67.5	DJ
General purpose glass	16.7	47.8	ADJ
Polyethylene	88.9	100.0	EFGIK
Nylon	60.5	71.4	EFGIL
Mini-vial™	100.0	37.7	CEFGIKMN
Sterilin mini-vial	51.2	24.9	CEFGIKN
3 Dram mini-vial	58.6	39.9	DIJN
Quartz	65.0	77.6	DHJ

Table 3. Intercomparison of various liquid scintillation vials — Carbon-14

	Relative E^2/B	Relative $EV/\sqrt{B}$	Disadvantages
Borosilicate glass	64.7	87.0	DJ
General purpose glass	42.9	73.2	ADJ
Polyethylene	100.0	100.0	EFGIK
Nylon	87.4	82.1	EFGIL
Mini-vial™	99.3	34.7	CEFGIKMN
Sterilin mini-vial	91.4	29.4	CEFGIKN
3 Dram mini-vial	85.2	50.4	DIJN
Quartz	89.0	93.3	DHJ

figures are given to maximise vial comparison.

It can be seen from the tables that for samples with little quench, the Mini-vial™[a] and polyethylene bottles appear to be superior; unfortunately, unquenched samples of this type are rare in bioanalytical research. For larger samples where more quenching is present, 20 ml vials are superior to any of the miniature vials.

Percent counting efficiencies in this study did not vary much with the type of counting vial, but backgrounds varied considerably. In applications where sample count rates are greater than 100 times background, background count rate need not be considered,[6] and economical general purpose glass bottles can be used. It is interesting to note that relatively high figures of merit are obtained for general purpose glass vials when counting beta-emitters other than tritium in moderately to highly quenched systems.

A WORD OF CAUTION IN USING THE EXTERNAL STANDARD RATIO METHOD OF QUENCH CORRECTION

It is not convenient to prepare a quenched standard calibration set in screw cap, polyethylene or polypropylene vials, as the permeability of these containers renders the set useless after several days. Many laboratories use commercial quenched standard sets in 20 ml flame-sealed glass bottles to prepare an external standard ratio calibration curve which is then applied to other types of counting vials, such as nylon, polyethylene and miniature vial combinations. Figure 2 shows the gross error which can be introduced in the experiment in such cases. A commercial quenched standard set in 20 ml flame-sealed glass borosilicate

---

a Trademark Nuclear Associates.

vials was compared with the internal standard method to determine actual counting efficiencies.

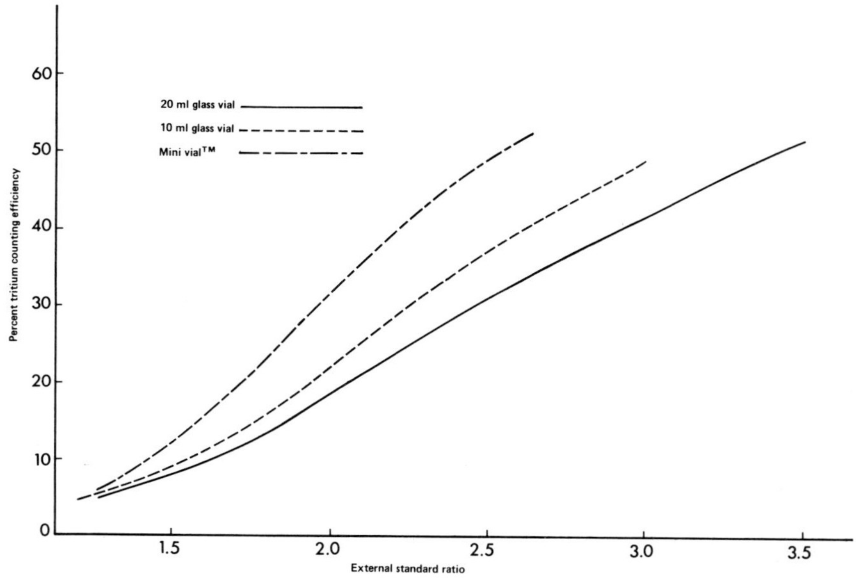

Fig. 2. External standard channels ratio calibration curves for vials of different composition and shape.

We have found the external standard ratio method of quench correction to be impractical for any counting bottle other than 20 ml glass vials. This is not surprising since one would not expect to obtain the same external standard gamma interaction with plastic and glass, and hence a different Compton continuum or simulated beta spectrum will result. Despite claims to the contrary from one supplier, we have found this variability to be true of instruments from all commercial manufacturers.

We have shown in Fig. 3, however, that with the use of the sample channels ratio method[7] of quench correction quite satisfactory results can be obtained using a commercial glass quenched standard set to construct a quench correction curve for many different types of counting containers.

For those who have access to on-line computers, Glass[8] has suggested an excellent method of simultaneously computing and comparing both sample channels ratio and external standard ratio. We strongly recommend this approach.

WASHING AND RE-USING COUNTING VIALS

When one considers the cost of washing scintillation vials, i.e. technician time, counter time and the chance of a contaminated bottle affecting the precision of data, re-using standard counting bottles is false economy. Polyethylene, polypropylene and nylon vials should never be re-used, and caps should be discarded after a single use.

Rummerfeld and Goldman[9] have discussed the economics of vial washing versus the accuracy required for the experiment. Harris and Friedman,[10] Kushinsky and Paul,[11] Radin[12] and Drosdowsky and Egoroff[13] have described various scintillation vial washing devices. A word of caution, however; in several of the

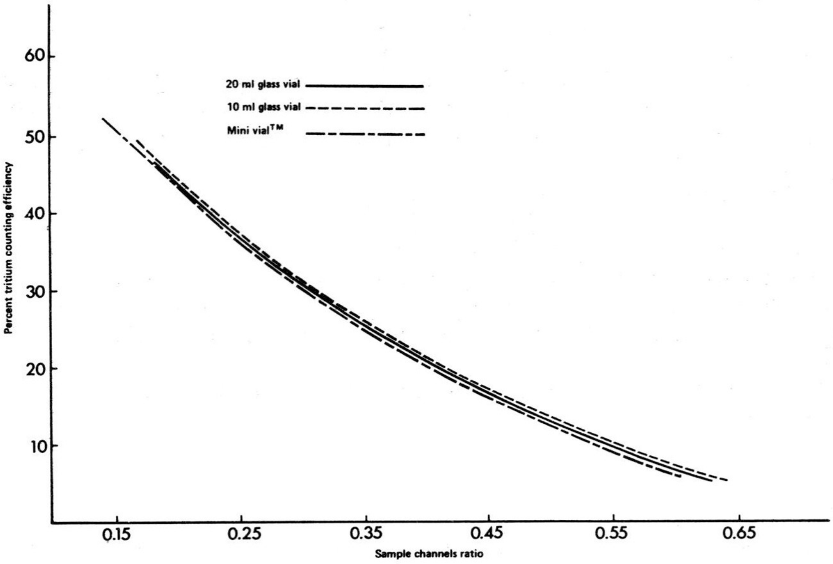

Fig. 3. Sample channels ratio calibration curves for vials of different composition and shape.

studies mentioned above radioactive steroids were employed to test the effectiveness of the washing device. Since steroids are known to be fairly soluble in non-polar aromatic solvents and tend not to adsorb to glass,[14,15] those contemplating re-use of counting vials should not presume that similar results will be obtained with other types of samples. Lipids, cations, multivalent anions, proteins and basic amino acids are particularly prone to glass adsorption.

Washing and re-using counting bottles is a risky business unless each vial is filled with fresh scintillant and counted prior to the addition of the radioactive sample.

SUMMARY

A comparison of common counting vials has shown that the optimum counting vial for a particular experiment depends upon volume of sample, counting efficiency or quench and background count rate. Eight commonly used vials including three miniature vials are intercompared with respect to two figures of merit. Users are warned regarding the re-use of scintillation vials and the application of the external standard ratio method of quench correction to vials of different construction. The sample channels ratio technique of quench correction is shown to be independent of vial material or shape.

REFERENCES

1 K. Painter, 'The Choice of Counting Vial For Liquid Scintillation: A Review' Sydney International Symposium on Liquid Scintillation Counting, 1973.
2 E. Schram, Organic Scintillation Detectors, Elsevier, New York, 1963.
3 F.E. Kinard, Rev. Sci. Instr. 28, 293 (1957).
4 C.A. Ziegler, D.J. Chleck and J. Brinkerhoff in Liquid Scintillation Counting (Eds. G.C. Bell and F.N. Hayes), Pergamon Press, New York, 1958, p. 185.

5 R. Loevinger and M. Berman, Nucleonics 9, 26 (1951).
6 A.C. Kuyper, J. Chem. Educ. 36, 128 (1959).
7 E.T. Bush, Analyt. Chem. 35, 1024 (1963).
8 D.S. Glass, in Organic Scintillators and Liquid Scintillation Counting, (Eds. D.L. Horrocks and C.T. Peng), Academic Press, New York, 1971, p.803.
9 P.S. Rummerfeld and I.H. Goldman, Intern. J. Applied Radiation and Isotopes 23, 353 (1972).
10 J.E. Harris and L. Friedman, Anal. Biochem. 30, 199 (1969).
11 S. Kushinsky and W. Paul, Anal. Biochem. 30, 465 (1969).
12 Norman S. Radin, Anal. Biochem. 55, 637 (1973).
13 M. Drosdowsky and N. Egoroff, Anal. Biochem. 17, 365 (1966).
14 G.J. Litt and H. Carter, in The Current Status of Liquid Scintillation Counting (Ed. E.D. Bransome), Grune and Stratton, New York, 1970, p.156.
15 E.A. Evans, The Radiochemical Centre, Amersham, personal communication.

DISCUSSION

R. Dierckx: Clean vials are very important for accurate counting; polyethylene vials are especially sensitive in this respect. It is enough to take a clean vial in your hand to change the counting efficiency before and after by a few percent. We clean our vials with trichlorethylene. What do you recommend for cleaning the vials?

K. Painter: Cleaning methods depend entirely on the types of samples being counted. There are three publications in the literature which enthusiastically recommend various cleaning devices, but at least two of the papers describe the use of radioactive steroids which tend not to adsorb to glass. The type of cleaning agent to be used varies with the sample being counted.

B.W. Fox: We have found that ultrasonic cleaning provides an excellent means of cleaning vials used for a variety of sample preparation methods. Have you experienced the formation of a noxious gas resembling formaldehyde when using polypropylene vials? We suspect that it occurs when the Mahin and Lofberg peroxide–perchloric acid procedure is used in these vials. The result of this gaseous emission can be corrosion of the belts and damage to the light collecting system.

K. Painter: Polypropylene is not used to any extent in the States; I have not observed this phenomenon. All digestions, whether acidic, basic or oxidative in nature, should be carried out in glass vials.

D.E. Bowyer: With respect to trying to wash vials, it is obviously important to remember that the washing-up solution must be chosen so as to remove the chemical entity absorbed and does not relate to the isotope used as label.

P. Johnson: You made the point early in your talk that one advantage of the narrow mini-vial over larger vials was that the photomultiplier tube faces in the counter could be closer together, thus reducing light scatter. Therefore, what is the point of using a mini-vial inserted within a larger vial?

K. Painter: The point is that, with the reduced diameter of the mini-vial, photons have a shorter path length within the sample, which is advantageous in a high-quenching situation. Fewer photons will be lost due to quenching (not scattering) if the path length through the quenching medium is reduced.

P. Johnson: Many of us with biomedical applications have enough trouble dissolving components in much more than 10 ml and the small-volume vials would therefore be disadvantageous. I am not sure that you have really brought out the advantages of the mini-vials — can you re-iterate them?

K. Painter: There is one big advantage — cost. Lower volumes of scintillator mean lower costs. Counting efficiencies are higher and backgrounds are significantly lower, resulting in superior figures of merit. The table shown previously which intercompares the various counting vials demonstrates this point.

M. Ergin: It seems that standard glass vials are most suitable for Carbon-14 counting. In counting natural Carbon-14 activity would you like to comment on the contribution of the vials to the background activity, due to Potassium-40, etc? Does it change from vial to vial?

K. Painter: In our experience the greatest variation in background count rate with glass vials is from different lots, either from the same supplier or from different suppliers. I would not recommend glass vials for counting natural Carbon-14 activity, but rather teflon, nylon or polyethylene in that order.

SHORT COMMUNICATION

Colour Quenching in Liquid Scintillation Coincidence Counters

F.E.L. ten Haaf

N.V. Philips' Gloeilampenfabrieken, Nuclear Applications Laboratory, Eindhoven, The Netherlands

I should like to point out an interesting aspect of the quenching model I put forward two years ago.[1] Briefly, its essential features were as follows. In this highly simplified model the three-dimensional sample is replaced by a one-dimensional one, i.e. by the line joining the photocathodes of the photomultiplier tubes (see Fig. 1). In the absence of optical quenching the pulse height at the output of each

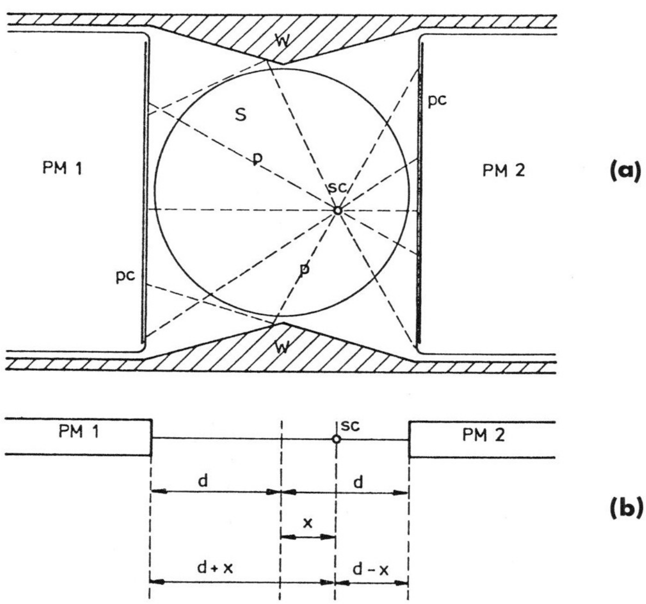

Fig. 1. (a) Schematic cross-section through liquid scintillation detector chamber with photomultipliers (PM) and their photocathodes (pc), reflective chamber walls (W), sample (S), arbitary scintillation event (sc) and photon paths (p). (b) Simplified one-dimensional representation as discussed in text.

photomultiplier, as a result of an arbitrary scintillation event (sc), is represented by:

$$h_1(E) = h_2(E) = \tfrac{1}{2}\, qcE \qquad (1)$$

where E is the energy of the β-particle, c is the conversion factor for an unquenched sample and q represents the pulse height reduction factor due to chemical quenching.

In the presence of colour quenching there is an additional attenuation of the pulse due to light absorption, which depends on the photon path length. Consequently, the pulse heights at the output of the photomultipliers will now generally be different. If the one-dimensional 'sample' has a 'diameter' 2d, the distance of the scintillation event to the centre of the 'sample' is x, and the optical attenuation constant is μ, we find:

$$h_1(E,x) = \tfrac{1}{2} \exp\left[-\mu(d+x)\right]\left[qcE\right] \qquad (1a)$$

and:

$$h_2(E,x) = \tfrac{1}{2} \exp\left[-\mu(d-x)\right]\left[qcE\right] \qquad (1b)$$

In coincidence counters these two pulses are usually added linearly in a pulse summation circuit prior to analysis. The pulse height at the output of this circuit is then given by:

$$h_s(E,x) = h_1 + h_2 = \frac{\exp\left[-\mu(d+x)\right] + \exp\left[-\mu(d-x)\right]}{2}\left[qcE\right]$$

This can also be written as:

$$h_s(E,x) = \cosh(\mu x)\left[\exp(-\mu d)qcE\right] \qquad (2)$$

Equation (2) shows that, unless $\mu = 0$, the pulse height for a given energy E depends on the position x of the scintillation event in the sample. From this we may conclude that chemically quenched samples and optically quenched samples must produce differently shaped pulse height spectra in coincidence counters provided with linear summation.

However — and now I come to the point I want to make — if a logarithmic summation is applied we obtain a different picture:

Taking the logarithms of expressions (1a) and (1b) and adding yields the following answer:

$$\begin{array}{rl} & \log h_1(E,x) = -\mu d - \mu x + \log(\tfrac{1}{2} qcE) \\ + & \log h_2(E,x) = -\mu d + \mu x + \log(\tfrac{1}{2} qcE) \\ \hline & \log h_1 + \log h_2 = -2\mu d + 2\log(\tfrac{1}{2} qcE) \end{array} \qquad (3)$$

We see that the term containing x is cancelled out.

If the model were sufficiently representative of the actual conditions in a liquid scintillation counter, it would mean that, by summing the logarithms of the signals from both photomultipliers, we would obtain an instrument with a similar response to both chemically and optically quenched samples.

Unfortunately the model is too primitive for this purpose, but we may hope that considerations of this nature will prove useful when a more sophisticated model becomes available.

The objection has been made that logarithmic addition is, in fact, equivalent to a multiplication of the photomultiplier signals and that this is very far from producing a useful signal for analysis. I am inclined to disagree with this and prefer to look upon logarithmic summation as a way of obtaining the geometric mean of the individual photomultiplier pulses as opposed to the arithmetic mean that is obtained by linear summation.

Equation (3) can be transformed as follows:

$$\begin{aligned} \tfrac{1}{2}(\log h_1 + \log h_2) &= -\mu d + \log(\tfrac{1}{2}qcE) \\ \log(h_1 h_2)^{\frac{1}{2}} &= -\mu d + \log(\tfrac{1}{2}qcE) \\ \sqrt{h_1 h_2} &= \exp(-\mu d)\,(\tfrac{1}{2}qcE) \end{aligned} \quad (4)$$

This indicates that, according to the model, the geometric mean of the individual photomultiplier pulses is proportional to E and independent of the position of the scintillation event in the sample.

The similarity in the response of the model to chemical and optical quenching, if logarithmic summation is applied, is even more striking if we replace q by the approximation:

$$q = \exp(-k\gamma)$$

where γ is the chemical quencher concentration and k is a constant of proportionality.

Equation (4) then becomes:

$$\sqrt{h_1 h_2} = \tfrac{1}{2}\exp(-\mu d - k\gamma)\,[cE]$$

REFERENCE

1 F.E.L. ten Haaf, in Liquid Scintillation Counting (Eds. M.A. Crook, P. Johnson and B. Scales), Heyden, London, 1972, Volume 2, p.39.

SECTION II

QUENCH CORRECTION TECHNIQUES

Chapter 5

Standardisation in Liquid Scintillation Counting

Elizabeth Bush Mueller

706-208 S.W. 16th Avenue, Gainesville, Florida 32601, U.S.A.

The evolution of the commercial liquid scintillation counter can be seen as a response to some of the needs of its users as perceived by the manufacturers: for greater sensitivity of detection, especially of low-energy emitters; for greater precision, with long-term stability and reliability; for greater sample handling capacity, with versatility in accepting different kinds of samples; and for reduction in user time, attention and expertise required. As with many laboratory instruments this last need, for 'automation', came with the transformation from a specialist's research tool into a routine analytical instrument processing constant streams of samples. In addition to straightforward data-processing functions, capabilities for data-interpretation and decision-making have been incorporated into many models. The ideal instrument is often conceived of as a black box into which a sample in the required form can be inserted and from which the desired answer will then emerge without further manipulation by the operator. Such a box has a single control switch with two positions, 'on' and 'off', and no operating manual.

Efforts to substitute counter automation for user education have been accompanied, perversely, by an awesome proliferation of external counter controls. Each must be correctly set by the operator if the output data are to have meaning. Primarily this has resulted from the simultaneous need to satisfy the other goals, providing versatility while optimising the sensitivity, precision and counting time for each sample. In some cases one set of user choices has apparently been supplanted by another set which is presumed by the designers to be easier to comprehend. It is seriously questionable as to what extent this has succeeded in reducing the expertise required to extract accurate information from these instruments.

Conspicuous among the parameters which are being removed from direct operator control are the individual discriminator and gain settings. One is no longer forced to visualise the pulse height spectrum and its relationship to these parameters in order to select the counting channels. I want to present some of the reasons why I believe that a user — at least one who has occasion to count more than one type of sample — cannot afford to be unaware of the pulse height spectrum, or to be ignorant of the basic processes of liquid scintillation counting. There are some problems with which the instrument is not yet equipped to cope. These are problems in the determination of counting efficiency, and they are at the same time problems of sample preparation.

I daresay that we might have achieved the goal of the black box if we had been willing to abandon the capability of counting many different forms of sample. I do not think this would have been a good trade-off, in the present state of sample preparation technology, because it would require too great a sacrifice either in sample size or counting efficiency or in convenience of preparation for many kinds of sample materials. But the alternative to limiting ourselves to one form of sample is to be able to provide an appropriate standard for each different sample form. We must also be able to recognise those forms of sample which are not acceptable because it is simply not possible to determine their counting efficiencies by any other method.

The meaning of sample 'form' as used here should become more explicit shortly. It may be defined as that collection of characteristics which must be reproduced in the standard with which the sample is to be compared. Some sample properties are not relevant to this comparison and they do not determine 'form'. Unfortunately, it is apparent that many users lack an adequate understanding of the required relationship between the samples and the standards.

In order to identify the determinants of form we may start with a simple expression for the magnitude of the electrical pulse produced in the counter by an event in the sample. This will be written for beta particles and their pulse height spectra explicitly, because the liquid scintillation counter is most often used to count betas. The implications of the following discussion for the counting of other high-energy radiations should be readily apparent, however.

$$\text{Pulse height} = \underset{1}{E_\beta}\ \underset{2}{f_E}\ \underset{3}{e_E}\ \underset{4}{f_p}\ \underset{5}{e_p}\ \underset{6}{f_{pe}}\ \underset{7}{A_{PM}}\ \underset{8}{A_{amp}}$$

1 E_β = energy with which β is emitted,

2 f_E = fraction of β-energy deposited in scintillator solution,

3 e_E = efficiency of conversion of energy $E_\beta f_E$ to photons of energies to which the photocathodes are sensitive (photons/keV),

4 f_p = fraction of photons collected at the photocathodes,

5 e_p = efficiency of conversion of photons to photoelectrons (pe/p),

6 f_{pe} = fraction of photoelectrons collected at the first dynode,

7 A_{PM} = gain of the dynode string,

8 A_{amp} = gain of the remainder of the circuit before pulse height analysis.

For simplicity in the following discussion it will be assumed that the counter has two identical photomultiplier tubes and employs linear amplification.

It may be noted that the expression does not contain a separate term for each energy-conversion step in the liquid scintillation counting process. Since it will be convenient to consider certain of these steps together, the corresponding variables have been combined, giving the seven terms labelled 2 through 8. Each of these terms may therefore be the product of several variables.

The expression is not linear in beta energy. Thus the shape of the pulse height spectrum is not the same as that of the beta spectrum. One source of this difference is the uncertainty associated with each of the variables in the pulse height equation: while an average value may be assigned to each variable, for any particular sample and counter, that value is just the mean of a distribution of values which the variable takes for successive beta disintegrations. The final result is a distribution of pulse heights for any given beta energy, and a total spectrum which is the sum of these distributions. The relative breadth and the asymmetry of the pulse height distribution increase with a decreasing mean number of events (or participating quanta) per step of the liquid scintillation process. These statistical factors increasingly modify the shape of the spectrum as the beta energy decreases or the sample quenching increases, yielding fewer photoelectrons per disintegration.

Another source of discrepancy between the shapes of the beta- and the pulse height-spectra is the energy-dependence of Factors 2, 3 and 4 of the equation. Since each has its own form of energy-dependence, each will affect the shape differently: that is to say, different kinds of quenching produce different spectral shapes. Nevertheless, the pulse height spectrum should change in a reproducible way with any single variable affecting pulse height. That fraction of the spectrum falling within the fixed range of pulse heights selected by any counting channel should vary reproducibly with any other such fraction, as long as only one variable is allowed to change. Any portion of the pulse height spectrum should also vary reproducibly with the fraction of the total number of disintegrations which is not counted. This is the basis for the channels ratio method of standardisation. It is also a necessary condition for the method which measures the net counting rate of some portion of an external standard spectrum in order to determine what percentage of the sample disintegrations has been observed.

If two factors which affect the shape of the pulse height spectrum in different ways are allowed to vary independently, no unique relationship will be observed between chosen portions of a spectrum. The limits placed upon the form of the samples and their standards must be those which ensure that spectral shapes vary identically with efficiency.

Let us look in more detail at some of the factors in the equation in order to see how they affect the spectral shape.

FACTOR 2

The fraction of the beta energy deposited in the scintillator will be reduced if any of the following occurs: escape of electrons at the periphery of the sample, energy loss in electron backscattering from the vial wall, electron energy loss due to self-absorption in undissolved sample material, or absorption losses in any other component of a second, non-scintillating phase which might be present in the sample. The wall losses should be negligible (less than 1%) in an homogeneous solution, if the beta energy does not exceed that of Carbon-14. This is not true for Phosphorus-32, however, whose maximum beta range is roughly two-thirds the radius of the conventional sample vial. Wall losses will be very significant if the labelled sample compound is adsorbed onto the wall or is precipitated onto the bottom of the vial. Counting geometry is then reduced from 4π to 2π. (In this case a reduction in Factor 4 of the equation also results, since light emitted close to the walls or bottom of the vial is less likely to be detected.)

Absorption losses may also be expected in a sample precipitate. If the loss is not severe, as may be the case for a high-energy emitter in a precipitate which is well penetrated by scintillator solution, any efficiency loss may be compensated to some extent by other effects of the precipitation, such as reduction of quenching by removal of the precipitated compound from solution, or perhaps a slight enhancement of light output due to light-scattering by the precipitate, with a consequent reduction of light-trapping in the vial. The resulting pulse height spectrum, however, cannot be expected to have the shape of one produced by a dissolved sample material. Beta energy losses in backscattering and absorption are not linear with energy, and their effects on the shape of the pulse height spectrum are not the same as those of quenching of the scintillator solution.

If the labelled compound is partly in solution and partly in another phase, its pulse height spectrum will be a composite of that which is characteristic of the solution and that which has been degraded in energy. As an exercise one might try to treat this in the manner of a doubly labelled sample with superimposed spectra, and solve for the activity in each phase. A standard for each phase would be required. If solution efficiency were allowed to vary, the spectra would be required to vary in a fixed relationship as a function of quenching only, with the efficiency of the undissolved phase a known function of the efficiency of the solution. This would be possible only if the spectrum of the undissolved phase did not vary simultaneously with any other parameter, such as sample thickness. Alternatively, if solution quenching were held constant, the thickness of the undissolved phase could be allowed to vary. Efficiencies might be determined from the external standard ratio, in the first case, and from the sample channels ratio, in the second case (in which the solution efficiency would be known and constant). If, however, the solution efficiency and the value of Factor 2 vary independently, in general the overall counting efficiency cannot be determined from a combination of these ratios.

The requirement for two distinct standards, each of which is known to be present in one phase but not in the other, might be met. But the other requirements severely restrict the practical application of such a procedure. Ordinarily, when a second phase forms in a sample, neither the solution quenching nor the distribution of the solid phase is well enough controlled.

The prospects for accurate counting of a single undissolved species with negligible solubility in the scintillator solution are much better. One frequently used technique involves the precipitation onto a filter of a very small mass of sample. Usually the counting efficiency is determined by preparing a similar sample of a compound of known activity. Since most of the compounds which one wishes to measure will not be available in standardised preparations one is usually forced to compare one's unknowns with a standard sample prepared by precipitation of a different compound. There are certain obvious hazards in this method, such as the determination of chemical yields, problems of adsorption losses in manipulating very dilute solutions, and the possibility of partial leaching of either the sample or standard into the scintillator solution. But in addition there are less obvious problems, such as those which have been explored by Gill,[1] by Furlong,[2] and by Bransome and Grower.[3] When using absorptive filters, such as paper, they found different counting efficiencies for molecules of different sizes, suggesting that these penetrate to different depths in the filter, and that the escaping betas suffer different absorptive energy losses. Gill found a significantly different ratio in the counting rates of large and small molecules according to whether they were precipitated onto paper filters or glass fibre filters. He advanced the very reasonable explanation that the large molecules stay on the surface in either case, but may suffer from self-absorption losses, while the small molecules show higher efficiencies when on the surface of the non-absorptive glass filter, but suffer from absorptive losses when embedded in the paper. Meyer and Keller[4] also found different efficiencies on filter discs for the two compounds they wished to compare in their experiment: the precursor nucleoside triphosphate and the product DNA. Both Hydrogen-3 and Carbon-14 were used in all these experiments.

One straightforward way of resolving this problem is to determine the activity of one sample of each species by an independent method in a preliminary experiment. Since samples of this kind often weigh a few milligrammes at most, it should usually be possible to get them into a true solution either by digestion in a basic solubiliser or by combustion, and to standardise using the usual solution standards. One may then proceed to establish the counting efficiency for each species on the kind of filter one wishes to use. One may also establish the way in which that efficiency varies with sample weight over the range of weights to be anticipated.[4] It is often possible to derive a sample channels ratio curve to measure efficiency as sample thickness varies (for samples of less than infinite thickness). Thickness should be uniform across the sample support, and all variables other than sample weight should be held constant.

Partial leaching of the sample into the scintillator would result in a faulty method, as already discussed. It is easy enough to determine whether this is occurring by removing the filter from the vial and recounting the solution. One would normally expect that the scintillator used would consist only of toluene and fluor, since this provides high efficiency and minimum solubility for most of the compounds it is necessary to count in the solid phase. (If they are toluene-soluble to any appreciable extent they are probably best counted in solution.) Yet it is surprising to find reports in the literature of filters placed in dioxane or surfactant solutions for counting, with little or no evidence as to whether the samples have been eluted. It is not safe to assume that samples can be completely removed from filters in this way, and careful experimentation should be carried out to establish it. This is illustrated by some work of de Bersaques,[5] who found incomplete, extremely slow, or non-reproducible elution from paper in several scintillator solutions. Similar problems are encountered when thin layer chromatogram media are immersed in scintillator solution.

Finally, there is another variable associated with counting on filters, one which properly belongs with the discussion of the fourth factor in the pulse height equation. The efficiency with which the scintillations are detected will depend upon the location of the filter in the vial, and upon its opacity. The seriousness of these problems with filters is, of course, a function of particle energy. When counting high-energy betas, such as those of Iodine-131 and Phosphorus-32, efficiency has been reported to be quite insensitive to optical or self-absorption effects.[6,7] This undoubtedly depends upon the choice of a suitably wide counting channel, in which efficiency may remain quite constant despite small pulse height shifts.

The foregoing discussion has been concerned with small solid samples, but larger weights — up to one or more grammes — can also be successfully measured under properly controlled conditions. The best efficiencies are usually achieved by dispersing the solid particles in a gel scintillator. Fortunately, in this case it appears that reproducible counting efficiencies can be expected if the average particle size and the size distribution are held within the limits obtainable by normal chemical techniques for precipitation. This permits standardisation by comparison with similar samples of compounds of known activities. Such standard compounds must be like the unknowns in physical form (particle size and optical transparency) and in beta absorption coefficient. Counting efficiency varies with the total mass of the suspended sample material, due in part to variations in Factor 4 of the equation, such as absorption and scattering of light by the sample particulate. A common problem in obtaining reproducible efficiencies is that of controlling sample geometry, or spatial distribution. Too thin a gel permits settling out of the sample, and excessive viscosity makes uniform dispersion difficult. If reproducible geometry can be achieved, the sample channels ratio can often be used to correct for changes in efficiency due to varying self-absorption and transparency. The standard set is prepared using a compound of like physical form, varying only that property which is allowed to vary in the samples (usually weight).

The most popular of all types of two-phase samples is undoubtedly the emulsion. It is popular because there are available some very efficient and inexpensive emulsifying agents which can incorporate relatively large amounts of polar compounds into the best alkylbenzene solvents. Because the technique is so versatile, and because it often provides a high figure of merit (the product of the sample weight and the counting efficiency), it is extraordinarily tempting. A number of detailed studies have been done, especially on Triton surfactants in toluene or xylene, to find compositions suitable for the incorporation of various sample solutions. For the Triton systems and for a number of others there are at least three identifiable kinds of mixtures one can produce by varying the relative proportions of surfactant, alkylbenzene, and aqueous solution: one type appears to have the properties of a true solution, being fluid and clear; one is quite apparently an emulsion, being fluid and hazy or cloudy; and one is a gel, viscous and clear-to-hazy. This latter may be used to suspend solids, as previously mentioned. The following discussion will concern the counting of aqueous and other polar liquids.

We may eliminate from consideration any formulation whose properties are not stable with time and with the changes in temperature the sample must undergo before counting. This probably eliminates most of the very cloudy fluid emulsions, as well as those gels which have a tendency to settle slowly, showing separation of a liquid after a day or so. Accurate assay of unstable samples cannot be expected because the radioactive sample may be undergoing a change of phase and/or the solution efficiency may be changing with time.

Assuming a stable composition, how are we to determine counting efficiency in these preparations? Which ones are true solutions, and which ones must be treated as heterogeneous? The literature of liquid scintillation counting is marked by considerable confusion in the terminology it applies to these mixtures, and little information as to their actual molecular arrangements. The lack of such information, however, is not an obstacle to correct standardisation. Only a knowledge of those properties which can be measured by liquid scintillation counting is necessary. Visual examination of samples can supply clues which are valuable in the prediction or interpretation of counting rates.

The following <u>operational</u> definition of a true solution is suggested as appropriate and useful for classifying samples in liquid scintillation counting: <u>a solution is a mixture of components in which any one of those components is counted with the same efficiency, if tagged with a given radionuclide</u>. Thus, if a mixture containing Triton, toluene and water exhibits the same counting efficiency for tritiated water as for tritiated toluene, it behaves as a solution. It can then legitimately be compared with solution standards, providing, of course, the samples and standards have the same values for Factors 3 and 4 of the pulse height expression. By this definition, all true solutions have the same value for Factor 2 (except for any difference caused by varying the surface-to-volume ratio, which is trivial except for high-energy emitters). This criterion can be applied to both the fluid and the gel samples. It can also be applied to the cocktails employing basic solubilisers, such as Hyamine hydroxide, NCS and Soluene. The microstructures of these mixtures are not known, but they are irrelevant to counting if the mixtures behave as true solutions according to the definition.

Standardisation is much simpler for true solutions than for other kinds of samples. The solution is to be preferred unless there is an important advantage to be gained in convenience or figure of merit by using a heterogeneous sample. The heterogeneous preparations of surfactants may be expected to present the same problems as do the partially precipitated samples already discussed: if the radioactive material is distributed in an unknown way between the phases, and if both the self-absorption and quenching factors are allowed to vary, the counting efficiency is not easily determined. If it is necessary to use such an emulsion it is most prudently limited to the counting of sample material containing a single labelled compound. A standard of this same compound should be available for efficiency determination by preparing duplicate standard samples, or by internal standardisation with care that the phase distribution of the sample is not altered by the addition of standard.

If solution behaviour can be established for a preparation, by comparing its counting efficiencies for appropriate standardised lipophilic and hydrophilic compounds, a channels ratio method may be used for counting samples. For reasons to be discussed in connection with Factors 3 and 4, it is best to prepare a special standards set, using the same surfactant system quenched with unlabelled material like the sample material to be assayed. Experience in the author's laboratory with a variety of surfactant systems, including proprietary preparations sold exclusively for liquid scintillation counting, indicates that both the gelled and the fluid compositions are more likely to behave as solutions if optically clear. Unfortunately, this is not an infallible criterion, nor is it well-defined: there is often a gradual transition from clarity to haziness and translucence as composition is varied. The user is advised to test any composition range he expects to use by counting appropriate internal standards.

To sum up the methods of determining counting efficiency for non-homogeneous samples, it is necessary, first of all, to have available a compound of known activity which is very similar to the unknown. The similarity should be in physical characteristics, if the sample is insoluble, and in chemical characteristics, if the sample is soluble in some non-scintillating phase. Comparison with the standard can then be carried out by one of three methods: preparing a duplicate standard sample, internal standardisation (for soluble emulsion samples), or use of a sample channels ratio curve, if one pertinent parameter can be varied while others are held constant.

As previously discussed, there are many non-homogeneous samples for which the determination of counting efficiency is not feasible. It is necessary to be able to recognise such samples when they occur accidentally. Bush[8] has discussed various methods for detection of phase separation, including a 'double ratio' technique, which requires a set of standard quenched samples known to be homogeneous. An interesting extension of this technique has been described by Glass,[9] who routinely counted all samples in this fashion. By monitoring both the sample channels ratio and the external standard ratio, he was in effect providing an independent check on the expected output of each of the three counting channels. He was thus able to detect discrepancies from various sources, such as chemiluminescence, contamination by another nuclide, and counter drift or other malfunctioning, in addition to revealing phase separation in samples. In his laboratory, liquid scintillation counting data were reduced by computer. This represents an imaginative use of the capability inherent in automatic data processing to overcome the threat that it presents in alienating the experimenter from his data. I anticipate that we shall see an expansion in the use of software to perfect the black box counter.

The need to monitor and stabilise instrumental functioning has been met by the manufacturers with improved hardware. At least one counter automatically monitors itself for changes in gain and makes any adjustment necessary to return itself to a calibrated state, so that channel efficiencies and ratios should be stable indefinitely. A start has been made towards monitoring the sample as well: I believe there are at least two counters with devices designed to recognise and discriminate against chemiluminescence or other processes which produce single photons at a rate high enough to cause significant accidental coincidence counts. We may hope that future automation will include ways of utilising that information in the signal which reveals how counting efficiency is related to sample form.

FACTOR 4

Among the variables that determine what fraction of emitted photons reaches the photocathode are absorption in and total internal reflection at the vial wall, both varying with vial material. Reflection losses are greatest at the external surface of a vial with a clear, smooth wall, like the glass vial, and are decreased by roughening the outside wall or by using a diffusing wall material such as polyethylene. Light trapping may also be broken up by a scattering phase within the sample. The increases in absolute counting efficiency achieved by these means are usually only a few percent. The pulse height spectra of scattering samples have slightly different shapes than those of clear samples, and produce slightly different channels ratio curves.

The position of the photon source in the vial affects the probability for escape of the photon, and for its chance of ultimately reaching the phototube face at a favourable angle. Thus the value of Factor 4 depends upon the scintillator volume and the location of the radioactive source in that volume (whether distributed throughout or on a surface somewhere within it). The internal dimensions of plastic and glass vials are sufficiently different to produce observable efficiency changes due to this geometry factor. The pulse height spectrum of the external standard is even more dependent on scintillator volume, which determines the probability for interaction in the scintillator of a gamma ray, as a function of gamma energy. For all these reasons, the same kind of vial should be used for the standard samples as for the unknowns. If polyethylene vials are used, the standards should be prepared at the same time as the unknowns, since the vials will swell and the samples change in efficiency as time passes. Samples containing light-scattering materials, such as Cab-O-Sil or other gelling agents, should not be compared with clear standards. Sample volume should be held within the limits determined for any given counter to yield a single calibration curve.

Finally, light absorption in a coloured sample not only decreases the number of photons escaping but changes the way in which the shape of the pulse height spectrum varies with efficiency. In 1971 ten Haaf[10] presented a simple model for colour quenching which, while it was only one-dimensional, established a basis for understanding the difference in shape of the spectra of colour-quenched and chemically quenched samples. The colour-quenched spectrum is broader because the dispersion of pulse heights is greater for events of any given energy. In addition, according to this model, coincidence losses vary with energy in different ways for the two types of quenching.

Since the first study of the channels ratio method was published in 1960[11] we have known that different calibration curves were required for coloured and colourless samples. Modern counters provide considerably better spectral resolution, and one now finds that for precise work different curves are needed for different colours, and for various admixtures of chemical and colour quenching. Most coloured samples are also chemically quenched to some degree.

When the two quenching factors vary independently in a set of samples, no single channels ratio curve will fit all the points. Fortunately, in many cases one type of quenching predominates, so that reasonable accuracy can be obtained with a single set of standards. These should be prepared so as to duplicate the samples as nearly as possible. For example, in our laboratory we found it necessary in assaying muscle tissue in various solvent systems to prepare a separate set of standards for each system, using not only the same solvent and the same kind of muscle from an unlabelled animal, but a piece of muscle which had been washed until it had approximately the same colour as the sample to be assayed.

A recent innovation called 'colour restoration' is now available on one commercial counter. Its function is to obviate this problem in choosing standards. For the calculation of sample channels ratios, pulse height analysis is performed on a spectrum consisting of the smaller of the two pulses produced in coincidence by the two tubes, rather than on the spectrum of summed pulses. The shape of this 'lesser pulse height spectrum' was found to be less sensitive to the kind of quenching than was that of the summed pulse height spectrum. Very few results with this system have yet been published. While both the manufacturer's literature[12] and an independent report[13] indicate that the 'lesser spectra' are not identical for chemically and colour-quenched samples, a very substantial reduction is reported in the discrepancy between the channels ratio curves for the two types of samples. For tritium samples, which have relatively short quenching ranges and thus lower maximum colour intensities, the discrepancy between the two curves may be within normal curve-fitting errors over the usable range of efficiencies.

FACTORS 6 AND 7

These variables have been factored out explicitly in writing the expression for pulse height because they are employed by different manufacturers in their systems of instrumental response to quenching. As the average pulse height decreases with quenching, it is highly desirable to 'follow' the spectrum with the channels in which one is measuring it, so as to maintain optimal discrimination between spectral regions. Either the channel discriminator settings or gain can be varied to keep the spectrum in a channel. These procedures ensure that all pulses corresponding to a chosen beta energy range will be counted, but they cannot maintain a constant efficiency. Total efficiency drops with quenching because of losses below the threshold for detection, while efficiency in any given range of the spectrum may change with quenching because of change in the spectral shape. The spectrum of a quenched sample cannot be made identical with that of an unquenched sample by increasing the gain. For pulses produced by small numbers of photoelectrons (such as those typical of Hydrogen-3 and Carbon-14 counting) a change in the mean number of photoelectrons per pulse results in a significantly different pulse height distribution than does an equivalent change in mean pulse height produced by Factors 7 or 8.

Some counter manufacturers have provided for automatic monitoring of the pulse height and for appropriate adjustment of some channel parameter(s) to follow the desired portion of the spectrum as it is shifted by quenching. Monitoring is performed by making a measurement of an external standard ratio for each sample as it arrives at the counting station. The measurement is always taken at the same gain and in fixed channels. The observed value of the ratio determines the adjusted channel settings at which the final sample counting will be done. The efficiency under these counting conditions must be determined from some empirical relationship which has been established by counting a quenched standard set.

In one type of counter the system gain is adjusted by changing the high voltage applied to the photomultiplier dynode string (Factor 7).[14] The voltage on the first dynode remains constant. Discriminator levels remain fixed. The efficiency is taken from a calibration curve of the external standard ratio measured at a constant gain versus efficiency at the adjusted gain.

In another type of counter the discriminator levels are varied, but not as a continuous function of the external standard ratio. Ten discrete settings are provided, and over the narrow range of quenching for which the spectrum falls into one of these windows an attempt is made to keep the spectral shape constant. This is done by varying Factor 6, through application of a magnetic field external to the photomultiplier tube. The change in the statistical distribution of electron numbers at the first dynode caused by a small decrease in Factor 6 closely resembles that which results from a reduction in the number of photons arriving at the photocathode. The efficiency at each of the ten settings is determined by counting standards.[15]

FACTOR 3

I should like to return now to Factor 3 of the pulse height expression. This is the one whose value is affected by chemical quenching and fluor self-quenching. In the past we have not had to concern ourselves much with the particular kind of Factor 3 quencher when using the sample channels ratio, because they were all observed to produce the same kind of change in the sample pulse height spectrum. With the advent of external standardisation the chemical nature of the sample became important because the electron density of the sample determines the gamma absorption cross section. This not only determines the counting rate for the external standard, but may also affect the shape of the electron spectrum produced by the gammas. The cross section for gamma interaction affects the geometric distribution of electrons produced within the sample volume, and this will affect the pulse height spectrum. For these reasons, more stringent limitations on composition and volume must be observed when comparing samples and standards by external standardisation.

Progress, however, has not stopped here. We have arrived at the point where we may now have to consider the effect of sample chemistry on the internal sample pulse height spectrum. I believe that this is the result of improved photocathode efficiency, and the resulting improvement in spectral resolution. Earlier such improvements had similar consequences in the counting of coloured samples: in the early 1960's we were not able to distinguish a chemically quenched from a colour-quenched spectrum for tritium samples, principally because for those small pulses the statistical broadening at the photocathode obscured such differences. Now that our tritium efficiency is about twice as great, we can see those differences in the tritium pulse height spectra, just as we could then in the Carbon-14 spectra.

Five or six years ago in our laboratory we began to notice unexpected differences between toluene and dioxane scintillators. The apparent activity of a tritiated water sample, as determined by a sample channels ratio method, depended upon whether it was counted in dioxane or in toluene—ethanol. The same difference was observed for a sample of tritiated toluene. In each case the comparison was made with a sealed standard set containing toluene quenched with acetone. The most quenched of these standards contained a rather large volume percent of acetone. Finally, by carefully preparing complete quenched sets of the three types of solutions we discovered that we had three slightly different channels ratio curves. This posed a real and unexpected problem in standardisation, as well as a challenge to our curiosity.

In seeking an explanation, we first noticed that the indices of refraction of dioxane and acetone were lower than that of the glass vial, while the toluene index was higher. Small differences in light-trapping within the various samples seemed possible. We already had a simple model for colour quenching which qualitatively explained the shape of the pulse height spectrum in terms of dependence of pulse height on the position of an event in the vial, although it was not developed as quantitatively as ten Haaf's model. It seemed reasonable that light-trapping differences might affect the shape of the pulse height spectrum in an analogous way. We therefore investigated a number of solvents with different refractive indices. We also put solvents which gave different channels ratio curves into vials in which light-trapping was reduced, such as polyethylene, nylon and frosted glass vials. We found similar differences between the solvents in all vials. We were not able to correlate these differences with the refractive indices of the solvents.

In the original comparison of toluene and dioxane solutions we had observed deviations of up to about 4% in apparent d.p.m. as determined by the sample channels ratio: i.e. if the toluene samples were used as the standards, an efficiency taken from their channels ratio curve would lead to a calculated d.p.m. for a dioxane sample which differed by as much as 4% from the true d.p.m. In expanding our range of solvents, we found considerably larger deviations. We also found a dependence on quenching agent in some cases. Figure 1 shows some typical deviations. These are plotted horizontally, and represent the error in d.p.m. which results from using a sample channels ratio curve of a particular set of standards to determine efficiencies of different kinds of samples. In this case the standards contain butyl-PBD in toluene and are quenched with carbon tetrachloride. The scale shows negative deviations to the right of zero simply because most of the deviations we observed were negative (as are those due to colour), and in plotting graphs we were accustomed to placing the origin at the left. We therefore adopted this convention. The true (internal standard) efficiency of each sample is plotted vertically. Each curve on the graph represents a quenched sample set in which one component is varied to change the counting efficiency while others are held constant. The standards are represented by a vertical line at zero error. The distance of any curve from this axis can also be expressed as the relative efficiency difference at a given channels ratio between the channels ratio curves for the 'standards' and the 'samples'. Such a graph thus shows how various channels ratio curves diverge or otherwise deviate as a function of quenching.

Note that in Fig. 1 the units of the abscissa are comparable in magnitude with the experimental error. The uncertainty in the internal standard activity was believed to be less than $\pm$ 1% (at the 68% confidence level), as was that of the channels ratio. The error in determining the position of the channels ratio curve is somewhat larger, however. The data were usually reduced by a computer program which fitted a curve to the standards data points. It was necessary to use care in the spacing of these points in order to avoid gyrations in the curve, particularly at the ends. If the lines on the graph zigzag in unison, this is assumed to be due to the shape of the standards curve. When we were in doubt about the position of the standards curve we either prepared new standards, used one of the sample sets as standards in order to get a better spacing of points, or hand-fitted a curve for comparison with the computer-derived curve. When the deviations we were measuring were comparable in magnitude with experimental error we often replicated the samples or repeated the counting using a different channels ratio. The remaining figures present smoothed curves which may be the averages of such composite data; in Fig. 1 actual data points are shown to indicate their

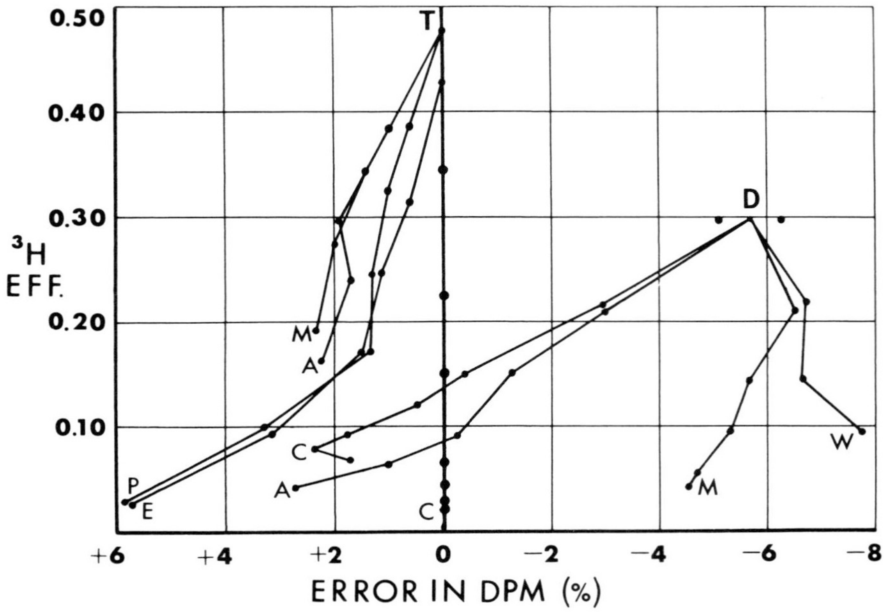

Fig. 1. Relative error due to deviations in channels ratio curves for ^3H samples as a function of quenching agent and of solvent
Standards: Butyl-PBD in toluene + CCl_4, 0–2.8% by vol.
Samples: (T) Toluene + butyl-PBD 6.1 g/litre, (M) CH_3OH to 40%, (A) CH_3COOH to 30%, (E) $(C_2H_5)_2NH$ to 4%, (P) C_6H_5OH to 11.5%, (D) Dioxane + butyl-PBD 25 g/litre, (A) CH_3COOH to 12%, (C) CCl_4 to 1.4%, (M) CH_3OH to 53%, (W) H_2O to 15%.
Quencher concentrations in volume percent. For all figures in this chapter, the fluor concentrations given are those in the final counting samples, and all samples were labelled with 1% toluene-^3H or -^{14}C.

precision. The two points representing unquenched dioxane samples are indicative of the reproducibility for replicate samples. There was a small difference in their channels ratios in this case, although their efficiencies in a wide channel were essentially identical. The values for these two points were averaged and this average was used in drawing the curves.

These data show that the nature of the chemical quenching agent may have quite a significant effect on the shape of the pulse height spectrum. These and a limited number of results with other solvents suggest that such effects are larger in aliphatic than in aromatic solvents. However, the extent to which curves for different quenchers will diverge depends upon factors other than whether the solvent is aromatic or aliphatic, such as the amount of quencher present at a given efficiency. This of course depends on the inherent solvent efficiency and its resistance to quenching as well as on the fluor and its concentration. In any case, one should be careful in choosing a quenching agent for a standards set. For instance, it appears from Fig. 1 that carbon tetrachloride might be a particularly bad choice for many toluene samples, since it deviates from the other quenchers which were examined.

Figure 2 shows another characteristic difference between aromatic and aliphatic solvents. In these samples there is no quencher, but the fluor concentration is

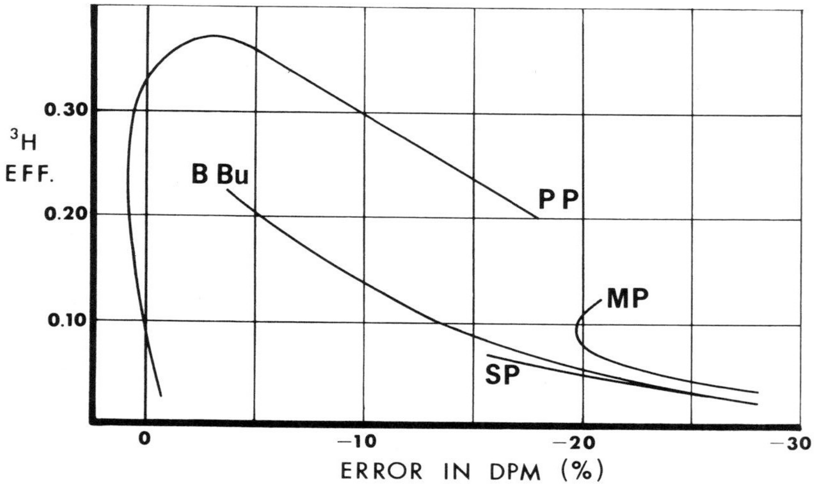

Fig. 2. Relative error due to deviations in channels ratio curves for ^{3}H samples as a function of fluor concentration, and of solvent
Standards: 7 g/litre butyl-PBD in toluene + CCl_4.
Samples: (PP) PPO 0.25–100 g/litre in phenylcyclohexane, (BBu) Butyl-PBD 0.2–5 g/litre in bicyclohexyl, (MP) PPO 1.25–30 g/litre in mineral oil, (SP) PPO 8–20 g/litre in spinacane.

varied. (The particular fluor used is irrelevant to this comparison). The fluor concentration was extended approximately to the limit of solubility in each solvent. Efficiency was still increasing with concentration except in the case of phenylcyclohexane, in which the fluor was much more soluble (to 100 g/litre) and showed marked self-quenching. Three features of the phenylcyclohexane curve were typical of those for the aromatic solvents we examined: at low fluor concentrations there is not much difference between the channels ratios of samples which have low fluor concentrations and those which are quenched with carbon tetrachloride (the standards for this graph); near the concentration of maximum efficiency (about 8 g/litre in this case) ratios begin to differ from those of the standards; the deviation then increases as efficiency drops with increasing fluor concentration. The aliphatic solvents all showed strong deviations from the aromatics at low fluor concentrations. If the fluors can attain high enough concentrations in the aliphatic solvents their curves also show increasing deviations after reaching maximum efficiency: this is beginning to happen in the mineral oil at about 6 g/litre, but at the maximum concentration of 30 g/litre the efficiency has not yet begun to drop. Figure 3 shows a similar curve for dioxane containing butyl-PBD, which reaches a concentration of 100 g/litre. Here the pulse height spectrum does deviate sharply at high fluor concentration, although the fluor suffers only a small loss in efficiency due to self-quenching.

Figure 3 also shows the effects on the channels ratios of varying the concentrations of four different fluors in toluene. A small deviation from the carbon tetrachloride-quenched standards appears at low fluor concentration (low efficiency). A small difference between fluors is just measurable at low concentrations: the long-wavelength emitters are slightly displaced from the short. In trying to verify this we repeated the experiment a number of ways, and I believe that this is a real difference. In dioxane these same four fluors had slightly

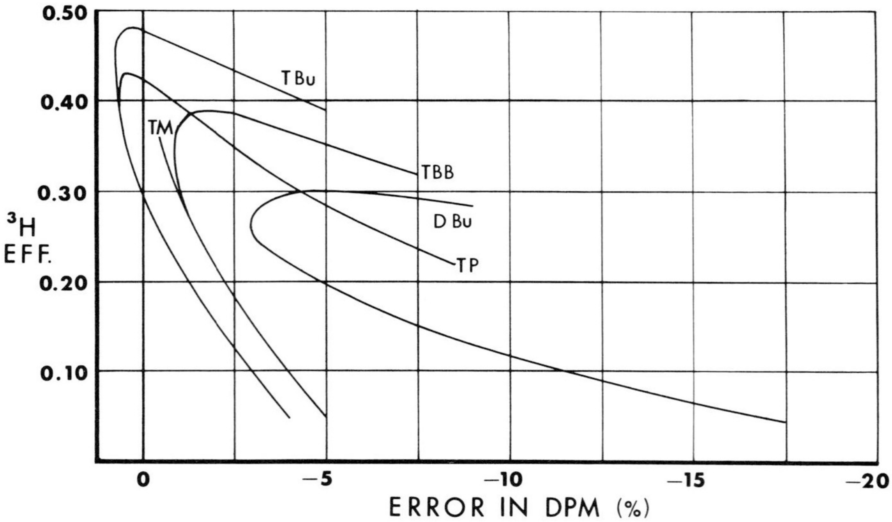

Fig. 3. Relative error due to deviations in channels ratio curves for ^{3}H samples as a function of fluor and its concentration, and of solvent
Standards: Butyl-PBD 7 g/litre in toluene + CCl_4.
Samples: (TBu) Butyl-PBD 0.1–100 g/litre in toluene (maximum efficiency at ~7.5 g/litre), (TP) PPO 0.075–100 g/litre in toluene (maximum efficiency at ~5 g/litre), (TBB) BBOT 0.15–25 g/litre in toluene (maximum efficiency at ~8 g/litre), (TM) Dimethyl-POPOP 0.1–2.5 g/litre in toluene (maximum efficiency at 2.5 g/litre), (DBu) Butyl-PBD 0.3–100 g/litre in dioxane (maximum efficiency at ~25 g/litre), minimum error at ~5 g/litre).

different curves but they were not ordered by wavelength. A very interesting thing happened when the counter was operated in the unsummed mode, i.e. when pulse height analysis was performed on the output of one phototube only while using the output of the other tube simply to operate the coincidence detector. The curves for the long-wavelength fluors were then displaced relative to those for PPO and butyl-PBD, so that they were closer to the axis. A definite separation then appeared in dioxane, and a reversal of position for the two curves at low concentration in toluene.

Pulse summation might be expected to affect the relative difference between the shapes of the pulse height spectra of coloured and colourless samples. Although none of these samples was coloured in the usual sense, there is significant self-absorption of fluor emission at all these concentrations, the amount differing with the fluor. Light absorption losses outside the solution are also wavelength-dependent.

In an analogous experiment with Carbon-14 labelled samples counted in the summed mode, the channels ratio curves of the same four fluors were not resolved. It might have been expected that a greater difference would be observed than with Hydrogen-3, since the counting efficiency was higher for a given fluor concentration, and lower concentrations were reached. (These samples were not counted in the unsummed mode.) Figure 4 shows typical Carbon-14 curves for varying fluor concentration in toluene and in dioxane. Although the maximum fluor concentrations were the same in these samples as in the Hydrogen-3 sets, the effect

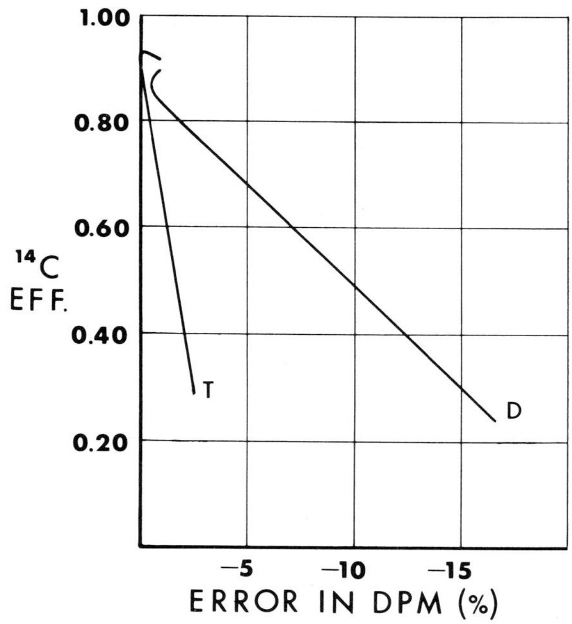

Fig.4. Relative error due to deviations in channels ratio curves for ^{14}C samples as a function of fluor concentration, and of solvent.
Standards: Butyl-PBD 7 g/litre in toluene + CCl_4.
Samples: (T) Butyl-PBD 0.01–100 g/litre in toluene. (D) Butyl-PBD 0.01–25 g/litre in dioxane.

of self-quenching is hardly evident in these curves. Little or no loss of efficiency was measured at high concentration, although the sample channels ratios did show sizeable changes indicative of decreasing pulse heights. This is explained by the fact that efficiency was measured in an integral channel at high gain, and that it changed only very slowly with channels ratio at high efficiency. The flatness of the channels ratio curves would tend to obscure deviations between them. A different choice of channels would be needed to measure differences in the spectra at high concentration.

It is tempting to attribute the changes in shape of the Hydrogen-3 pulse height spectra at high fluor concentration to self-absorption, since these changes are in the same direction as those due to colour. It is also very difficult to explain the large deviations at low fluor concentration in aliphatic solvents on the same basis! Changes in self-absorption might also result from the effects of solvents and quenchers on the emission and absorption spectra of the fluors. Such spectra were examined, and some small shifts were seen. We were unable to correlate these with the counting results on the basis of any simple assumptions. However, our data were certainly not sufficient to prove that differences in solution transparency did not contribute to the observed differences in the shapes of the pulse height spectra.

In an effort to get independent information on the relevance to these phenomena of geometry factors, i.e. the factors responsible for the change in shape of the colour-quenched spectrum, we looked at some different sample configurations. We sealed solutions labelled with tritium and containing varying fluor concentrations into capillary tubes which could be suspended in an upright position inside an ordinary counting vial. The inside diameter of the tubing was 1 mm, and the depth of filling corresponded to a 20 ml filling of the vial. The tubes were then surrounded with 20 ml of various unlabelled solutions. They were counted together with an ordinary quenched standard set of bulk-labelled solutions, and the channels ratio curve of that set was used to calculate their activities. These were compared to the known activities. By this means we hoped to get some measure of self-absorption effects on efficiencies and channels ratios for sources at various locations in the vial.

When the tubes were centred in the vials and surrounded by solutions of their own respective compositions their percent deviations in d.p.m. followed the same order as for the same compositions in ordinary bulk samples, and were close to being numerically the same in many cases. When the tubes were surrounded by pure toluene, approximately the same deviations were observed, although they were slightly smaller in a few cases. The pulse height spectrum of the 1 mm thick source thus had very nearly the same characteristics as the spectrum of a 20 ml bulk source of the same solution.

The efficiency for a tube sample surrounded by its own scintillator was considerably higher than the efficiency for 20 ml of the same solution labelled in bulk. This would be expected when the tube is centred, since light collection from the centre of the vial is relatively good. When tubes containing PPO solutions at concentrations of 0.075–4 g/litre were transferred from vials containing pure toluene to vials containing their own scintillators, respectively, there was very little change in their counting efficiencies. It appears that these 1 mm sources must be considered as 'thick' with respect to self-absorption at their own concentrations. When they were placed in a solution of 100 g/litre PPO their counting rates dropped by 20—45%, and there was a small loss in relative d.p.m. as would be expected for a 'colour' effect.

The tubes were also counted at other positions in the vials, surrounded by a (yellow) coloured solution. The changes in efficiency and calculated d.p.m. were in general agreement with what we expected from our simple model of colour quenching. We concluded from these experiments with capillary tubes that fluor self-absorption along a multiplicity of light paths can cause a shift in the pulse height spectrum like that due to colour, but that this does not adequately explain the differences in the shapes of the pulse height spectra of the unquenched, 'colourless' solutions.

We considered the possibility that the scintillation time decay curves might be different in these solutions, and that variation in the ratio of fast and delayed decay components might lead to variable discrimination between large and small pulses. The expected magnitudes of the fast and delayed components, however, made it seem unlikely that this effect would be observable.

No temperature dependence was observed for the deviations over the range accessible in the counter (-10 to +20°C). This was true for a non-viscous solvent, toluene, as well as for a viscous one, benzyl alcohol, whose viscosity varied significantly over this range. The benzyl alcohol was quenched with a variety of strong quenchers in small volumes, and diluters of low viscosity in large volumes. The pulse height spectra of these tritiated solutions showed only small variations, as was typical of aromatic solvents. In our dioxane samples the observed deviations were not correlated with solution viscosity.

The data presented here were obtained with a Nuclear-Chicago Mark I counter. Similar results were seen when samples were counted in a (higher efficiency) Mark II. The differences in the channels ratio curves for Hydrogen-3 samples were observable with any choice of channels. They could also be seen with a

Barium-133 external standard ratio. At large pulse heights, as for high efficiency Carbon-14 samples, errors in efficiency determination due to differences in curves may be minimised in the same way as for spectral changes from other causes (such as quenching or phase separation): by counting in a channel in which efficiency is nearly constant and not sensitive to spectral shape.

Our investigation was terminated without having produced an explanation of the observed variations in pulse height spectral shape among colourless solutions. We found no clear relation with 'geometry factors'. I am reporting this exploratory work not only because of the practical implications of these phenomena in standardising, particularly as counter resolution improves, but in the hope of stimulating further investigation by others. My present feeling is that a complete quantitative description of the optics of the liquid scintillation counter is long overdue. I hope that Dr. ten Haaf and others are continuing their work in this area. We may not know to what extent these effects are optical or geometric until we get an adequate model.

REFERENCES

1 D.M. Gill, Intern. J. Applied Radiation and Isotopes 18, 393 (1967).

2 N.B. Furlong in The Current Status of Liquid Scintillation Counting (Ed. E.D. Bransome, Jr.), Grune and Stratton, New York, 1970, p.201.

3 E.D. Bransome, Jr. and M.F. Grower, Anal. Biochem. 38, 401 (1970).

4 R.R. Meyer and S.J. Keller, Anal. Biochem. **46**, 332 (1972).

5 J. de Bersaques, Intern. J. Applied Radiation and Isotopes **19**, 166 (1968).

6 E.D. Sebring and N.P. Salzman, Anal. Biochem. 8, 126 (1964).

7 H. Horwitz, J.G. Kereiakes, E.L. Saenger and T. Selkirk, J. Nuc. Med. **4**, 426 (1963).

8 E.T. Bush, Intern. J. Applied Radiation and Isotopes 19, 447 (1968).

9 D.S. Glass, in Organic Scintillators and Liquid Scintillation Counting (Eds. D.L. Horrocks and C.T. Peng), Academic Press, New York, 1971, p.803.

10 F.E.L. ten Haaf, in Liquid Scintillation Counting (Eds. M.A. Crook, P. Johnson and B. Scales), Heyden & Son, London, 1972, Vol.2., p.39.

11 L.A. Baillie, Intern, J. Applied Radiation and Isotopes 8, 1 (1960).

12 B.H. Laney, 'Quench Correction', Searle Analytic Inc., Des Plaines, Ill. 60018.

13 C. Ediss, A.A. Noujaim and L.I. Wiebe, Intern. Symposium on Liquid Scintillation Counting, Sydney, Australia, August, 1973.

14 D.L. Horrocks, Biomedical Technical Report TR–560, Beckman Instruments, Inc., Fullerton, Calif., 1973.

15 Liquid Scintillation Counting Marketing News LSC–33, Packard Instrument Co., Downers Grove, Ill., Feb. 16, 1970.

DISCUSSION

B.W. Fox: A mixture of a hydrocarbon solvent and an aqueous solution will in most cases be two phases. If a surfactant is added, it only disperses these two phases as a colloid or emulsion. Thus if any mixture contains surfactant it must be regarded as heterogeneous and not a true solution. Most commercial mixtures appear to be based on mixtures which are situated between the solvents, detergent margin and the centre and these are fortuitously similar in counting efficiency for tritiated toluene and tritiated water standards. At lower detergent, the toluene efficiency is higher than water, and at low toluene levels, the efficiencies from H_2O– ^3H are higher (although the absolute efficiencies are much lower).

E. Mueller: It was to avoid such a controversy that I offered an 'operational' definition of a solution — if the efficiences of ^3H-toluene and $^3H-H_2O$ are the same, the mixture behaves as a solution from the point of view of counting.

ACKNOWLEDGEMENT

The author's experimental work was done in the laboratories of Nuclear-Chicago Corporation (now Searle Analytic Incorporated) and Amersham-Searle Corporation. Significant contributions were made by D. L. Hansen and M. J. Gezing. Thanks are also due to J. Floeckher for assistance in data collection.

Chapter 6

Fundamental Approaches for the Assessment of Chemical and Colour Quenching in Backgrounds and Samples

P.E. Stanley [a]

Department of Agricultural Biochemistry, Waite Agricultural Research Institute, The University of Adelaide, Glen Osmond, South Australia 5064

INTRODUCTION

In this presentation two quite different aspects of quenching will be discussed. Both give rise in certain circumstances to considerable errors in the final calculation of disintegration rate, but nevertheless many workers are unaware of or ignore them. The topics are (1) the change in the background counting rate at various degrees of quenching, and (2) the very different shapes of pulse height spectra observed for colour and chemically quenched samples and the consequent problem of the assessment of quenching using the pulse height shift procedure.

CHANGES IN BACKGROUND WITH QUENCHING

This subject is of special interest when low activity samples with various levels of quench are being counted or when the blank (as distinct from the background) from, for example, an enzyme assay is several times the unquenched background. If external standard values are being used to estimate quenching, then the subtraction of the corrected background is readily achieved, but the background counts depend on the fluor and quenching agent as well as the instrument.[1] Here it will be shown that it is possible to subtract the corrected background when the channels ratio procedure is employed to assess quenching. In this case the ratio itself is of course affected when dealing with samples having low activities of up to five to ten times the true background. A computer program will be described which can be used to calculate the appropriate background for not only single-label experiments but also for those involving double labels when the two ratio channels are set in the pulse height range for the higher energy isotope.

The problem which the program solves is illustrated in Table 1. It can be seen that not correcting for background or subtracting a constant background count

---

[a] Present address: Department of Clinical Pharmacology, The Queen Elizabeth Hospital, Woodville, Adelaide, South Australia.

gives rise to a gross error in the true channels ratio, which is of course used to assess the counting efficiency of the sample. The results in Table 2 show that

Table 1.

	Channel 1	Channel 2	Channels ratio
	Using a constant background		
Background (c.p.m.)	30	20	
Gross (c.p.m.)	70	45	1.55
Sample (c.p.m.)	45	25	1.60
	Using a quenched background		
Background (c.p.m.)	37	15	
Gross (c.p.m.)	70	45	1.55
Sample (c.p.m.)	33	30	1.10

Table 2.

	Channel 1	Channel 2	Channels ratio
	Using a constant background		
Background (c.p.m.)	168	70	
Gross (c.p.m.)	205	135	1.52
Sample (c.p.m.)	37	·65	0.57
	Using correct quenched background		
Background (c.p.m.)	133	107	
Gross (c.p.m.)	205	135	1.52
Sample (c.p.m.)	72	28	2.57

similar errors are encountered when the background count is somewhat higher. This may be the case for an analytical procedure where the blank is high. Thus there is a quenching of the true background as well as in the blank. These may well be affected in different ways and it is clear that if any reliance is to be placed on the assessment of the efficiency using the channels ratio procedure, it is mandatory that the corrected background be introduced into the calculation.

Figure 1 shows the change in the counting rate for coloured and chemically quenched samples in three different energy channels. Figure 2 illustrates the effect when samples have a large 'background' made up of the true background plus the blank. These results were obtained by placing samples in vials which were specially selected for their similar backgrounds in a wide tritium window. Each vial was counted for 20 min during each of ten cycles and then vials were selected which had means within ± 0.8 c.p.m. of a running mean. Workers undertaking low activity measurements should make a similar selection of vials. A wide tritium window is a better and generally more stringent setting for double-isotope work with Hydrogen-3 and Carbon-14.

The computer program is written in Fortran Extended and runs on a CDC 6400 instrument. The code is structured so that the correct channels ratio is attained using an iterative procedure and depends on the input of data from a quenched set of radioactive standards and a set of backgrounds quenched in precisely the same fashion. Each set should have at least six standards. Figure 3 illustrates the flow chart for the program. The program is written to allow the user maximum flexibility; thus on the first card the input format is punched and on the next one the number of standards and the channels to be used for the ratio. Default values of six standards and channels 2 divided by 3 are provided. The program then

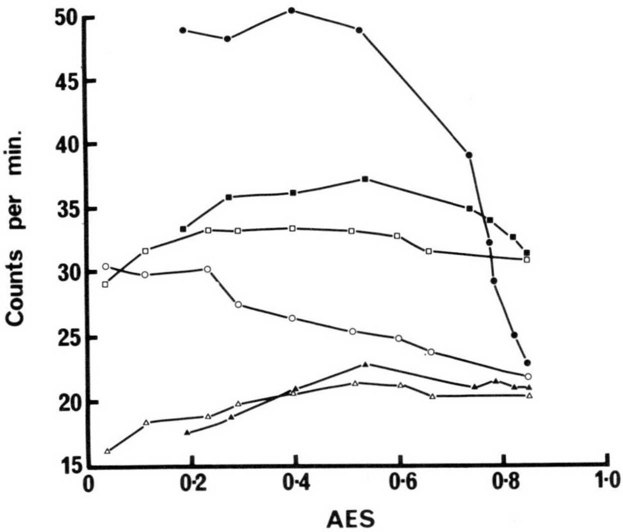

Fig. 1. Background counts per minute for colour and chemically quenched samples. Methyl orange in ethanol and carbon tetrachloride were the quenching agents. Readings were made on a Packard Tricarb 3375 operated at 20°C and fitted with EMI QB9635 photomultipliers. They are the means of ten 20-min counting periods for each sample. AES = automatic external standard ratio.

○ Chemical quench. P.H.A. setting 20–1000 at 45% (wide tritium window)
● As above but colour quenched.
□ Chemical quench. P.H.A. setting 50–1000 at 6% (20:1 Carbon-14 window)
■ As above but colour quenched.
△ Chemical quench. P.H.A. setting 130–1000 at 6% (Carbon-14 above Hydrogen-3 endpoint).
▲ As above but colour quenched.

calculates counts per minute for both sets of standards and ratios of the labelled standards, and the ratios are then associated with the identically quenched background for each channel. A sample is then read and the channels ratio, CR1, is obtained from it. If the sample is of medium to low activity, CR1 is probably not the correct value for assessing the counting efficiency. This value is tested as follows. It is used to obtain, by linear interpolation, backgrounds for each channel. If CR1 is larger than the biggest ratio the backgrounds appropriate to that ratio are used, and similarly for the lowest ratio. These backgrounds are then subtracted from the gross counting rate in each channel for the sample and a new ratio, CR2, is derived. If CR1 and CR2 differ by no more than 0.2%, then CR2 is used to obtain backgrounds and the result is printed prior to reading the next sample. However, if CR1 and CR2 are not that close, then the program sets CR1 equal to CR2 and the process starts again and is continued for up to 40 times, after which an informative message is printed. Under certain circumstances, CR1 and CR2 oscillate between the same two values and if this occurs a weighting factor is included so that the new CR1 is not equal to CR2 but falls at a definite point in between the two values. This assumes that there is some validity in the values of both CR1 and CR2. In this way a proper convergence is attained. The program terminates with a blank card. The user is protected in the case of

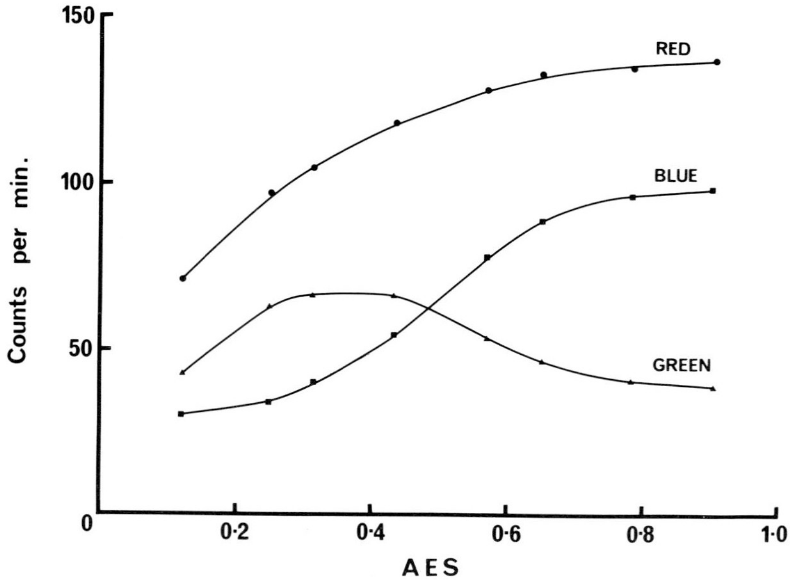

Fig. 2. Background plus Carbon-14 blank counts per minute. The quenching agent was carbon tetrachloride. Instrument as in Fig. 1.
RED P.H.A. setting 50–1000 at 6% (20:1 Carbon-14 window).
GREEN P.H.A. setting 50–150 at 6%.
BLUE P.H.A. setting 150–1000 at 6%

non-convergence and when samples have an apparently lower counts per minute than the most quenched standard. Thus, in the latter case, one of both channels giving rise to negative net counts per minute causes the program to reassign to that data positive values and iterations are allowed to continue for up to five occasions, after which time an informative message is printed and the next sample is read in.

The program works well in practice and if the user counts the samples to satisfactory statistical limits it usually gives a result for each sample rather than printing the default message. The program is readily adapted for double-label work when the two ratio channels are set above the endpoint of the lower energy isotope. However, the dynamic range for assessment of quenching by this procedure is somewhat limited[2,3] since as quenching increases the ratios usually either tend to a value of one or zero, depending on the way in which the discriminators are set. Recently Hansen and Carroll[4] extended the useful range of this procedure by setting the lower discriminators for the upper energy isotope below the endpoint for the lower energy isotope. The value of this procedure is shown in Fig. 4 but it means that the lower energy isotope increasingly contaminates the ratio channels as the discriminator is set lower and Carbon-14 is counted with better efficiency, thus allowing a bigger dynamic range for assessment of quenching. Hansen and Carroll obtain the correct ratio and hence efficiency by an iterative procedure. By employing their procedure together with the program described here, it should be possible to obtain good assessments of quenched background thus:

1. set channels ratio CR3 = 100. 0;
2. use the spillover procedure of Hansen and Carroll[4] to obtain a channels ratio CR4;
3. is CR3 within 0. 2% of CR4? If yes, use CR4 to get backgrounds and then exit. If no, go to 4;
4. more than 50 iterations? If yes, print default message and exit. Otherwise, go to 5;
5. using CR4, obtain new background and thus a new ratio CR3. Go to 2.

Work is continuing to complete this part of the program and also to assess whether the use of differences in successive background determinations makes a better and quicker termination of iterations than the difference between CR1 and CR2 and also CR3 and CR4.

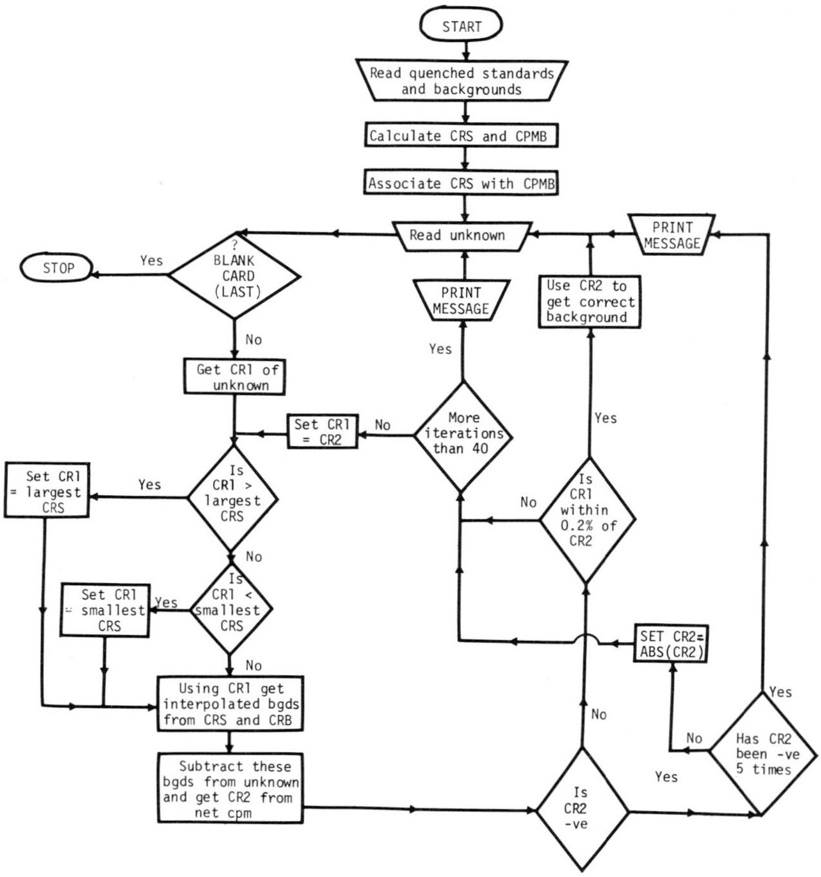

CRS = Channels ratio of standards

CPMB = CPM's of background standards

CR1 and CR2 - Channels ratio calculated by programme

Fig. 3. Flow diagram for the computer program which calculates the correct value for the quenched background when using the channel ratio procedure.

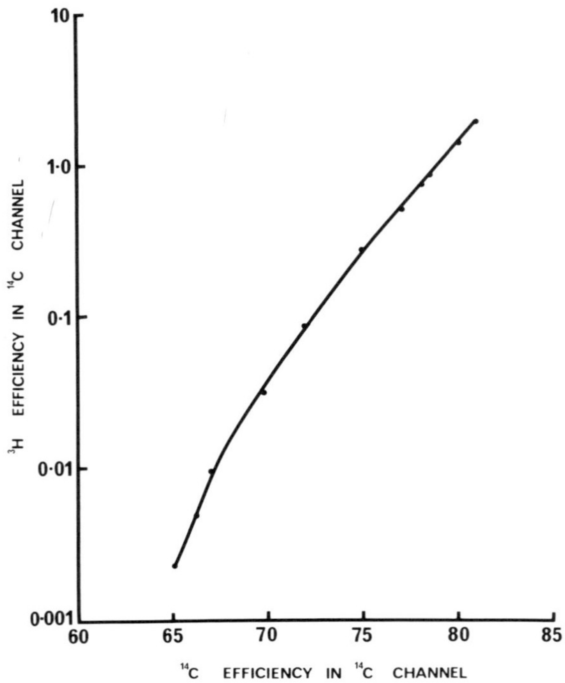

Fig.4. Spillover of Hydrogen-3 into the Carbon-14 channel as a function of Carbon-14 efficiency. Data were collected for an oxygen quenched sample using the instrument mentioned in the legend to Fig.1. The lower discriminator of the carbon channel was set to give around 5% Hydrogen-3 efficiency and then gradually raised. This illustrates the value of setting this discriminator some way into the tritium spectrum, as there is considerable gain in Carbon-14 efficiency (and quenching range accessible by channels ratio) for only a marginal change in the spillover of Hydrogen-3.

Some typical results obtained using the program are presented in Table 3.

Table 3. An example of the results produced by the program.

Channel	Gross c.p.m.	Ratio 1/2	Actual net c.p.m.	Ratio 1/2	Calculated net c.p.m.	Ratio 1/2
1	78.7		56.0		54.6	
		1.81		2.58		2.59
2	43.5		21.7		21.1	
3	20.2		6.4		5.7	

A NEW APPROACH TO THE PROBLEM OF CHEMICAL AND COLOUR QUENCHING

The pulse height spectra for chemically and colour quenched samples of medium-high energy β-emitters are different in shape. This difference is significant with Carbon-14 (see Fig. 5) but is barely discernible with the low energy emitter tritium, since it is closely associated with the number of photons produced by the β-particles. Thus, for a mean β-energy (50 keV) from Carbon-14 about 800 photons are produced,[5] while for Hydrogen-3 (mean 5 keV) the value is much lower, around 50. Since about one-fifth of this number finally results in the production of a photoelectron, it follows that for tritium a much smaller number of photoelectrons is involved; thus the difference between the effects of the two types of quenching is small. However, for Carbon-14 when larger numbers of photons are involved, there is more scope for statistical variation at all stages in the process and the difference in spectra occurs because of this.

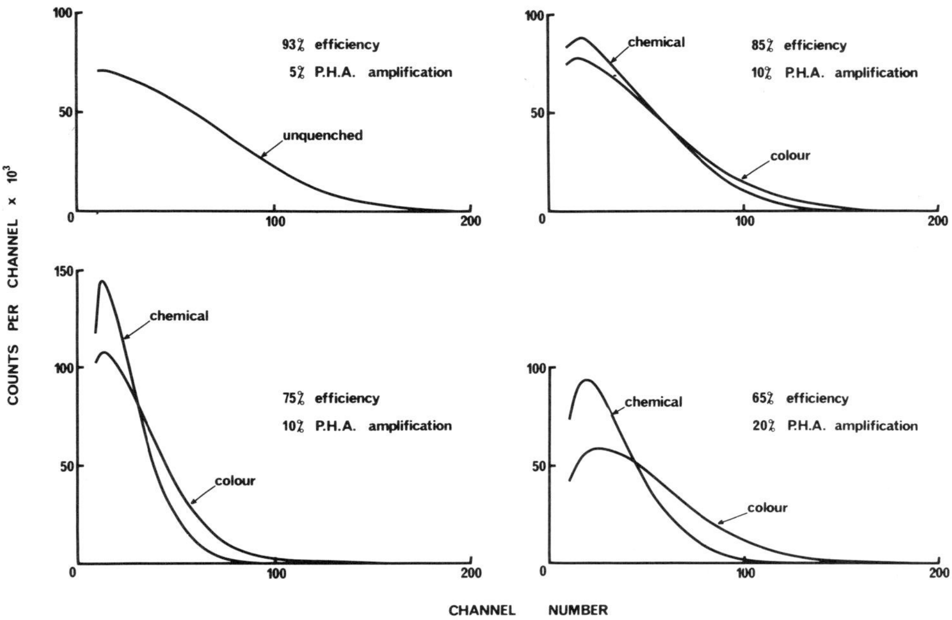

Fig. 5. Pulse height spectra for quenched Carbon-14 samples. Carbon tetrachloride and methyl orange were the chemical and colour quenching agents respectively. The 'unquenched' sample was air saturated. Quoted efficiencies were obtained in coincidence at integral settings on a Packard Tricarb 3390–544 fitted with RCA 4501 V4 bialkali photomultipliers and operated at $20 \pm 0.5^{o}C$ (3 to ∞ at 100% amplification). Pulse height spectra were obtained using a Packard Spectrazoom 200 channel multichannel analyzer connected by a coaxial cable via a 1K resister (for pulse stretching) to the test point at the rear of the RED pulse height analyzer. Coincidence was established by a pulse from the RED ratemeter socket. The spectra were conveniently fitted into the 200 channel analyzer by varying the pulse height analyzer gain (P.H.A.).

It is worth while to explain briefly the fundamental differences between colour and chemical quenching. Chemical quenching results in a diminution of the photon output from a scintillation and the effect is quite independent of the position in the vial where the event occurs. However, colour quenching involves the absorption of emitted photons by the sample itself. Thus, as the path length taken by a photon increases, the likelihood of it being absorbed increases exponentially. Taking a hypothetical β-particle producing a scintillation at the centre of the vial and the photons (of similar wavelength) being emitted in the plane of the vial base, then the probability that each photon will reach one photomultiplier is about the same. However, for a scintillation occurring very close to one photomultiplier, then photons emitted in its direction will have a much higher probability of reaching it than photons having to traverse the diameter of the vial to the other photomultiplier.

Thus, for example, given a total of 1000 photons reaching the photomultipliers, the pulse produced at the nearer photomultiplier may be the result of 900 of these and consequently a pulse height ratio of 9:1 is recorded for the photomultipliers. For the scintillation occurring in the middle of the vial, the ratio would be about 5:5. This, of course, is an over-simplified way of considering the process since photons are not of constant wavelength, do not travel in one plane and do not divide themselves equally to the left and right directions. It does, however, illustrate the point that the pulse height spectrum for colour quenched samples is the result of a highly probabilistic set of factors. It is therefore apparent that in attempting to apply the channels ratio procedure for β-emitting isotopes which are more energetic than tritium, it is necessary to use two standard curves, one for colour and the other for chemical quenching.

Thus it is not good practice to use the ratio procedure to assess the chemical and colour quenching in the same sample. Very recently Ediss et al.[6] have shown that reasonably effective corrections can be applied if the smaller of the two pulses is used to produce the spectrum for analysis. This is known as lesser pulse height analysis. The present work has had as its long-term aim the analysis of the shape of the pulse height spectrum so that a quantitative assessment of both colour and chemical quenching can be made in the same sample. To do this a mathematical model is being developed to study the factors that are important in influencing the shape of the pulse height spectrum starting with the β-particle itself. A much more detailed account of the present model was presented by Malcolm and Stanley at a recent symposium.[7] Only a brief synopsis will be given here.

Monte Carlo techniques are used in simulating various stages of the liquid scintillation counting process. As a result of this, the amount of central processor time (CP) needed for each scintillation for Carbon-14 is considerable, i.e. approximately 1 s on a CDC 6400 computer. Thus, to obtain a reasonable pulse height spectrum, at least 10,000 s (nearly 3 h) of CP time is required. The program, written in Fortran Extended, requires less than 40^8 K core.

The simulation is initiated by the computer generating a hypothetical β-particle in a random three-dimensional position within the vial. The β-particle is then randomly assigned an energy according to the Fermi probability function for Carbon-14. Currently 13 photons are then assigned per keV of the β-particle. Earlier in this symposium Dr. Horrocks showed that the number of photons produced per keV of energy was energy dependent below 300 keV and so it will now be necessary to modify the model to include this correction. If chemical quenching is to be simulated, a fixed fraction of photons for each scintillation is discarded at this stage. The photons produced are then randomly assigned different wavelengths according to the fluorescence spectrum of PPO in toluene and are emitted in random directions throughout the vial. From the measured absorption spectrum of colour quenched samples, the photons of appropriate wavelength are then either absorbed or not, prior to reaching the vial wall. This is accomplished using an exponential probability function and the random number generator. The next stage in the simulation is to determine the angle of incidence of the photon at the vial—air interface. If it is greater than the critical angle, then the photon will be totally reflected internally and thus lost.

The current version of the model produces a value of approximately 50% loss of photons by this process, the refractive index of toluene and the glass used for making vials each being approximately 1.50. Evidence for the considerable loss of light in the vial walls by the process of internal reflection is given in Fig. 6. The sample elevator apparently does not reflect much of this light towards the photomultipliers, since covering the elevator with black paper diminishes the counting efficiency by only around 1%.

It is presently assumed that the detector chamber is perfectly reflective and there are no transmission losses in air. Moreover, photons are not reflected back into the vial but go directly to the right-hand photomultiplier if their initial path is to the right of the line perpendicular to the photomultipliers or to the left-hand one if the initial path is to the left. Thus photons arrive at each photomultiplier and the photocathode is then modelled by taking account of its quantum efficiency for different photon wavelengths. Again using a probability function and the random number generator, a photon of a given wavelength either produces a photoelectron or does not. Coincidence is modelled simply by requiring that at least one photoelectron has been generated at each photocathode during the simulation of a β-event and scintillation. Summation of the coincident pulses simulated the summation amplifier.

While work is still progressing on this model, it currently produces spectra for coloured and chemically quenched Carbon-14 samples which are qualitatively

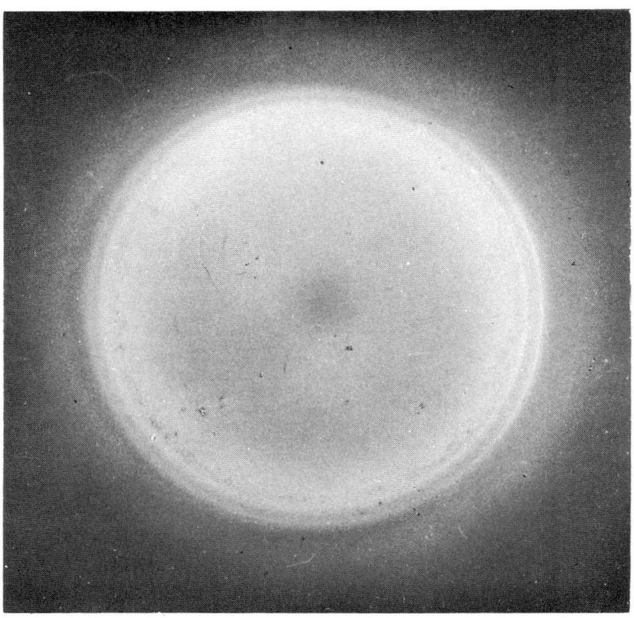

Fig. 6. A photograph of the bottom of a scintillation vial filled with a Carbon-14 sample, taken by its own light. Notice the lighter annulus at the rim of the vial. This is thought to be caused as follows. A photon which gets totally internally reflected in the upper vial walls by virtue of their circular symmetry will occasionally be reflected towards the base. Here there is an abrupt change in symmetry, especially near the rim, and thus more light escapes at this point.

similar to those observed in practice, and the integral counting efficiencies, down to 65%, are within a few per cent. It is of interest that satisfactory results for tritium are obtained with about 11 photons per keV instead of the 13 for Carbon-14. The reason for this has been unclear but the values given by Dr. Horrocks for the scintillation efficiencies of 0.044 and 0.052 may well explain this result. Further additions to the model will include information on the optics of both the detector chamber and the vial and the perturbations of pulses known to occur in the photomultiplier.[8]

ACKNOWLEDGEMENTS

The author wishes to acknowledge Mr. P.J. Malcolm, with whom the mathematical model has been developed. The financial assistance of the Australian Wheat Industry Research Council is gratefully acknowledged, as is the encouragement given by Professor D.J.D. Nicholas throughout this work.

REFERENCES

1 R.J. Herberg, in Organic Scintillators and Liquid Scintillation Counting (Eds. D.L. Horrocks and C. T. Peng), Academic Press, New York and London, 1971, p.783.
2 R.W. Hendler, Anal. Biochem. **7**, 110 (1964).
3 R. Ninomiya, Intern. J. Applied Radiation and Isotopes **14**, 541 (1963)
4 D.L. Hansen and C.O. Carroll, Intern. J. Applied Radiation and Isotopes 22, 677 (1971).
5 J.W. Hastings and G. Weber, J. Opt. Soc. Amer. **53**, 1410 (1963).
6 C. Ediss, A.A. Noujaim and L.I. Wiebe, in Liquid Scintillation Counting : Recent Developments (Eds. Philip E. Stanley and Bruce A. Scoggins), Academic Press, New York and London (in press).
7 P.J. Malcolm and P.E. Stanley, in Liquid Scintillation Counting : Recent Developments (Eds. Philip E. Stanley and Bruce A. Scoggins), Academic Press, New York and London (in press).
8 Y.K. Akimov, Scintillation Counters in High Energy Physics, Academic Press, New York and London, 1965.

DISCUSSION

B.E. Gordon: How do you square the 50% total internal reflection loss with only 5% increase in efficiency of Hydrogen-3 counting?

P.E. Stanley: The important thing is to recall that all you need to detect an event is a coincidental pulse from each photomultiplier. If more photons are collected at the photocathode then this increases the pulse height but not the efficiency. What you are really increasing when using sand blasted vials is the number of few photon events which enable coincidence since the photons are not so readily totally internally reflected.

B.W. Fox: Was the photograph of light emission which you showed done with an optically flat vial base — it looked remarkably like the unprecise moulding on our own vials?

P.E. Stanley: Yes, the variation may be due to this source. What I attempted to demonstrate was that the lower corners of the vial represent a non-symmetrical area and photons being totally internally reflected on the vertical vial walls would stand a better chance of leaving the vial at this point.

B.W. Fox: Has anyone considered vertical moulding on vials, much like some modern lighting techniques, to increase the light output from the vial?

P.E. Stanley: No, except for Dr. Gordon's work in attempting to define the light output by various surfaces.

Author's note added in proof

The efficiencies for Carbon-14 counting mentioned in Fig. 5 are probably about 3 percent too high due to what is thought to be after-pulsing in the photomultipliers. The shapes of the spectra are unchanged since counts in the first few channels, where after-pulses would be apparent, are not included. Normal pulse height analyser settings for Carbon-14 would, to a large extent, eliminate after-pulses which have only a small pulse height yet may be resolved by the pulse height analyser from the main scintillation pulse.

Chapter 7

Colour Quenching by Carotenoids

P.M. Bramley,[a] B.H. Davies and Averil F. Rees

Department of Biochemistry and Agricultural Biochemistry,
University College of Wales, Aberystwyth, Wales, U.K.

INTRODUCTION

The carotenoids are a group of lipid-soluble pigments which occur in all plants and algae and in many fungi, bacteria and animals. Their basic structural feature (see Fig. 1), on which all their properties and functions depend, is the polyene chain, the sequence of conjugated double bonds, which makes up the backbone of the molecule. The number of conjugated double bonds may be as high as 15 or as low as 3, so that the carotenoids can be purple, red, orange, yellow or even colourless. The light absorption properties of the polyene chromophore are modified by the nature of the end groups; some 40 different end groups have been detected in the 350 or so naturally occurring carotenoids that are currently recognised.

Studies on the biosynthesis of these compounds, particularly in micro-organisms, have been in progress in this laboratory for a number of years. An important feature of the experimental approach is, of course, the use of suitable radioactive substrates (labelled with Carbon-14 and/or Hydrogen-3) as precursors of the carotenoids in whole organisms or in enzyme systems. Although their solubility characteristics are such that the carotenoids dissolve without difficulty in the simplest and most efficient of scintillation fluids, the very nature of these pigments dictates that such samples interfere with the basic liquid scintillation counting process by causing colour quenching. The use of liquid scintillation counting for the assay of radioactivity in numbers of carotenoid samples which are at different concentrations and which have different absorption spectra obviously requires a versatile, accurate and preferably rapid procedure for colour quench correction.

METHODS OF CORRECTING FOR COLOUR QUENCH BY CAROTENOIDS

The ease with which carotenoids are photodegraded by ultraviolet radiation enables

---

[a] Present address: Department of Biochemistry, Royal Holloway College, Englefield Green, Surrey, U.K.

this procedure to be used prior to counting for the elimination of colour quenching.[1] This method has certain disadvantages in that it is destructive and is time-consuming because of the luminescence which results from the ultraviolet bleaching and which has to be allowed to decay before counting can commence. Much the same criticisms can be made of other methods (e.g. bleaching with benzoyl peroxide and light[2]) which have been used in the carotenoid field. The availability of liquid scintillation counters equipped with automatic external standards has enabled the exploitation of the external standard channels ratio (ESCR) method in correcting both for chemical and for colour quenching. While separate calibration procedures are necessary for the respective types of quenching, it has been shown in studies using a limited number of widely differing coloured samples (e.g. β-carotene, scarlet red, oracet blue, chlorophyll, road tar, etc.) that the same relationship between counting efficiency and ESCR applies to samples of all colours.[3,4]

In establishing quench correction curves in order to use this method for counting Carbon-14 and Hydrogen-3 in carotenoid samples, we chose ten different carotenoids (Fig. 1) which, between them, have a wide variety of absorption maxima covering the entire wavelength range of phosphor emission and photomultiplier sensitivity (Fig. 2). Each part of the spectral range of the instrument can be covered by selecting the appropriate carotenoid. In spite of their differences in chromophore, all ten carotenoids are based on the same 40-C (tetraterpenoid) skeleton; seven are hydrocarbons (carotenes) and, of these, three (β-carotene, γ-carotene and lycopene) are isomers. Because they are structurally so similar and are basically hydrocarbon in nature and are therefore not chemical quenchers, these ten pigments provide the opportunity for making an extremely detailed and controlled study not only of the validity of the ESCR method of colour quench correction, but also of the quenching phenomenon itself.

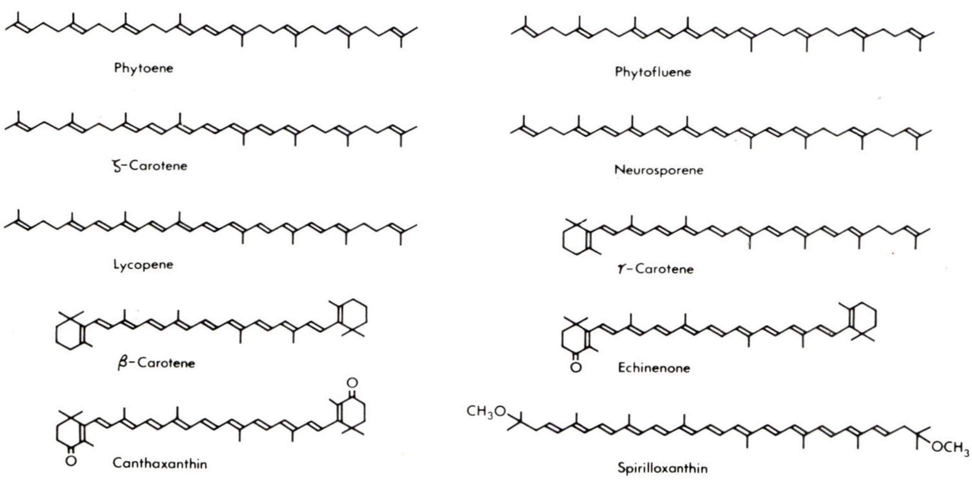

Fig. 1. Structures of the carotenoids used in these studies.

MATERIALS AND METHODS

Instrument

Measurements of radioactivity were made with a Nuclear Enterprises 8310

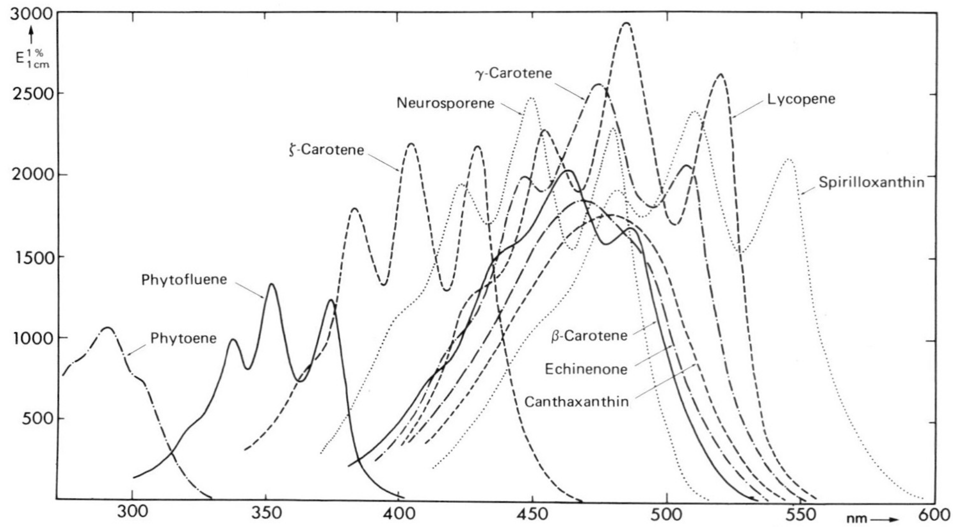

Fig. 2. Absorption spectra of carotenoids in toluene.

Automatic Liquid Scintillation Spectrometer[a] equipped with a Caesium-137 50 μCi external standard. The discriminator settings were as follows: counting channel 1, E = 0.6 V, ΔE = 4.4 V; counting channel 2, E = 5.0 V, ΔE = 5.0 V; counting channel 3, E = 0.6 V, ΔE = 9.4 V; external standard channel 2, E = 9.80 V, ΔE = 10.02 V; external standard channel 3, E = 8.00 V, ΔE = 4.32 V. The counting efficiencies in the counting channels, as determined using the manufacturer's standards, were: Carbon-14, 17.4% (1), 72.7% (2), 90.7% (3); Hydrogen-3, 44.9% (1). Each experimental vial was counted in triplicate for 10 min, the counting modes used being βESOL and βES for Carbon-14 and Hydrogen-3 respectively. External standard channels ratios and apparent counting efficiencies were calculated with an on-line Olivetti Programma 101 computer.

Experimental samples

The counting efficiencies of Carbon-14 were assayed by preparing, for each of the ten carotenoids, a series of ten counting vials, each containing a different accurately measured amount of the carotenoid (0 to 10 μg or, in the case of phytoene, 0 to 25 μg) and 10 ml of a scintillation fluid containing 3.0 g PPO[b] and 0.5 g dimethyl-POPOP[c] per litre of toluene. The scintillation fluid also contained a concentration of (1-Carbon-14)n-hexadecane[d] equivalent to 18731 d.p.m. per vial. The counting efficiencies of Hydrogen-3 were assayed by an identical procedure except that each vial contained an equal amount (equivalent to 34056 d.p.m.) of (1,2 Hydrogen-3)n-hexadecane.[e]

---

a Nuclear Enterprises Ltd., Sighthill, Edinburgh 11, Scotland, U.K.

b 2,5-Diphenyloxazole; Packard Instrument Co. Inc., La Grange, Illinois, U.S.A.

c 1,4-Bis(4-methyl-5-phenyloxazol-2-yl)benzene; Packard Instrument Co. Inc., La Grange, Illinois, U.S.A.

d Specific activity 1.10 μCi/g; Radiochemical Centre, Amersham, Bucks., U.K.

e Specific activity 2.00 μCi/g; Radiochemical Centre, Amersham, Bucks., U.K.

Carotenoids

The carotenoids were obtained as follows: phytoene, phytofluene, ζ-carotene, neurosporene, γ-carotene, lycopene and β-carotene were isolated from normal and inhibited cultures of a number of mutants of the fungus Phycomyces blakesleeanus,[5,6] spirilloxanthin was obtained from cultures of the purple, non-sulphur, photosynthetic bacterium Rhodospirillum rubrum,[7] while samples of echinenone and canthaxanthin were a generous gift from F. Hoffmann-La Roche & Co. Ltd., Basel, Switzerland. All the carotenoids were purified to spectral purity by standard methods of column and thin-layer chromatography.[8]

Absorption spectra

Absorption spectra (Fig. 2 and Table 1) were determined in toluene using a Unicam SP800 Recording Spectrophotometer, the wavelength scale of which was calibrated using a holmium oxide filter. Quantitative measurements were made in one of a matched pair of 1 cm path-length silica cuvettes and specific extinction coefficients ($E^{1\%}_{1cm}$) were calculated (Table 1) from those quoted for solutions in other solvents after quantitative comparisons of carotenoid spectra in toluene and in the appropriate solvent. Samples were dispensed into the counting vials as accurately pipetted aliquots from standard solutions and were evaporated to dryness prior to dissolving the residue in 10 ml scintillation fluid containing the appropriate labelled species of n-hexadecane.

Table 1. Absorption characteristics of carotenoids in toluene.

Carotenoid	Absorption maxima (nm)			$E^{1\%}_{1cm}$ at λ_{max}*
Phytoene	(277)	290.5*	302	1040
Phytofluene	337	354 *	374	1140
ζ-Carotene	383	406 *	431	2200
Neurosporene	425	451 *	480	2480
Lycopene	457	484 *	519	2950
γ-Carotene	450	474 *	507	2580
β-Carotene	(434)	463 *	490	1960
Echinenone		468 *	(498)	1830
Canthaxanthin		479 *		1910
Spirilloxanthin	478	508 *	542	2400

QUENCH CORRECTION CURVE

After recording in triplicate the counting efficiencies of some 200 standard samples and the corresponding values of ESCR, it was possible to plot four curves representing the relationships between the efficiency of isotope counting in the different channels and the ESCR (Fig. 3). The results of previous workers[3,4] are confirmed in so far as the nature of the carotenoid is irrelevant, for points for all the pigments fitted the same lines. Any slight deviations of the individual points (all <2%) could satisfactorily be accounted for on the basis of pipetting errors.

These calibration curves have been used for the routine counting of Carbon-14 in carotene samples as part of our biosynthetic studies. They have consistently given results comparable with those obtained following a bleaching procedure; the latter has now been abandoned in this laboratory.

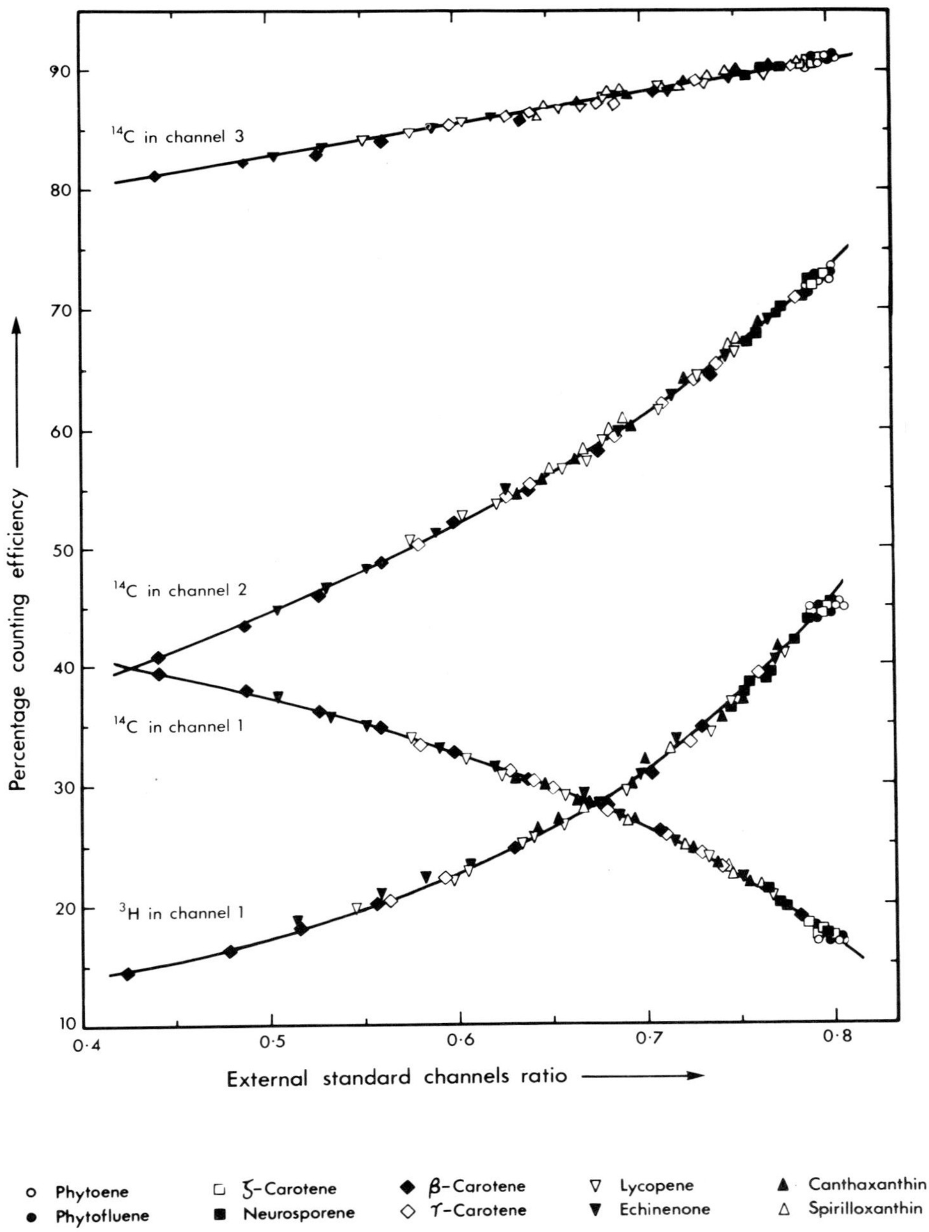

Fig. 3. Relationship between external standard channels ratio and counting efficiencies for Carbon-14 and Hydrogen-3 colour quenched by various carotenoids (see text for conditions)

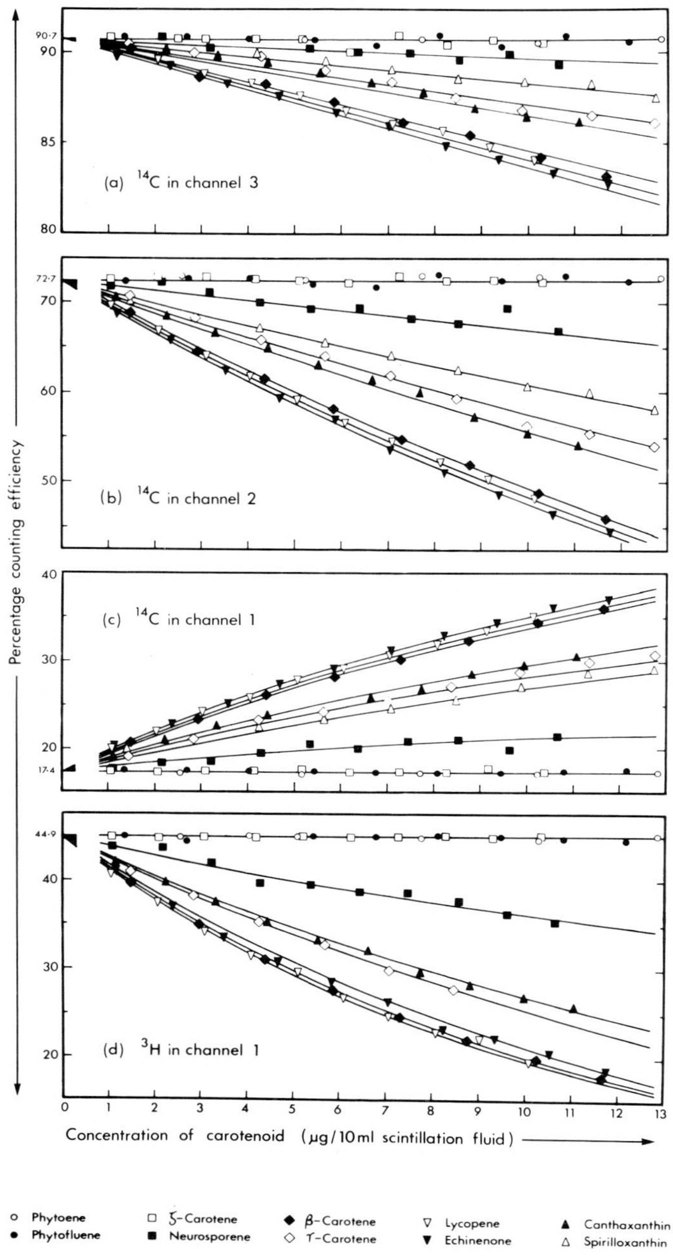

Fig. 4. Variations in counting efficiency with carotenoid concentration (see text for conditions).

COLOUR QUENCHING BY DIFFERENT CAROTENOIDS

The results upon which the quench correction curves are based were studied in more detail. When the percentage counting efficiencies were plotted against the concentration of carotenoid in each vial, the values corresponding to each individual pigment gave rise to a different line (Fig. 4). This confirms that, as would be expected, some carotenoids are more efficient quenching agents than others.

It has often been assumed[9] that the counting of phytoene and phytofluene, which are both colourless but absorb ultraviolet radiation (Table 1), would require some correction procedure, especially phytofluene, whose strong fluorescence (maximum excitation 369 nm, maximum emission 490 nm, both values in ethanol[10]) might be expected to lead to some spurious results. The presence of neither of these polyenes showed any effect on the recorded count rates of standard samples. As the sensitivity of the EMI 9635QB Photomultipliers[a] (with which the NE 8310 scintillation counter is fitted) is maximal at about 380 nm[11] and the fluorescence emission spectrum of the scintillator mixture (in toluene) has maxima at 410 (infl.), 430 and 455 nm,[12] it would be expected that the conjugated heptaene, ζ-carotene (λ_{max} in toluene 383, 406 and 431 nm), and the conjugated nonaene, neurosporene (λ_{max} in toluene 425, 451 and 480 nm), would show the highest degree of colour quenching per unit concentration. This was not the case. Of the coloured carotenoids, these two polyenes showed the least quenching, although neurosporene was a stronger quenching agent than ζ-carotene. Rather, the most efficient quenchers were lycopene (λ_{max} in toluene 457, 484 and 519 nm) and echinenone (λ_{max} in toluene 468 nm), both of which absorbed maximally in a region where both photomultiplier sensitivity and phosphor emission are at some 60% of their highest values.

True comparisons of quenching efficiencies, however, cannot be made on the basis of concentration alone, for this takes no account of the different extinction coefficients of the carotenoids. Further comparisons were therefore made. In the first, values for percentage quench determined from the above data at carotenoid concentrations calculated to give extinctions of 0.25 at their wavelengths of maximal absorption were plotted against the wavelengths of maximal absorption (Fig. 5). This indicated quite clearly that, on the basis of matching peak wavelengths, those carotenoids which absorb maximally at about 465 nm, well away from the expected region, are the most efficient quenchers of this liquid scintillation system.

Since this anomaly may have arisen through considering peak wavelengths alone instead of spectral overlap integrals, a second approach was used. In this, values were found for the percentage quench shown by those graphically determined concentrations of the different carotenoids which absorbed a standard total amount of light over their entire absorption range, and these were plotted against the corresponding mean wavelengths of light absorption, again determined graphically for each carotenoid. There was, however, no significantly better correlation between the apparent wavelength of maximal quenching (465 to 470 nm) and the wavelengths of maximal phosphor emission and photomultiplier response.

THEORETICAL APPROACHES TO COLOUR QUENCHING BY CAROTENOIDS

Two theoretical attempts were made to correlate the experimental data with expected behaviour. The first utilised the integral of Ross,[13] where the percentage

---

[a] EMI Electronics Ltd., Electron Tube Division, Hayes, Middlesex, U.K.

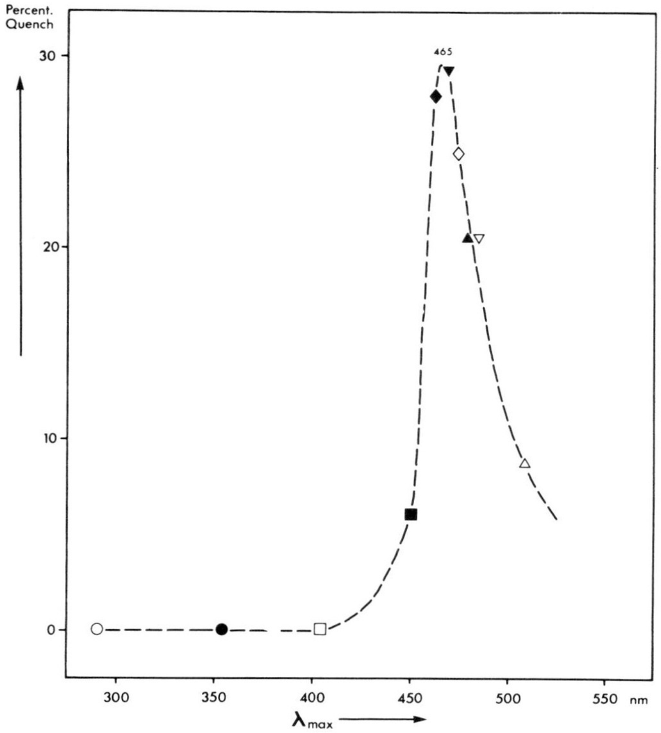

Fig. 5. Percentage quench of Carbon-14 counts in counting channel 2 by carotenoid solutions of E_{max} 0.25 plotted against λ_{max} (carotenoid symbols as for Figs. 3 and 4).

colour quench (Ω) of a sample is given by:

$$\Omega = \int_{\lambda_1}^{\lambda_n} (K)\ (A)\ d\lambda$$

where K is the quenching coefficient, A is the absorbance of the solution and λ is the wavelength. This approximates to:

$$\Omega = (K_{\lambda_1})\ (A_{\lambda_1}) + (K_{\lambda_2})\ (A_{\lambda_2}) + (K_{\lambda_3})\ (A_{\lambda_3})\ \ldots + (K_{\lambda_n})\ (A_{\lambda_n})$$

The second method, suggested to us by Dr. J.B. Birks, likens the system at wavenumber $\bar{\nu}$ (cm^{-1}) to (1) an emitter (phosphor) of fluorescence $F(\bar{\nu})$, (2) a detector with response $S(\bar{\nu})$ and (3) a filter of transmittance $T(\bar{\nu})$, equivalent to an absorbing solution of molar extinction coefficient ε and molar concentration c in a scintillation vial of internal diameter t, where:

$$\log_{10} T(\bar{\nu}) = -\varepsilon(\bar{\nu})ct/2$$

The factor 't/2' is an approximation, since for one of the pair of photomultipliers the length of the absorbing path varies from zero to t, while for the other it varies from t to zero. The quenching factor (Ω) is given by:

$$\Omega = \frac{\int F(\bar{v})T(\bar{v})S(\bar{v})d\bar{v}}{\int F(\bar{v})S(\bar{v})d\bar{v}}$$

Unfortunately, neither calculation gave useful results. This is presumably due to the complex and persistent multi-peaked absorption spectra of the carotenoids, especially those of the acyclic carotenes which show pronounced fine structure. Such spectra do not lend themselves to the use of simple integrals in an experiment with only a limited number of samples.

SUMMARY AND COMMENTS

When viewed solely from the point of view of the carotenoid specialist, the data presented above justify the use of the ESCR method of correcting for colour quenching by a wide variety of these pigments. One set of correction curves is applicable to all the compounds of this class that we have examined, and it is unlikely that any others are sufficiently powerful chemical quenchers to introduce inaccuracies into the results and so necessitate an alternative approach. As long as the limitations of the technique, with regard to very high degrees of quench[14] and to doubly labelled samples,[9] are appreciated, the method is as accurate and is certainly more convenient than other procedures available.

Attempts to correlate the degree of colour quenching by the individual carotenoids with their light absorption characteristics led to the unexpected, if not anomalous, observation that those pigments which absorb maximally in the region 350 to 450 nm (phytofluene, ζ-carotene and neurosporene) are not necessarily the most powerful quenchers in the system examined. Indeed, phytofluene consistently shows no quenching, while the effect of ζ-carotene is virtually negligible. The latter observation contrasts with a report that the conjugated heptaene is a strong quencher.[2] Perhaps the reason for the discrepancy lies in the different instrumentation used or in the difficulty of separating ζ-carotene from β-zeacarotene and β-carotene.[15]

Although losses in efficiency in counting carotenoid samples are due solely to colour quenching, the use of a series of different carotenoid pigments to investigate the relationship between colour quench and wavelength did not give as satisfactory a correlation between experimental observation and theoretical prediction as has been achieved by other workers. Iwakura and Kasida[16] used a series of 14 dyes, which were very different in their individual structures and which required methanol for solubilisation, in order to carry out a similar study. That a better correlation between their experimental data and theory was obtained than in our case must be attributable, at least in part, to the relative simplicity of the absorption spectra of the dyes chosen for their study.

ACKNOWLEDGEMENTS

Our grateful thanks are due to F. Hoffmann-La Roche & Co. Ltd., Basel, Switzerland, for the generous provision of samples of synthetic echinenone and canthaxanthin, to Dr. J.B. Birks of the University of Manchester for his suggestions on the theoretical treatment of colour quenching, and to the Science Research Council for a Research Studentship (P.M.B.) and a Research Grant for studies on carotenoid biosynthesis.

REFERENCES

1 R.J.H. Williams, G. Britton, J.M. Charlton and T.W. Goodwin, Biochem. J. **104**, 767 (1967).

2 W.M. Walter, Jr. and A.E. Purcell, Anal. Biochem. **16**, 466 (1966).

3 A. Noujaim, C. Ediss and L. Wiebe, in Organic Scintillators and Liquid Scintillation Counting (Ed. D.L. Horrocks and C-T. Peng), Academic Press, New York and London, 1971, p. 705.

4 J.F. Lang, in Organic Scintillators and Liquid Scintillation Counting (Ed. D.L. Horrocks and C-T. Peng), Academic Press, New York and London, 1971, p. 823.

5 P.M. Bramley, Ph.D. Thesis, University of Wales, 1973.

6 B.H. Davies, Pure Appl. Chem. **35**, 1 (1973).

7 B.H. Davies, Biochem. J. **116**, 101 (1970).

8 B.H. Davies, in Chemistry and Biochemistry of Plant Pigments (Ed. T.W. Goodwin), Academic Press, New York and London, 1965, p. 489.

9 G. Britton and T.W. Goodwin, in Methods in Enzymology, Volume XVIIIC (Ed. D.B. McCormick and L.D. Wright), Academic Press, New York and London, 1971, p. 654.

10 R.J. Davis, M.Sc. Thesis, University of Wales, 1967.

11 'EMI Photomultiplier Tubes', EMI Electronics Ltd., 1970.

12 B.H. Davies and P.M. Bramley, unpublished work.

13 H.H. Ross, in Radioisotope Sample Measurement Techniques in Medicine and Biology, International Atomic Energy Agency, Vienna, 1965, p. 409.

14 J.L. Spratt, in Liquid Scintillation Counting, Volume 2 (Ed. M.A. Crook, P. Johnson and B. Scales), Heyden and Son, London, 1972, p. 245.

15 R.J.H. Williams, B.H. Davies and T.W. Goodwin, Phytochem. **4**, 759 (1965).

16 T. Iwakura and Y. Kasida, in Radioisotope Sample Measurement Techniques in Medicine and Biology, International Atomic Energy Agency, Vienna, 1965, p. 447.

DISCUSSION

E.B. Mueller: Have you tried counting carotenoid solutions with no fluor or with only primary fluor? Are the (non-quenching) carotenoids possibly acting as fluors or wavelength shifters, and this accounts for their not quenching?

B.H. Davies: No, but we have seen no shift in the emission spectrum of samples of PPO + Me_2 POPOP in the presence of carotenoids, and they are present at such small concentrations that they would not be expected to act as fluors.

Chapter 8

New Method for Obtaining the Quench Correction Curve in Liquid Scintillation Counting

M. A. Reunanen

Dept. of Chemistry, University of Turku, Finland

E. J. Soini

Wallac OY, 20810 Turku 81, Finland
Åbo Akademi, Dept. of Physics, 20500 Turkuso, Finland

INTRODUCTION

The use of a liquid scintillation counter to perform quantitative analysis requires that it be calibrated in absolute units. In addition, some scheme for making data corrections must be incorporated since the effects of quenching vary from sample to sample. No longer is there any problem in obtaining an instrument with a built-in computer, or feeding the counting data into such a computer. Instead, the difficulty presently encountered lies in the calibration of the scintillation counter. No real progress has occurred in this area, and the same old manual methods are still being employed.

Application of quench correction curves is usually required in order to make valid quantitative measurements with a liquid scintillation counter. Such curves are obtained by measuring a series of samples which have identical amounts of known radioactivity but increasing levels of quenching. Generally, the user either purchases a commercially available set or makes these standard quenched samples in his laboratory. Commercial sets are expensive, and the lifetime of these samples is not necessarily very long. The calibration can be made by pipetting an additional amount of quenching agent into the sample vial after every measurement. The preparation of standard samples demands the utmost care and accuracy in laboratory procedures, and whenever the composition of the scintillation liquid or sample is different, the corresponding calibration samples must be provided. Yet too many research workers calibrate their instruments against improper standards. Some even ignore the entire quench correction because they find the calibration procedures and associated data handling too difficult. An accurate, absolute radioactive standard is a requisite in this approach, however, and the maximum care is required to achieve reliable results. Storage of a standard solution requires close supervision to prevent changes in the standard sample or errors in its use. Solutions kept in containers in the liquid form can evaporate whenever the container is opened. Also, volumetric dosing or weighing of the solution can be a source of errors.

PRINCIPLE

In order to simplify the calibration of a liquid scintillation counter, a new method has been developed, which permits counters to be readily manufactured and made commercially available to the largest number of users. This new approach is intended for counters having either external standardisation or channels ratio standardisation functions. With this system the user is able to freely select the scintillation solution, sample vial and volume of sample to suit his own particular needs.

Two basic components are employed within this system: (a) a standard capsule for the radioactive labelled compound (internal standard), and (b) an evaporation device for automatically adding the quenching agent.

The capsule, a small glass cup, contains an accurate amount of radioactive material which readily dissolves in the solvent. The compound containing the isotope is secured to the inner wall of the capsule in the form of a thin film, permitting the safe handling of capsules with either the hands or pincers (Fig. 1). Capsules are packed in plastic, press-thru strips, each strip containing 10 capsules, for example.

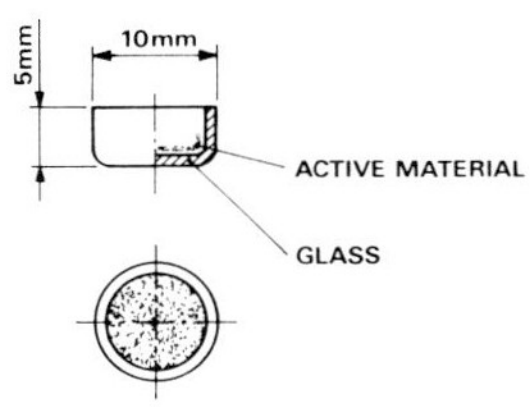

Figure 1

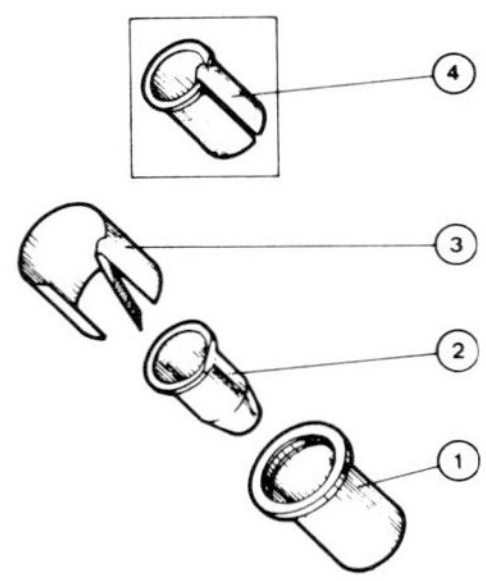

Figure 2

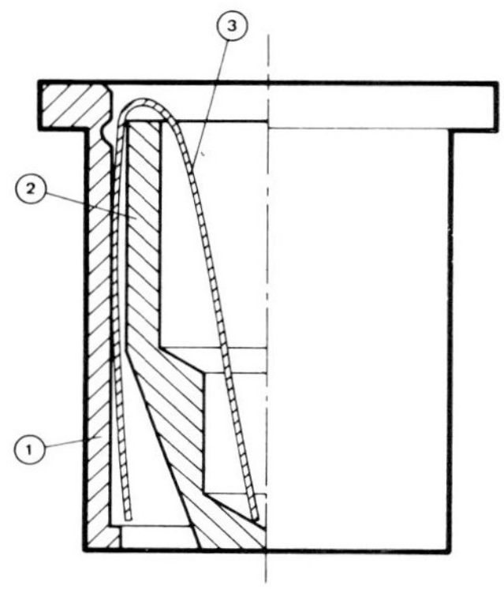

Figure 3

Fig. 1. The internal standard capsule.

Fig. 2. Construction of the evaporation device. 1, Outer body. 2, Quench agent container. 3, Chromatography paper. 4, Paper fixed around container.

Fig. 3. Construction of the evaporation device. 1, Outer body. 2, Quench agent container. 3, Chromatography paper.

The active material, dissolved in a suitable solvent, is dosed into the capsules by using an automatic pipetting machine. The solvent is then dried. For organic solvent systems, ^{3}H-tristearate and ^{14}C-tristearate have been used as labelled compounds. ^{3}H-sucrose and ^{14}C-sucrose have been used for water solvent systems. The initial activity has been 400,000 d.p.m. for the Hydrogen-3 capsules and 100,000 d.p.m. for the Carbon-14 capsules.

The capsule can then be inserted into the sample vial by simply pressing the plastic, press-thru strip. Once the capsule has been added to the vial, it must be turned so that the active film is facing upwards. This is accomplished by agitating the vial. The scintillation solution can be added immediately to systems with organic solvents. Tristearate dissolves immediately, and activity is homogeneously distributed throughout the scintillation solution by the motion of the vial. However, when using scintillation systems with water soluble samples, the capsule must first be dissolved in a small amount of water in the counting vial, the quantity of water being approximately the same as the water content of an actual sample. Finally, the scintillation solution is added.

The plastic evaporation device (Figs. 2 and 3) contains a small container for the quenching agent and a filter-paper evaporation element. Carbon tetrachloride, a quenching agent which evaporates quite rapidly, is pipetted into the container, being absorbed into the filter paper and passing to the evaporation surface outside the container. The shape of the evaporation surface is such that the evaporation rate increases as a function of time, permitting complete quenching of the sample within a reasonable period. The quenching agent is distributed evenly by convection in the vial, provided the viscosity of the scintillation solution is substantially low. Carbon tetrachloride has been chosen as the chemical quencher; no colour quencher has been selected yet.

200 μl of quenching agent is generally the quantity pipetted into the evaporation device. The amount of quenching liquid used is not critical, and the pipette accuracy can be ±20%. After being filled, the evaporation device is immediately placed on the vial in place of the lining, and the vial is closed with its normal cap. Finally, the vial is moved to the liquid scintillation counter for repeated counting. The evaporator automatically increases the quenching level of the sample, and the counter begins its tabulation.

An automatic liquid scintillation counter must be set so that it measures the sample repeatedly, printing out the channels ratio and c.p.m. values after each count. The channels ratio can be determined by using either the 'channels ratio' method or the 'external standard channels ratio' method. Counting should be continued until a desired quenching level is exceeded. In practice, it will take approximately 15 min to get observation points from 50% to 5% efficiency with 10 ml tritium samples. The total number of measuring points per curve depends upon both the counting time set for one measurement and the sample volume. For example, the values of the preset count and sample volume shown in Table 1 produce approximately 30 measuring points within a Hydrogen-3 efficiency range of 50% to 5%.

Table 1. Recommended Preset Parameters:

Counting preset count	Standardisation preset count	Sample volume (ml)
10, 000	10, 000	2
15, 000	15, 000	3
20, 000	20, 000	4
25, 000	25, 000	5
50, 000	50, 000	10
75, 000	75, 000	15

Since the counting of sample activity does not occur simultaneously with the measurement of the external standard channels ratio, a mean value of two consecutive external standard ratios must be calculated. This mean value represents, with sufficient accuracy, the simultaneous observation of the external standard ratio and the count (Fig. 4).

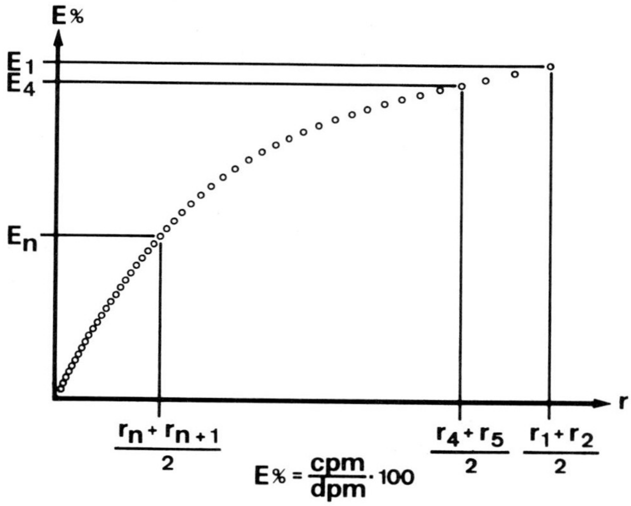

Fig. 4. Mean values of two consecutive external standard ratios are to be calculated. This mean value represents the simultaneous observation of external standard ratio with counting result.

The quench correction can be performed by either a computer contained within the liquid scintillation counter or with a separate calculator. The absolute activity, A, of a sample is obtained from Eqn. (1), where N is the measured c.p.m. and E is the counting efficiency:

$$A = \frac{N}{E} \tag{1}$$

It is convenient and practical to feed values of E into the computer according to Eqn. (2), which is an equation representing the quench correction curve. The coefficients, a_0, a_1, a_2, a_3, have been obtained by using the least square polynominal method, given the observation points (r, c.p.m.):

$$E = a_3r^3 + a_2r^2 + a_1r + a_0 \tag{2}$$

A sufficient number of observation points can be selected within the quenching range where the actual measurements of the sample will be taken. If there is little change in the level of quenching of the unknown samples, the quench correction can be represented sufficiently by two points lying on the straight line described by Eqn. (3). Of course, it is assumed that the curve is quite linear in that region:

$$E = a_1r + a_0 \tag{3}$$

Coefficients a_1 and a_0 can easily be determined manually. However, the coefficients of Eqn. (2) should be calculated with a computer, using an adequate number of observation points. All commercial computers, and the best of the desk calculators, are furnished with ready-made library programs for '3rd-order curve-fitting'.

Finally, the absolute activity of an unknown sample can be determined with a built in computer, or separate computer, using the previously determined coefficients along with Eqns. (1) and (2).

ACCURACY AND RELIABILITY OF THE METHOD

The determination of the absolute activity of standard capsules is based on a comparative measurement with a Tritium-standard No. 4947 ($C_6H_5CH_3$) and Carbon-14 standard No. 4222 ($C_{16}H_{34}$) of the National Bureau of Standards. The Bureau lists the accuracy for Tritium-standard No. 4947 as ±1% which is the estimated maximum inaccuracy, which was assigned to the tritiated water standard of NBS, the standard deviation in the comparative measurement being less than 0.1%.[1]

The accuracy of Carbon-14 standard No. 4222 is listed as $tS_m + \Delta = 3\%$, where $tS_m = \pm 0.5\%$ is the computed standard error at the 99.73% confidence level (3.23σ) of comparison measurements and $\Delta = \pm 2.5\%$ is the linear sum of the systematic errors.[2]

The estimated limits for absolute accuracy of the standard capsules set by comparison measurements with reference standards of NBS are for tritium capsules ±1.2% and for carbon capsules ±3.2%.

Deviation in activity between different capsules depends upon the accuracy of the automatic pipetting machine. Experience has shown that the machine in use doses the isotope with a standard deviation of ±0.2%.

The stability of the activity of the capsules depends upon the purity of the carrier compound and ambient conditions. The tristearate used during the preparation of capsules was found to contain so many light components that chromatographic purification had to be performed prior to dosing. Stability was also improved when the non-active carrier compound was added, resulting in a low specific activity of the capsule. With regard to ambient temperature, the capsules have been found to be stable for several months in an environment between room temperature and -20°C. The long-term stability is still being evaluated, so results are not yet available. Dosing of the quenching agent with the evaporation device is satisfactory for toluene, xylene and dioxane-based scintillation systems. The equilibrium obtained in scintillation solutions has been so good that no significant differences have been observed when comparing quench curves with manually prepared quencher additions. With increasing viscosity of the scintillation solution, however, the quenching agent becomes less evenly distributed, and the resulting curves begin to diverge from each other. This means that the quenching agent must be added manually for samples with high viscosity as emulsion samples. Figures 5, 6, 7, 8, 9 and 10 illustrate quench correction curves obtained for different scintillation solutions, either done manually or with automatic evaporation.

The preceding discussion has presented a method for determining quench correction curves in liquid scintillation counting. The main significance of this technique is that it simplifies and expedites practical procedures in preparing the correction curves. The observation points obtained by means of the evaporation device accurately correlate with results of other methods. In addition, this evaporator approach produces many points over the entire quenching range and, depending upon the type of sample and investigation, any desired number of these points can be selected.

As a final note, the use of an internal standard capsule might be widened to other fields of practical isotopic work, such as standards and means of control. This requires, however, that the manufacture of capsules is developed for industrial production and that a reliable quality control is established.

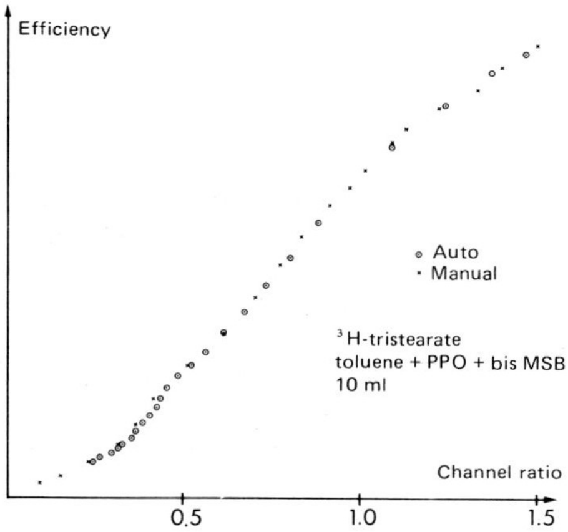

Figure 5

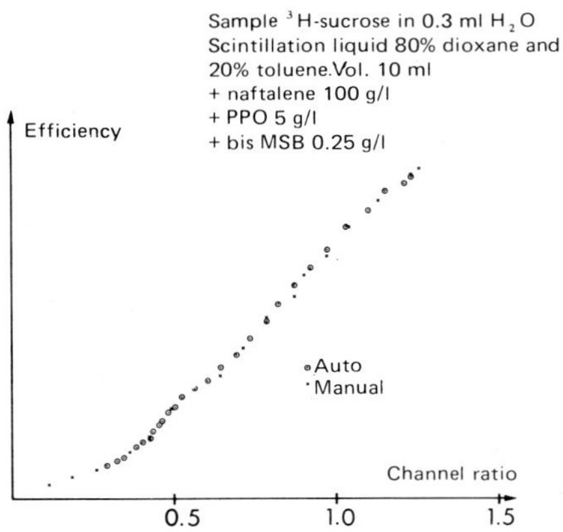

Figure 7

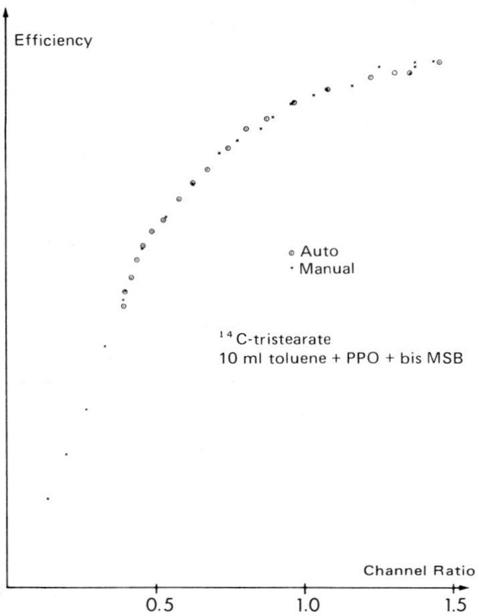

Figure 6

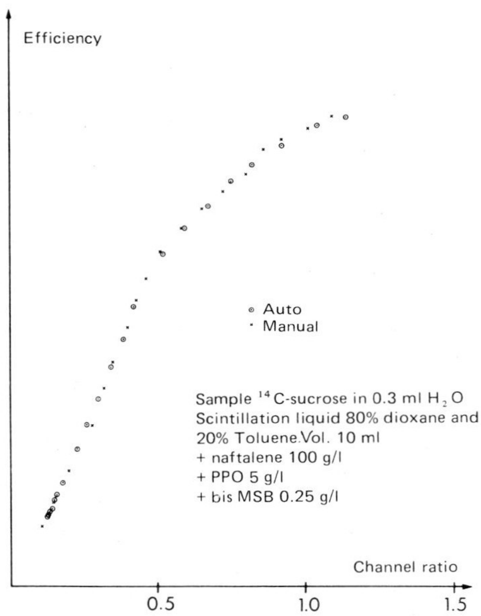

Figure 8

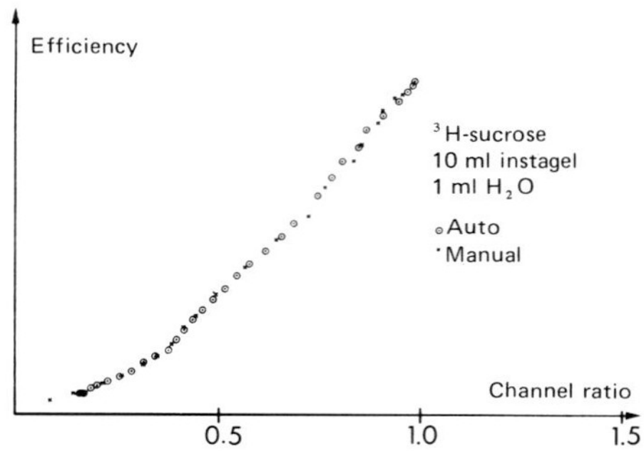

Figure 9

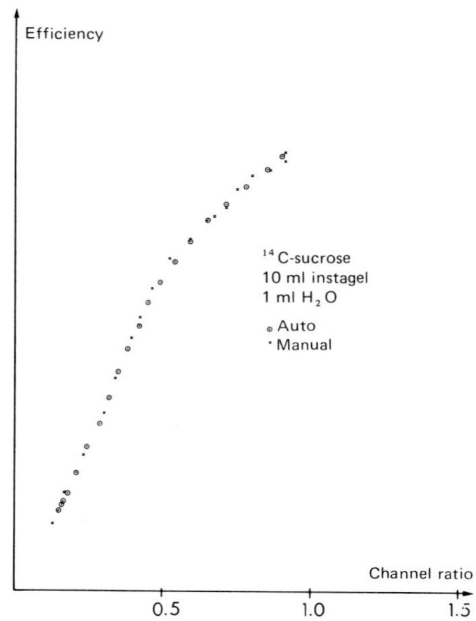

Figure 10

SUMMARY

A new method for quench correction in liquid scintillation counting has been described. The active internal standard, in dry form, is added in a glass capsule. The quenching agent is added by means of a special evaporating device while the counter is repeatedly tabulating the counting results of c.p.m. and channels ratio. The reproducibility of the standard capsules has shown a standard deviation of 0.2%. The absolute accuracy depends mainly on the accuracy of the reference materials used. The evaporation device is suitable when using low viscosity scintillation mixtures.

REFERENCES

1 S.B. Garfinkel, W.B. Mann, R.W. Medlock and O. Yura, Intern. J. Applied Radiation and Isotopes **16**, 27 (1965).

2 W.B. Mann, R.W. Medlock and O. Yura, Intern. J. Applied Radiation and Isotopes **15**, 351 (1964).

DISCUSSION

F.R. Jacobsberg: You suggested that your evaporator could be used for colour quench; presumably involving involatile materials. How does this work?

E. Soini: This is a problem. We have used a number of compounds but these will also involve chemical quenching.

P. Johnson: You intrigued people, I think, with your implication that you may have thought of ways of applying your method to colour quenching, but obviously you have not done so. As you realise, many of us who are using biological samples are unable to apply a single chemical quench curve, such as you obtain with carbon tetrachloride, in some of the quenching situations which we encounter. 'Your elegant method would obviously have greater general interest if it were applicable to such problems, but obviously it is restricted to volatile quenching agents such as carbon tetrachloride.

E. Soini: Yes, it is true that it is only applicable to volatile quenching agents; it is difficult to find suitable coloured quenching agents.

J.A. Heslop: How do you obtain the efficiency of counting and correlate this with an external standard channels ratio, as the counting efficiency is changing with time? What errors are introduced?

E. Soini: The efficiency is taken as the mean of two consecutive counts. As the rate of evaporation of quencher is slow, no large error is introduced.

Chapter 9

A Flexible FORTRAN IV Program for Scintillation Counter Data Processing with Optional Features for User's Specific Requirements

D.E. Bowyer and J.D. Pearson

Department of Pathology, University of Cambridge,
Tennis Court Road, Cambridge

INTRODUCTION

Nowadays most users of scintillation counters expect automatic processing of raw data in order to calculate the absolute activity in their samples and possibly to perform further analysis of results. The techniques which have been described include:

1. simple production of counts per unit time within a calculator in the counter (for example, continuously during counting as in the Packard 3375, which also computes the standard deviation of the counts, or at the end of counting as in the Nuclear Chicago Mk. II);
2. 'on-line' computation by either a small computer built into the counter (as in the Intertechnique SL40) or by a desk-top calculator such as the Olivetti Programma 101, or Diel calculator as used with the Tracerlab Corumatic 2000;
3. transfer of data to a large computer, followed by processing the data as a batch-job, either off-line or by direct access via an on-line time sharing terminal.

The main advantages of the large computer are that the data can be readily stored and edited, and, because of the speed and storage capacity of the machine, more sophisticated and extensive processing of the data can be performed.

The kinds of data reduction which users require depend upon the type of experiments they perform and, in any environment where there are a number of users of one machine, there may be a wide variety of demands. Descriptions of a number of programs, written for large computers have appeared in the literature. Most have been written in high level languages such as FORTRAN or ALGOL, and thus can be used with only minor modification at most computer installations. Many of these programs have been cited in the review by Spratt,[1] other and more recent examples are those of Johnson et al.;[2] Barber and Bourne;[3] Stanley;[4] Assailly et al.;[5] Ayrey and Evans;[6] Deterding;[7] Haissig and Schipper;[8] Yund et al.;[9] Cramer et al.;[10] Hansen and Carroll;[11] Forker and Wycoff;[12] Boeckx et al.;[13] O'Toole and Osburn.[14] These programs have usually been written for a particular application and with the needs of a few users in mind. They thus have the following disadvantages: (1) the format of the data expected by the program relates to one counter; (2) it is assumed that all of a given batch of data is to be processed in the same way; (3) the instructions concerning which samples are to be processed, methods of quench correction and degree of further data reduction are often not self-explanatory. The user is often required to provide instructions in a numerical, abbreviated form, a scheme which is not easy to remember nor provides a reliable system in a multi-user environment.

GENERAL FACILITIES OF THE PROGRAM

In order to overcome some of these disadvantages we have written a program in FORTRAN IV (for the IBM 360/370 G1 and H extended compilers and using list-directed READ routines), which provides very general facilities for processing scintillation counter data, according to simple self-explanatory instructions supplied by the user. The program which compiles in about 150K and runs in approximately 92K bytes of main store, performs the following functions:

1. reads data, from paper tape, cards, magnetic tape or disc, as required, in a format which is specified by the user;
2. translates users' instructions;
3. outputs a copy of instructions and raw data to facilitate data checking;
4. calculates and outputs counts per minute (c.p.m.) for each sample for each channel and also channels ratios if requested;
5. performs subtraction of background from samples. If determination of efficiency is by the External standard Channels Ratio (ECR) method (see below), quench corrected backgrounds may be determined by reference to a set of quenched backgrounds.[15,16] When the Sample Channels Ratio (SCR) method is used, quench corrected background subtraction is not used, but a mean background is used;
6. calculates disintegrations per minute (d.p.m.) for samples containing up to three isotopes. The following methods of computation can be selected:
   i. counting efficiencies for each isotope, which are assumed constant for a set of samples, are supplied by the user;
   ii. counting efficiencies, assumed constant for a set of samples, are computed from samples containing reference amounts of activity;
   iii. quench correction for a set of samples is applied by SCR or ECR methods. In the SCR and ECR modes, the curves relating counting efficiency to channels ratio are described by second order polynomials. The coefficients of the polynomials may either be calculated by the program from sets of reference standards or supplied by the user. Optionally, quenched coefficients which have previously been computed or supplied may be normalised by reference to a 'reference' standard or standards.[17] Using the ECR mode, d.p.m. for three mixed isotopes may be computed from counts in three channels. In the SCR mode, d.p.m. are calculated assuming no overspill of the lower energy isotope into the ratio channels used for the high energy isotope[18-20] and d.p.m. for a maximum of two mixed isotopes can be calculated;
7. d.p.m. for each isotope are corrected for decay during the time elapsed since the first sample began counting. The time elapsed during sample transfers, etc. in the sample changer is allowed for. Assumed values for individual counters are selected when the user specifies the format of the data input;
8. the mass of each labelled species in each sample is computed, if the specific activity of each isotope is provided;

9. additional subroutines can be called to perform further arithmetical procedures according to individual requirements. The basic program does not have to be altered because further subroutines communicate with values already computed via 'common blocks' of the program.

FORMAT OF INSTRUCTIONS

The instructions allow the user to subdivide the data file into groups by sample pot number, each group being considered as one data set. Each data set may be processed in different ways according to the users' requirements by instructions punched on corresponding groups of cards, called instruction sets.

Each of the types of instruction card has as its first argument a name or an alias which reflects the nature of the instructions it contains. This name is then followed by further arguments, which provide information required for the instructions to be executed. The cards are as follows:

Card	First Argument	Function of Instruction Card
1	'*'	separation card between instruction sets.
2	'OPTIONS' 'O'	defines the counter used and format of data, for example 'NC1' defines 'one-line' output format from a Nuclear Chicago Mk. II counter.
3	'TITLE' 'T'	allows the addition of a title for the currently selected data set on the card following.
4	'ISOTOPES' 'I'	defines the isotopes present in the currently selected data set and the channel optimised for that isotope. Options for channels are 'A' 'B' 'C' and for isotope names '3H' '14C' '32P' '35S' '45CA' '125I' '131I'.
5	'SAMPLES' 'S'	defines the position (from the pot numbers on the scintillation counter output) of the currently selected data set.
6	'BACKGROUND' 'BG'	defines the position of background pots for the currently selected data set. If quenched background correction using ECR is required, the location of background pots is provided on the 'MODE' card, and this card is ignored.
7	'MODE' 'M'	defines the way in which d.p.m. are to be calculated for the currently selected data set. Permitted options which are specified by the second argument on this card are: 'READ' — reads the counting efficiencies of each isotope in each used channel; 'CALC' — calculates the counting efficiencies of each isotope in each used channel from specified pots containing reference material; 'SCR' — Sample Channels Ratio method; 'ECR' — External standard Channels Ratio method.
8	'REFERENCE' 'R'	this card is optional and allows 'normalisation' of quench correction coefficients by reference to a 'reference' standard.

9	'ADD' 'A'	this card is optional and defines additional subroutines provided by the user.

In order to process a single data set, one instruction set, consisting of cards 1—7 must be present, cards 8 and 9 are optional. To process several data sets or one data set in several ways, further instruction sets are provided. Only instruction cards which are to be changed are added; the other cards may be omitted, in which case the values of the arguments on these cards, which were defined for the previous data set, remain unchanged. Figure 1 shows an example of two instructions for the calculation of d.p.m. by the ECR and then SCR modes on one set of data, and Fig. 2 the output format for d.p.m.

```
Card  Instruction

 1    '*' /
 2    'OPTIONS' 'NC1' /
 3    'TITLE' /
      TEST RUN OF QUENCH SET PLUS SAMPLES
 4    'ISOTOPES' '3H' 'A' '14C' 'B' /
 5    'SAMPLES' 176 187 /
 6    'BACKGROUND' /
 7    'MODE' 'ECR' 'S' 234 8 10000 0, 218 8 10000 0, 250 8 /
 8    '*' /
 9    'M' 'SCR' 'B/C' 'S' 234 8 10000 0, 218 8 10000 0, 250 8 /
10    '*' /

The arguments on the 'MODE' cards (cards 7,9 in the example)
signify:

      'B/C' - ratio of channel B over channel C
      'S'   - quench coefficients to be computed from standards,
              then for each isotope
            - location of first pot containing standard (e.g. 234)
            - number of standards in quenched set (e.g. 8)
            - d.p.m. in standard (e.g. 10000)
            - time elapsed since standard was calibrated (e.g. 0)
              then for backgrounds
            - location of first pot containing background
              (e.g. 250)
            - number of backgrounds in quenched set (e.g. 8)
```

Fig. 1. An example of two instruction sets for processing a data set by the ECR and SCR modes.

```
TEST RUN OF QUENCH SET PLUS SAMPLES
NO BG POTS SPECIFIED ON BG CARD: NO BG COUNTS HAVE BEEN SUBTRACTED,
UNLESS BG POTS OR COEFFICIENTS HAVE BEEN DEFINED ON THE 'MODE' CARD,
IN WHICH CASE A NOTE OF THESE APPEARS BELOW

BG COEFFICIENTS OBTAINED FROM 'MODE' 'ECR' CARD
CHANNEL A  -.355E+02  0.282E+02  -.366E+01
CHANNEL B  -.701E+01  0.754E+01  -.962E+00
CHANNEL C  0.0        0.0        0.0

THE THREE COEFFICIENTS OF THE QUENCH CURVE
C.EFF = QC1 + QC2*RATIO + QC3*(RATIO) **2
ARE GIVEN BELOW FOR EACH ISOTOPE IN EACH USED CHANNEL

3H    IN CHANNEL A    0.166E+02    0.892E+01   -0.959E+00
14C   IN CHANNEL A    0.283E+02   -0.744E+01    0.627E+00
3H    IN CHANNEL B    0.194E+01   -0.122E+01    0.194E+00
14C   IN CHANNEL B   -0.467E+02    0.518E+02   -0.572E+01

TEST RUN OF QUENCH SET PLUS SAMPLES

DPM OF EACH ISOTOPE FOR THIS SAMPLE SET

SAMPLE    POT     DPM      DPM
 NO.      NO.     FOR      FOR
                  3H       14C

  1      176.        0.     9921.
  2      177.        0.     9966.
  3      178.       28.     9992.
  4      179.        0.    10137.
  5      180.    10218.        0.
  6      181.    10232.        4.
  7      182.    10209.        1.
  8      183.    10190.        0.
  9      184.    10219.     9733.
 10      185.    10179.     9966.
 11      186.    10161.    10037.
 12      187.    10177.    10053.
```

```
BG CPM OBTAINED FROM 'MODE' 'SCR' CARD ARE:
CHAN A    18.14   CHAN B    7.56   CHAN C    2.89

THE THREE COEFFICIENTS OF THE QUENCH CURVE
C.EFF = QC1 + QC2*RATIO + QC3*(RATIO) **2
ARE GIVEN BELOW FOR EACH ISOTOPE IN EACH USED CHANNEL

3H    IN CHANNEL A    0.221E+02    0.986E+01   -0.160E+01
14C   IN CHANNEL A    0.208E+02   -0.587E+01    0.472E+00
3H    IN CHANNEL B    0.199E+01   -0.190E+01    0.457E+00
14C   IN CHANNEL B   -0.401E+01    0.476E+02   -0.758E+01

TEST RUN OF QUENCH SET PLUS SAMPLES

DPM OF EACH ISOTOPE FOR THIS SAMPLE SET

SAMPLE    POT     DPM      DPM
 NO.      NO.     FOR      FOR
                  3H       14C

  1      176.        0.     9959.
  2      177.        0.     9922.
  3      178.        0.     9984.
  4      179.        0.    10097.
  5      180.    10077.        0.
  6      181.     9892.        0.
  7      182.    11890.        0.
  8      183.    10917.        0.
  9      184.    10057.     9798.
 10      185.    10045.     9949.
 11      186.    10029.    10006.
 12      187.    10028.    10076.
```

Fig. 2. Output giving d.p.m. for the instruction set described in Fig. 1.

DISCUSSION

General

We have found, in a multi-user environment, that the processing of scintillation counter data by computer is made much simpler and more reliable for most users by the provision of a program which allows input of instructions in a simple and self-explanatory format. A similar approach has recently been implemented in a program in PLAN4 for a 'pseudo-on-line' system using the ICL1900.[21]

Our program as it is presently implemented allows the choice between the two commonest methods of quench correction, SCR and ECR. The comparison of results produced by the two methods permits detection of inhomogeneity in sample preparation as suggested by Bush.[22]

It would be desirable to add further subroutines to provide options for quench correction by the so-called parabolic curve-fitting channels ratio method,[23] and by the SCR method in the presence of overspill by the technique of Hansen and Carroll.[11] These routines could easily be added and would be invoked by new calling arguments on the 'MODE' card, e.g. 'PARA' and 'SCR1', without interference with the rest of the program. Implementation of the Hansen and Carroll techniques would also allow quench correction of backgrounds using SCR as suggested by Stanley,[24] again without alteration to other options.

Concerning error calculations

Scintillation counting finds most use in the measurement of isotopes used as markers of molecules in physical, chemical and biological experiments. The experimental errors associated with typical experiments derive from:

1. inherent variability in the experimental system, e.g. biological variation;
2. inaccuracies in conducting the experiment — e.g. pipetting, weighing, etc. up to and including the preparation of sample for scintillation counting;
3. error due to the random nature of radioactive decay — this is assumed to follow the Poisson distribution model, although will not do so if the counter is insufficiently stable — see Matthijssen and Goldzieher,[25] cited by Spratt;[1]
4. errors due to machine instability, irregularity of sample vials and vial positioning;[4]
5. errors in performing quench correction by external standard due to irregularity of sample vials and positioning of external standard.[4]

It is obviously desirable to attempt an estimate of errors associated with the most significant error terms and as far as counting is concerned it is advisable to estimate both errors due to counting statistics for each sample and the error associated with fitting quench correction curves; the latter also being compounded of errors due to pipetting of the standards and counting statistics. It is probably most informative if these errors are presented separately as ± 95% confidence limits and then an attempt is made to combine them to give an estimate of the overall 95% confidence limits for the counting process. In this way, the major source of error, viz. sample counting or curve fitting, can easily be identified. Furthermore, it is probably advisable to calculate the error which is apparent in repeated counting of each sample and to compare it with the theoretical counting statistics. In this way machine drift may be detected.[7]

Concerning errors in counting statistics

It is obviously straightforward to compute errors associated with the counts accumulated in each channel and to combine these with the errors in counting the background in each channel, to give a net channel count rate error.[1, 26]

Remembering that in combining errors, the following rules apply, where $\sigma^2(A)$ is the variance of A, and $\sigma(A)$ is the standard deviation of A.

If $$C = A \pm B$$

then $$\sigma^2(C) = \sigma^2(A) + \sigma^2(B) \qquad (1)$$

and $$\sigma(C) = \sqrt{\sigma^2(A) + \sigma^2(B)}$$

If $$C = A \times B \text{ or } C = A/B$$

then $$\frac{\sigma^2(C)}{C^2} = \frac{\sigma^2(A)}{A^2} + \frac{\sigma^2(B)}{B^2}$$

$$\sigma^2(C) = C^2\left[\frac{\sigma^2(A)}{A^2} + \frac{\sigma^2(B)}{B^2}\right] \qquad (2)$$

and $$\sigma(C) = C\sqrt{\frac{\sigma^2(A)}{A^2} + \frac{\sigma^2(B)}{B^2}}$$

Let sample gross count = S; background count = B; sample counting time = T_S; background counting time = T_B; sample gross count rate = SCPM; background count rate = BCPM; and let net count rate = NCPM.

$$SCPM = S/T_S \qquad (3)$$

and $$BCPM = B/T_B \qquad (4)$$

We wish to find the variance in the net count rate, and hence the standard deviation and 95% confidence interval.

Now $$NCPM = SCPM - BCPM$$

from (3) and (4) $$NCPM = S/T_S - B/T_B$$

Using (1) to combine variances

$$\sigma^2(NCPM) = \sigma^2(S/T_S) + \sigma^2(B/T_B) \qquad (5)$$

Now $\sigma^2(S/T_S)$ is simplified by combining variances as in (2).

Thus $$\sigma^2(S/T_S) = (S/T_S)^2\left[\frac{\sigma^2(S)}{S^2} + \frac{\sigma^2(T)}{T^2}\right]$$

but $\frac{\sigma^2(T)}{T^2} \ll \frac{\sigma^2(S)}{S^2}$ because the counter time is sufficiently precise. Therefore

$$\sigma^2(S/T_S) = (S/T_S)^2\left(\frac{\sigma^2(S)}{S^2}\right)$$

$$= \frac{\sigma^2(S)}{T_S^2}$$

Likewise $$\sigma^2(B/T_B) = \frac{\sigma^2(B)}{T_B^2}$$

Substitute in Eqn. (5). Hence

$$\sigma^2(\text{NCPM}) = \frac{\sigma^2(S)}{T_S^2} + \frac{\sigma^2(B)}{T_B^2}$$

For the Poisson distribution, $\sigma^2(S) = S$, $\sigma^2(B) = B$. Therefore

$$\sigma^2(\text{NCPM}) = \frac{S}{T_S^2} + \frac{B}{T_B^2}$$

$$= \frac{\text{SCPM}}{T_S} + \frac{\text{BCPM}}{T_B}$$

Hence also σ(NCPM) and ± the 95% confidence limit (CI) where CI = 1.96 σ(NCPM). In practice the 2σ error is often used, being the 95.45% confidence limit, which is slightly more conservative.

Concerning the error in interpolating a quench correction curve

Determination of d.p.m. by ECR and SCR involves quench correction curves, which are often represented by a polynomial relating counting efficiency to channels ratio. The errors associated with interpolation of such a curve arise from counting statistics for the standards and also because, whatever the order of the polynomial, it may not represent the real behaviour of the machine. This is so even within the range of standards, let alone in extrapolated regions beyond. Some authors consider that other functions, for example logarithmic, give a better representation of machine performance.[6,7] We will consider the error associated with fitting a second order polynomial although the treatment is general,[27] allowing handling of high order polynomials, and is performed in the program by an IBM supplied subroutine.

Let $$E = a_0 + a_1R + a_2 (R)^2 \qquad (6)$$

where E is the efficiency and R is the ECR.

Assume we have n standards and thus produce n values of E and R given by e_n and r_n. The coefficients of this curve are generated by the method of least squares, i.e. minimising the sum of squares of the observed values (= e) from the expected values (= E) predicted by Eqn. (6). That is, we choose the values of a_i

which minimise $$Q = \sum_{i=1}^{n} (e_i - E_i)^2$$

which is $$Q = \sum_{i=1}^{n} (e_i - a_0 - a_1r_i - a_2r_i^2)^2$$

The values of a_1, a_2 are given by the solution of the set of simultaneous equations

$$a_1 C_{11} + a_2 C_{12} = D_{e1} \qquad (7)$$

$$a_2 C_{21} + a_2 C_{22} = D_{e2} \qquad (8)$$

where

$$C_{11} = \sum_{i=1}^{n} (r_i - \bar{r})\,(r_i - \bar{r})$$

$$C_{12} = \sum_{i=1}^{n} (r_i - \bar{r})\,(r_i^2 - \overline{r^2})$$

$$C_{21} = \sum_{i=1}^{n} (r_i^2 - \overline{r^2})\,(r_i - r_i)$$

$$C_{22} = \sum_{i=1}^{n} (r_i^2 - \overline{r^2})\,(r_i^2 - \overline{r^2})$$

$$D_{e1} = \sum_{i=1}^{n} (e_i - \bar{e})\,(r_i - \bar{r})$$

$$D_{e2} = \sum_{i=1}^{n} (e_i - \bar{e})\,(r_i^2 - \overline{r^2})$$

and

$$a_0 = \bar{e} - a_1\,\bar{r} - a_2\,\overline{r^2} \qquad (9)$$

Now in any subsequent determination of E from a single value of R we may determine an error in E from

$$\begin{aligned} \sigma^2(E) &= \sigma^2(\bar{e}) + (E - \bar{e})^2\,\sigma^2(a_1) \\ &\quad + (E^2 - \overline{e^2})\,\sigma^2(a_2) \\ &\quad + 2\,(E - \bar{e})\,(E^2 - \overline{e^2})\ \mathrm{CoV}\,(a_1\, a_2) \end{aligned} \qquad (10)$$

where

$$\sigma^2(\bar{e}) = \frac{s^2}{n} \qquad (11)$$

and s^2 is the residual variation of the regression.

Fortunately in solving Eqns. (7) and (8) for a_1 and a_2 if the matrix

$$D = \begin{vmatrix} C_{11} & C_{12} \\ C_{21} & C_{22} \end{vmatrix}$$

is inverted to give

$$1/D = \begin{vmatrix} C^{11} & C^{12} \\ C^{21} & C^{22} \end{vmatrix}.$$

then the terms C^{zz} are proportional to the variances and covariances of the coefficients a_1, a_2 as follows:

$$\sigma^2(a_1) = s^2 \; C^{11}$$

$$\sigma^2(a_2) = s^2 \; C^{22}$$

$$\text{CoV}(a_1 \; a_2) = s^2 \; C^{12}$$

$$= s^2 \; C^{21}$$

Hence from Eqns. (10) and (11)

$$\sigma^2(E) = \frac{s^2}{n} + (E - \bar{e})^2 \, s^2 \, C^{11} + (E^2 - \overline{e^2})^2 \, s^2 \, C^{22} + 2\,(E - \bar{e})\,(E^2 - \overline{e^2})\, s^2 \, C^{12} \quad (12)$$

In the program, this error calculation is performed for each determination of counting efficiency and the 95% confidence interval (CI) calculated from

$$CI = t_{n-2}\,\sigma(E)$$

where the value of t, Student's t for n - 2 degrees of freedom, is calculated by the program.

Concerning the combination of errors of counting statistics and interpolation of quench correction curves

Single isotope case. In this case

$$DPM = \frac{NCPM}{E}$$

Thus

$$\sigma^2(DPM) = (DPM)^2 \left[\frac{\sigma^2(NCPM)}{(NCPM)^2} + \frac{\sigma^2(E)}{(E)^2}\right]$$

This case is straightforward, but notice that if the net background count rate is determined from a quench corrected background, then $\sigma^2(BCPM)$ is determined as in Eqn. (12) and

$$\sigma^2(NCPM) = \frac{SCPM}{T_S} + \sigma^2(BCPM)$$

Double isotope case. In this case the combination of errors by the methods already discussed becomes more complicated. Assume the ECR method is used and that in counting the sample overspill occurs from the low energy sample channel, A, to the high energy sample channel, B. Let E_{ij} be the counting efficiency for isotope i in channel j. Then

$$(CPM_A) = E_{1A}\,(DPM_1) + E_{2A}\,(DPM_2)$$

$$(CPM_B) = E_{1B}\,(DPM_1) + E_{2B}\,(DPM_2)$$

hence

$$DPM_1 = \frac{E_{2B}(CPM_A) - E_{2A}\,(CPM_B)}{E_{1A}\,E_{2B} - E_{2A}\,E_{1B}} \qquad (13)$$

$$DPM_2 = \frac{E_{1B}\,(CPM_A) - E_{1A}\,(CPM_B)}{E_{2A}\,E_{1B} - E_{2B}\,E_{1A}} \qquad (14)$$

The values of $\sigma^2(E_{ij})$ are computed for the four quench correction curves as usual for each sample and the net count rate errors for CPM_A and CPM_B are also known. Hence by the use of Eqns. (1) and (2) for combinations of errors, $\sigma^2(DPM_1)$ and $\sigma^2(DPM_2)$ could be computed.

From inspection of Eqns. (13) and (14), however, it is clear that the expression for σ^2(DPM) is arithmetically cumbersome and does not apparently have a general form. The computation would probably best be performed by the use of two subroutines for combining errors. This has not so far been implemented.

General consideration. The computational untidiness associated with estimation of propagated errors could probably be overcome by the use of special techniques[28] such as (1) rounded interval arithmetic, (2) unnormalised arithmetic, (3) normalised floating point arithmetic with an index of significance, (4) automatic controlled precision arithmetic. Certainly some compilers have been written for rounded interval arithmetic in high level language such as Algol 60.[29] It is generally held, however, that measurement of propagation of errors by these techniques may yield excessively pessimistic estimates.

REFERENCES

1 J.L. Spratt in Liquid Scintillation Counting, Vol. 2, Heyden, London, 1972, p.245.

2 P. Johnson, P.A. Rising and J.T. Rising, in Liquid Scintillation Counting, Vol. 2, Heyden, London, 1972, p.267.

3 H.E. Barber and G.R. Bourne in Liquid Scintillation Counting, Vol. 2, Heyden, London, 1972, p.279.

4 P.E. Stanley in Liquid Scintillation Counting, Vol. 2, Heyden, London, 1972, p.285.

5 J. Assailly, C. Bader, J.-L. Funck-Brentano and D. Pevel, in Liquid Scintillation Counting, Vol. 2, Heyden, London, 1972, p.293.

6 G. Ayrey and K.L. Evans in Liquid Scintillation Counting, Vol. 2, Heyden, London, 1972, p.303.

7 J.H. Deterding in Liquid Scintillation Counting, Vol. 2, Heyden, London, 1972, p.313.

8 B.E. Haissig and A.L. Schipper, Analyt. Chem. 42, 1456 (1970).

9 M.A. Yund, E.W. Yund and F.C. Kafatos, Biochem. Biophys. Res. Commun. 43, 717 (1971).

10 C.F. Cramer, M. Nicholson, C. Moore and K. Teng, Intern. J. Applied Radiation and Isotopes 22, 17 (1971).

11 D.L. Hansen and C.O. Carroll, Intern. J. Applied Radiation and Isotopes 22, 677 (1971).

12 F.L. Forker and D. Wycoff, Anal. Biochem. 45, 107 (1972).

13 R.L. Boeckx, D.J. Protti and K. Dakshinamurti, Anal. Biochem. 53, 491 (1973).

14 J.J. O'Toole and J.O. Osburn, Intern. J. Applied Radiation and Isotopes **19** 821 (1968).

15 B. Scales, Anal. Biochem. **5**, 489 (1963).

16 P.D. Klein and W.J. Eisler, Analyt. Chem. **38**, 1453 (1966).

17 B.E. Gordon, W.T. Shebs and R.U. Bonnar, J. Amer. Oil Chemists' Soc. **44**, 711 (1967).

18 L.A. Baillie, Intern. J. Applied Radiation and Isotopes 8, 1 (1960).

19 E.T. Bush, Analyt. Chem. **35**, 1024 (1963).

20 R.W. Hendler, Anal. Biochem. **7**, 110 (1964).

21 B. Seaton, Biochemical Soc. Trans. **1**, 517 (1973).

22 E.T. Bush, Intern. J. Applied Radiation and Isotopes **19**, 447 (1968).

23 B.D. Caddock, P.T. Davies and J.H. Deterding, Intern. J. Applied Radiation and Isotopes **18**, 209 (1967).

24 P.E. Stanley, this symposium, 'Fundamental Approaches for the Assessment of Chemical and Colour Quenching in Backgrounds and Samples', p. 65.

25 C. Matthijssen and J.W. Goldzieher, Anal. Biochem. **10**, 401 (1965).

26 G.E.A. Wyld, in The Current Status of Liquid Scintillation Counting (ed. E.D. Bransome), Grune and Stratton, New York, 1970, p. 69.

27 O.L. Davies (ed.), Statistical Methods in Research and Production, 3rd Edn., Oliver and Boyd, London, 1967, p. 238ff.

28 E. Hansen (ed.), Topics in Interval Analysis, Clarendon Press, Oxford, 1969.

29 K. Nickel, in Topics in Interval Analysis (ed. E. Hanson), Clarendon Press, Oxford, 1969.

DISCUSSION

E.B. Mueller: Does your program average multiple countings of recycled samples?

D.E. Bowyer: Not at the moment but this option can be added and would, I think, be a desirable addition.

E.B. Mueller: I believe that the statistical variation in multiple countings is more important than the expected statistical error for a single counting.

D.E. Bowyer: I would not say more important in general, but certainly valuable for detecting erroneous counts due to machine malfunction, etc. during one counting period. The problem in everyday work in association with experimentalists, biologists at any rate, is persuading them to do repeated counting.

L.A. Wegner: Why do you restrict the polynomial fit to the second order? Using a modified Nuclear-Chicago program (which makes available fits of up to the order n 10) the experience of our laboratory is: performing the fitting routinely for three different types of Liquid Scintillation Spectrometers, in most cases the 'best' polynomial fit (with a minimum n) is achieved with values of n: $-4 \leq n \leq 6$.

D.E. Bowyer: You must be aware that the results produced by using higher-degree fits may not be correct in the regions between the points used for fitting.

L.A. Wegner: There may arise difficulties of this kind. But if one eliminates fits with 'oscillating' behaviour and only permits the use of functions with monotonic behaviour (inside of the interval of approximation) the difficulties mentioned will not occur.

D.E. Bowyer: I agree that the use of higher order polynomials will probably give a mathematically better fit to the experimental points as judged by the residual variation of the regression. The point at issue, however, is whether this computed curve is a better representation of the real behaviour of the machine. It is generally agreed that the oscillatory solution produced by a high order polynomial does not represent the real relationship between efficiency and, say, ECR. The problem is to find a balance between representation of machine behaviour and a mathematically defined best-fit, by least squares, to the experimental points. This, of course, depends upon the machine performance as determined, for example, by the vial variation, vial geometry and reproducibility of positioning of the external standard as discussed by Stanley {P.E. Stanley in Liquid Scintillation Counting, Vol. 2, (Eds. B. Scales, M.A. Crook and P. Johnson) Heyden and Son, London, 1972, p.285}. We believe that the counting conditions and range of quench correction standards should be chosen so that the quench correction curve shows, by graphing of repeated counts, neither maxima nor minima, thus avoiding the possibility of multiple solutions. The presence of a graphing routine for quench curves in a program is obviously very desirable for this reason. The usefulness of the computed curve as a representation of the machine behaviour may then be judged by testing for a significant difference between observed values produced by repeated (i.e. recycled) counting of standard(s) and the value predicted from the curve (F). This is simply achieved by an F test of the ratio of the variances of the curve fitting, $\sigma^2(E)$, to the variance associated with repeated counting, $\sigma^2(S)$. If there is no significant difference, then obviously the curve at that point is acceptable. If there is a significant difference, on the other hand, it would be necessary to compute a new curve, using a higher order polynomial, and to test at the point of interpolation again. We have not so far implemented this technique in our program, but it is obviously simple enough and assuming a reproducible, constant error for each counter, associated with repeated, recycled counting of a standard, the relevant values of $\sigma^2(S)$ for each machine could be invoked by the OPTIONS card which also defines data format and transfer times of the samples during sample changing.

SECTION III

SAMPLE PREPARATION : PART I

Chapter 10

Homogeneous Counting

B.E. Gordon

Koninklijke/Shell-Laboratorium, Amsterdam, The Netherlands
(Shell Research B.V.)[a]

INTRODUCTION

The rapid and astonishing advances in the technique of liquid scintillation counting have been due in large measure to the instrument designers. In the last 20 years such counters have gone from a single tube in a refrigerator-cum-glove box to the completely automatic, highly stable, computer coupled systems in use today. The speed with which new developments in electronics hardware and software logic have been integrated into liquid scintillation counters has been most gratifying. It is common practice for a user to insert several hundred samples into a counter on Friday and pick up his data computed as d.p.m., weight %, volume %, titer, etc. on Monday morning. Why, then, do we still get some wrong answers?

The explanation seems to be to a large extent due to the sample preparation methods, particularly of organic chemical or biological origin. It is simply that the great convenience and reliability of the modern instruments has led to the application of the methods to a bewildering variety of samples, many of which were and are unsuitable for analyses without careful preparation. As in most analytical techniques, the instrument has outstripped the sample preparative methods. This is also obvious from the literature where the bulk of recent publications is concerned with improvements in sample preparation, and very few with improved instrument performance.

A plenary lecture is supposed to review the current state of a subject, thus setting the stage for the working papers which follow. I wish to change that approach somewhat, but not completely, by presenting a selected review of the literature, not necessarily current, interspersed with personal opinions about the strengths or weaknesses of a particular technique. This will become most apparent in the preparation of biological samples which account for the bulk of samples today, although as a petroleum chemist I have rarely had to deal with such samples. However, I have followed the general literature over the years with a sense of increasing uneasiness over a situation developing to the point where the use of solubilisers, digesting systems, etc. seems to have gained a firm place in the radioanalytical arsenal as evidenced by the large number of papers on this subject and the steady availability of new reagent systems by research workers and by commercial houses. I hope to present a rationale for

[a] Home address: 1330 Brewster Drive, El Cerrito, California 94530, U.S.A.

the discontinuance of the use of such systems and a return to combustion as a sample preparative tool.

This chapter on homogeneous sample preparation will contain three sections — gases, inorganics and organics. Before I begin I suppose I should define a homogeneous sample (for liquid scintillation counting). This is a sample in which the composition of the volume traversed by the emitted radiation is, for all practical purposes, the same everywhere. Note that this, therefore, does not require true solution, i.e. homogeneity on the molecular level, but on a level considerably higher — perhaps $<0.1\mu$ radius for tritium. By this definition a cocktail homogeneous to one isotope may not be so to another because of different beta ray energies. In this sense true solution is a sufficient but not a necessary condition.

Considered in this way, gel counting with surfactants or cabosil is homogeneous if the droplet or particle sizes are small with respect to the beta particle energy. The reason for making this distinction is that homogeneous sample preparation does not restrict the user to true solutions (i.e. small molecular dimensions) but offers the option of using any techniques which disperses the sample in the counting system while avoiding absorption or adsorption losses.

GAS SAMPLES

This section is restricted to the analysis of gases based upon their solubility in the counting fluid rather than upon chemical reaction, as in CO_2 absorption by base.

Horrocks and Studier[1] in 1964 measured the solubility of some radioactive noble gases ($^{133/m}$Xe and ^{220}Rn) and found that if the counting vial had a small vapour space, the bulk of these gases would dissolve in the liquid. A factor expressing the ratio (cm^3 in liquid/cm^3 in vapour space) was five for Xenon-131, 32 for Radon-222 and 0.9 for Krypton-85. The latter value was too low for quantitative analysis. Later Horrocks[2] applied this to CO_2 in toluene where the ratio 4.4 was found, albeit at -20°C. Gordon et al.[3] applied the method to $H_2{}^{35}S$ in a toluene scintillator and Curtis et al.[4] applied it to tritium gas as well as Krypton-85. The method may be of interest for noble and other rather inert gases but it seems superfluous for reactive gases such as $^{14}CO_2$ which can be readily trapped in appropriate counting solutions containing bases like phenylethylamine. It is actively employed by us for counting hydrocarbon gases using a vial with a constricted neck with a small silicone rubber plug. The plug is vacuum tight and samples are injected by hypodermic syringe. It is the accepted method for the analysis of olefinic monomers used in polymerisation. Homopolymers made from these labelled olefins are also analysed and must have the same specific activity as the monomer before the work proceeds. Gaseous effluents from gas chromatography have also been trapped and counted. This is not strictly gas counting since compounds so analysed may be liquids or even solids at room temperature. On balance, gas analysis by liquid scintillation counting has only limited interest as indicated in the literature. The use of vacuum manifolds, sealed glass vials, etc. tends to limit its general popularity and I recommend the rubber plug plus hypodermic approach for most convenient operation. Calibrated hypodermics are used for highest accuracy and one can expect sampling errors less than 2% relative.

INORGANIC SAMPLES

Liquid scintillation cocktails are inherently inappropriate for ionic substances. However, the great convenience of this technique has led to considerable efforts in sample preparation to overcome the hostility of the usual scintillation solvents towards such materials. Three general approaches are: (1) solubility by use of more polar solvents such as ethers or alcohols; (2) alteration of the sample, particularly metals, to an oil soluble form; (3) avoidance of wall adsorption losses or precipitation by providing a very large surface for adsorption, viz. cabosil.

Carr and Parson[5] measured Calcium-45 as the chloride in a cocktail having 30% ethanol, Fairman and Sedlet[6] determined Lead-210 by conversion to the nitrate and solution in a dioxane scintillator. Methods to analyse both radio-isotopes of iron (^{55}Fe and ^{59}Fe) by liquid scintillation counting have employed precipitation as the hydroxide followed by conversion to the chloride, reduction with ascorbic acid and direct solution. Separation of the two isotopes was readily achieved because of the weak 5.9 keV X-ray of Iron-55 and the strong 0.4—1.5 MeV betas of Iron-59.

In this regard it is often the belief that the development of an aqueous compatible cocktail (e.g. Bray's solution, dioxane, ethanol/toluene) is adequate for aqueous solutions of salts. This is certainly a dangerous assumption since the behaviour of the salt may be quite different from that of the water. To avoid this error Bush[7] devised a double ratio technique for determining whether one was counting in true solution. The method involves plotting the external standard two channel ratio versus the sample two channel ratio. Departure from a standard curve previously developed is indication of adsorption or precipitation of the sample. The method has been successfully applied by Glass[8] to polymers and by Laurencot and Hempstead.[9] In addition, transfer of the sample to a fresh counting vial will indicate whether any activity is fixed to the walls of the vial, according to Bush.[7] However, it might not indicate the formation of a fine or flocculant precipitate which is also transferred. This useful technique really applies to homogeneous counting of which solution counting is one example. We have applied the Bush[7] technique to polyolefin counting and found that an obvious gel also fits the solution curve.

Conversion of inorganics to a more soluble, non-ionic form is a more desirable and analytically safer method of preparation. Nuclear fuel analysts have used the alkylphosphates of the heavy metals as the solutes. Elements such as promethium,[10] zirconium[11] and plutonium[12] are examples of organophosphate complexes. Other organic reagents include 2-ethyl-hexanoic acid for cadmium, lead, bismuth,[13] and caproic acid for nickel.[14]

Using the principle of 'if you can't beat them, join them', Blanchard and Takahashi[16] added cabosil to a toluene/ethanol counting solution to provide a very large surface for adsorption of labelled molecules. He showed that for an inorganic ($Na_2{}^{35}SO_3$) and several Carbon-14-labelled polymers a dramatic increase in the stability of count rate was achieved by a 3% w/v concentration of the silica. One would expect that this technique would show homogeneous counting by the Bush double ratio method. More energetic isotopes (e.g. Chlorine-36, Phosphorus-32) should certainly show such a response.

In all of the solution approaches to counting inorganics, whether salts or complexes, the concentration of the sample is of great importance. This is because the vial wall has a definite capacity for the labelled sample. For example, 20 ml of sample in a normal glass counting vial exposes about 3×10^{17} Å^2 of glass surface. Assuming no roughness factor and a simple value of 5Å^2 for the ionic area, we find that the glass has a capacity of 6×10^{16} molecules or 10^{-4} mmoles. For calcium chloride this corresponds to about 0.01 mg or a concentration of about 0.5 p.p.m. Thus a concentration of a few p.p.m. of a salt may result in a significant error in analysis. This in fact is not unique to inorganic materials. Hayes[17] in 1953 reported that 0.1 p.p.m. of benzoic acid in a toluene scintillator produced a counting efficiency of half that when 2 p.p.m. and higher concentrations were used. Eicholtz et al.[18] studied the adsorption of radioisotopes on glass and plastic surfaces from aqueous solutions. Not only glass adsorbed these materials (cations) but, surprisingly, polyethylene, polypropylene and silicone coated glass as well, but to varying degrees. The general conclusion was that the addition of carrier was the most effective means of reducing the adsorption. This, of course, suggests that carrier-free concentrations (usually in the p.p.b. region) of nuclides must be avoided. The problem then arises as to how much carrier to add since the problem of incipient precipitation always exists albeit more for ionic materials than organic complexes. The use of the Bush[7] double ratio method is certainly indicated in these cases to keep one out of serious trouble in the analysis. One must also bear in mind that some precipitation and/or adsorption reactions may appear hours later so that data on count rate versus time is essential to establish that not only the short-term but the long-term condition of the sample is stable and reproducible. This is best done by the use of standards having the same molecular structure as the sample. Standardisation of inorganic samples may be done by any of the accepted techniques, i.e. internal standard, external standard or sample channel ratio. Use of a standard labelled cation or anion in another chemical form from the sample is hazardous unless one clearly establishes that both have exactly the same distribution during the counting period.

ORGANIC SAMPLES

This section will deal with those samples which by the nature of their physical or chemical properties are unsuitable for direct homogeneous liquid scintillation counting. Those types which are readily soluble in the various liquid scintillation solvents, such as alcohols, hydrocarbons, ethers, carbonyls, esters, carboxylic acids, etc. need little further comment. The principle of carrier dilution should, however, be applied even to these samples of high polarity and low concentration simply because adsorption is always a threat. In addition, the permeability of polyethylene to aliphatic molecules (i.e. alkanes) is well known and one should exercise care when storing or counting such samples for long periods. Dioxane scintillator or Bray's cocktail are particularly appropriate solvents for those materials which are water soluble, highly polar, or both.

Samples of intense colour, low solubility or persistent chemiluminescence (e.g. recycle oil from refinery catalytic cracker) must be converted to an amenable form and combustion seems the only certain technique. In biological work, unlike petroleum or petrochemical studies, the use of an alternative route, i.e. digestion and/or solubilisation, has gained wide acceptance and I wish to deal with this technique before discussing the various combustion techniques now available.

The function of the solubiliser, whether it be acidic or basic, is to reduce a sample composed of high molecular weight compounds to a variety of fragments which by action of the powerful solvent ability of the reagent are then prevented from precipitation or agglomeration. Neither the chemical composition of the fragments nor their molecular weight distribution are known. It is sufficient that a visually clear solution is achieved which, when added to an appropriate scintillator cocktail, does not change in appearance or count rate with time. In the last 5—10 years a sizeable literature has grown around the use of such sample preparation treatments, the reason being that the large numbers of samples to be processed discouraged the use of the then existing combustion methods. However, for reasons advanced below, this procedure is fraught with danger and should be reconsidered as a method of analysis.

In 1958—59 we began studying the adsorption of various molecules at the oil/metal interface. At that time we came across the work of Binford and Gessler[19] who were studying the adsorption of different polyisobutylene polymers on carbon black from organic solvents. The polymers were of various molecular weights ranging from 5000 to over 300, 000 and of low polarity having good solubility in hexane. Their studies demonstrated that above a certain molecular weight adsorption from hexane onto the carbon black was irreversible and was not dependent on unsaturation but only on molecular weight. Smith et al.,[20] studying the adsorption of an oleophilic detergent calcium dinonylnaphthalenesulphonate Sulphur-35 at the gold/oil interface, found that the detergent was displaced from the gold surface by a polymeric alcohol (MW = 20, 000) or a polydodecyl methacrylate (MW = 50, 000). Further work using Carbon-14-labelled polymers showed that adsorption at the iron/oil interface was irreversible for polyalcohols, polyamine and polythiol of MW's from 20, 000 to 500, 000. Attempts to displace the adsorbed polymers with polar solvents or even pure monomer were unsuccessful. Increasing the temperature to 200°C for several days served only to remove some 20% of the amount adsorbed. Blanchard and Takahashi[16] reported the striking fact that acrylamide Carbon-14 in counting solution did not change in count rate with time but polyacrylamide Carbon-14 decreased. Glass[8] reported that an homogeneous-appearing polymer solution was shown to be non-homogeneous by the double ratio method.

The rationale for such behaviour of polymers may be advanced as follows: a polymer in a semi-coiled or uncoiled state in solution encounters a solid surface on which some of the repeat monomer units adsorb, albeit weakly. If the polymer then uncoils onto the surface, as observed by Binford and Gessler,[19] all or most of the adsorbing groups will be in a dynamic equilibrium between adsorption and desorption, i.e. some will be adsorbing and some desorbing, not, however, all at the same time. Assuming the heat of adsorption is low, 1—2 kcal/monomer unit, and that 500—1000 of the functional groups are exposed at any one time, it is clear that the probability that all of the adsorbing groups are off at the same time is exceedingly small. Stated another way, it would take several hundred kilocalories to desorb all the groups simultaneously, a requirement greater than C—C bond strength. It is also clear that as long as even one functional group of the polymer is fastened to the surface, the entire polymer is held close to the surface and counting geometry will be less than 4π. If, in the case of a non-polar polymer,[19] adsorption becomes irreversible at some molecular size (~50, 000), it clearly becomes irreversible with polar polymers at much lower molecular weights. This is the situation prevailing for the highly polar macromolecules originating from biological studies. Furthermore, if one bears in mind that the behaviour of such polymers is profoundly influenced by the nature of the solvent, the prior history of the polymer treatment, as well as the solid surface, one can see that predictability of surface behaviour is virtually impossible and analysis becomes uncertain. The quantitative effect arises from the calculations made for the capacity of the vial wall for inorganic ions. If, for example, a rather heavily coiled polymer of MW 10^5 has only a few groups of about $25Å^2$—$100Å^2$ cross section anchored to the surface, it is obvious that as much as a milligramme may be adsorbed on the glass wall from a counting solution. This means that sample sizes several times this order must be used to minimise the effect of these losses.

Because of macromolecule adsorption behaviour and the impossibility of correcting for it, it is my conviction that samples of this nature should always be burned and the products collected for counting. In our laboratory all polymers of the polyolefin type are burned, polymers of the polyaromatic type (polystyrene) are dissolved in amounts never less than 100 mg and usually 500 mg in 20 ml of a toluene scintillator. Such analyses are routinely checked by combustion (1 in 10) to make sure we are not fooling ourselves.

One is not out of the wood even if total degradation of proteins to amino acids is achieved. Thus Litt and Carter[21] showed serious adsorption losses of simple amino acids such as arginine and lysine depending upon the cocktail used and the state of the vial surface. For slightly more complex molecules labelled with Carbon-14, the effect of the vial surface varied with the nature of the compound. In one case plastic was superior to silanised glass; in another case they were similar. Generally, plastic vials were better, as expected.

Some workers have expended considerable effort to develop a general arsenal of methods to process the incredible variety of samples requiring liquid scintillation counting. Grower and Bransome[22] counted polyacrylamide gels following electrophoresis of tritium-labelled gel sections. They found that special slicing of the gel followed by hydrolysis of the gel was needed to obtain results within a few percent of the combustion value. Bray[23] studied a series of preparative methods for radioassay of aqueous samples. He screened several solubilisers as well as various cocktails and different kinds of samples. One must conclude from his careful study that each new sample type as well as each new labelled compound must be carefully checked under a variety of conditions to ensure that the final method used will yield accurate data. Kahlben and Rezvanie[24] described work involving seventeen scintillation cocktails, three solubilisers and two methods of oxidation. It was clearly demonstrated that with each system different problems were raised. These included chemiluminescence, varying efficiencies with sample volume, phase separation, etc. Little and Neary[25] showed that for each sample type and solvent an optimum fluor concentration exists. In other words, there is an optimum cocktail for each sample — a major problem for the user who wishes to operate under optimum conditions but has a variety of samples, some of virtually unknown composition. Kahlben and Rezvanie[24] concluded that it would be greatly to the benefit of workers in this field if only a limited number of sample preparation methods were used, viz. combustion or oxidation for solids and emulsion cocktails for liquids.

The situation, then, that we have got ourselves into is the development of a large variety of scintillation cocktails and a number of hydrolysing reagents to cope with an almost infinite variety of samples. In the effort to speed up sample preparation methods we have inadvertently introduced a number of other problems which need watching. Kahlben and Rezvanie[24] are indeed correct in calling for a simplification so that techniques and results can again become comparable and the user is assured of the accuracy (as opposed to the precision) of the analyses. In addition, the macromolecular adsorption problem described above raises the very real possibility that during an experimental study a small change in molecular weight, in molecular structure, or degree of unfolding of a polymeric chain can invalidate a sample preparation method carefully developed for the type of samples expected. I find it hard to believe that there is any programme using liquid scintillation counting which prefers a large amount of uncertain data to a smaller amount of accurate data. For organic samples of indeterminate composition the only way to ensure accuracy is by combustion.

COMBUSTION METHODS

Combustion of labelled organic samples preparatory to radioassay was in use years before liquid scintillation counting. Samples were burned then either in high pressure oxygen bombs, in combustion trains or in carius tubes with concentrated nitric acid. Carbon-14 was normally assayed as barium carbonate and Sulphur-35 as barium sulphate on planchets with thin window or windowless proportional counters. Tritium was assayed by converting the water from combustion to hydrogen gas over iron, zinc or magnesium and was introduced into a geiger or proportional gas counter or ion chamber for measurement.

These were tedious methods requiring considerable manipulative skill. The advent of liquid scintillation counting came as a great relief to those of us using the above techniques. Since then sample combustion methods have undergone radical changes in order to adapt them to the liquid systems found to be appropriate for scintillation counting.

Wet combustion

Samples may be wet or dry combusted. Wet combustion developed years ago in the petroleum industry, for trace metal analysis employs strong acids. The sample is totally destroyed and one must then somehow collect the $^{14}CO_2$ evolved. It is inappropriate for tritiated biological samples unless one condenses the acid and water vapour. Wet combustion as practiced by Mahin and Lofberg[26,27] is in fact not combustion in the conventional sense but a more drastic treatment of the biological samples than the alkaline solubilisers currently in vogue. They employed 60% perchloric acid/30% hydrogen peroxide at moderate temperatures to degrade the macromolecules to smaller, soluble molecules. A scintillation cocktail compatible with these acids (ethylene glycol monomethyl ether plus toluene) was added. The cocktail must be carefully chosen because of quenching and chemiluminescence problems with some systems (Bray's solution, gel systems). Wheeler and Strother[28] compared the NCS solubiliser with van Slyke oxidation to CO_2 and absorption in ethanolamine. It was found that ring-labelled compounds were not attacked. Van Slyke oxidation with hot chromic acid may be adequate for biological samples, may not be for many organic chemical types, and certainly is not for petroleum-based samples. Bartley and Abraham[29] oxidised paper chromatograms with potassium persulphate and absorbed the evolved CO_2 after acidification. It was effective for sugars and organic acids, including amino acids. However, earlier work on petroleum-based samples showed that persulphate was also ineffective for ring destruction, particularly cata-condensed aromatics (anthracene and higher). Smith[30] analysed organic samples for Chlorine-36 by total destruction with H_2SO_4/HNO_3 and distillation of the HCl into caustic. Counting was in a dioxane-based scintillator after neutralisation and buffering.

On balance it seems that wet combustion of organic samples can be an effective method if total decomposition is achieved. Anything less involves many of the uncertainties found with alkaline solubilisation. For Carbon-14 it may involve considerable manipulation and it is generally not employed for tritium. When used for non-volatile or easily isolated radionuclides (metals, halides),[27] total destruction of the sample can be successfully employed. The uncertainties of adsorption or precipitation must, however, be avoided through the judicious use of carriers.

It is clear from the foregoing that the virtue of complete destruction of a sample lies not only in the advantage of being able to count the sample homogeneously but, more important, in the conversion of the radionuclides into known chemical forms for which appropriate counting methods are available. On this basis the dry combustion methods hold the most promise, particularly for Carbon-14, tritium, and Sulphur-35.

Dry combustion

There are several methods of dry combustion of organic samples which have been reported in the literature. These may be divided into sealed systems and flowing systems. The former include high pressure bombs and atmospheric pressure oxygen flasks; the latter include combustion train systems and, recently, a flowing flask system.

The Parr oxygen bomb is an example of a sealed system. It comes in various sizes and operates with 20—25 atm oxygen. The sample is held in a platinum cup and ignited via an electrically heated filament. Acidic gases are trapped by placing several milligrammes of an aqueous caustic solution in the bottom of the bomb. Sample size capacity of a 500 ml bomb is about 1 g. Water of combustion can also be trapped by chilling the bomb before venting. The method is quantitative. Sheppard and Rodegher[31] used the system for Carbon-14 and tritium. The limitation is the high cost of a bomb and the rather long manipulation involved in setting up, venting and clean-up for each sample. Scott and Kennally[32] used much less costly sealed Vycor tubes containing oxygen slightly below atmospheric pressure. Samples ranging in size from 6—100 mg of carbon could be combusted in various size tubes by heating to 850°C for about 5 min. The tubes were chilled, opened and activity removed with either a dioxane or toluene/amine cocktail depending on whether water or carbon dioxide was collected. Yields were high, 96—98%, and the method well suited to large numbers of samples. One drawback was the need for glass manipulation and sealing. Explosion hazards are also present. Some 20 samples could be processed per day.

The Schoniger flask method for the analysis of various elements in organic samples lends itself admirably to the radioassay of Carbon-14 and tritium. The various modifications described in the literature include electrical ignition, infrared ignition, placement of absorbing scintillator in a cold finger before ignition or injection after ignition, and the use of simple side arm flasks with rubber stoppers as well as more complex three-neck flasks. In addition, combustion directly in a counting vial has been reported as a method for mass production.

Thus Olivierio et al.,[33] using a 2-litre filter flask and infrared ignition, injected, following combustion, a phenylethylamine-containing scintillator via rubber tubing attached to the side arm. Dobbs[34] overcame the quenching presence of halogen in the sample by adding pentene-2 to the absorbing scintillator solution injected into the flask after ignition. Conway et al.[35] demonstrated near quantitative recovery for $^{14}CO_2$ (97%) and ^3H_2O (98%) when using ethanol-in-toluene to absorb the water and ethanolamine plus cellosolve-in-toluene to absorb the CO_2. Gupta[36] described a novel technique for in-vial combustion of biological samples (2—3 mg) which is well suited for large numbers of samples. Vickers[37] improved the Gupta method by providing more certain ignition, larger (6 mg) samples and elimination of leaks. Ronucci et al.[38] determined simultaneously Hydrogen-3 and Sulphur-35 in samples by oxygen flask combustion and trapping the sulphur oxides and ^3H_2O in methanol/dioxane cocktail containing phenylethylamine.

Sample size limitations of the oxygen flask (usually ~100 mg sample plus 100 mg paper) were overcome by Dobbs[39] by the use of 5-litre flasks for samples of 2—3 g size, dry weight. He also extended the method to organic samples on inorganic substrates by adding powdered cellulose to ensure complete combustion. A significant advance in the oxygen flask method has been reported by Sher et al.[40] where sample combustion is aided by a stream of added oxygen and the combustion products are swept continuously into appropriate absorbers. Water adsorption effects are minimised by a steam purge. The method has been applied to Carbon-14 and Hydrogen-3 radioassay and seems to go a long way towards the automation of the combustion step of a sample preparation. Several studies of the performance of this unit have been reported[40,41] which demonstrate its quantitative nature and the absence of a tritium memory effect. Hunt and Gilbert[42] compared the combustion method with alkaline digestion on a variety of biological samples. They found lower recoveries in all cases with the digestion step.

The oxygen flask method is one of the most useful for combustion of organic samples. It lends itself well to scale-up to large numbers as well as large amounts of sample. It is suitable for virtually all radioisotopes. The major limitation seems to be the large amount of glassware clean-up and decontamination necessary for a high sample throughput. In this regard the Kaartinen—Packard[40] flow combustion flask is a significant step towards more efficient operation. A problem reported for the oxygen flask as well as other combustion methods involves the variable quenching by oxygen present in the counting solution.[43–45] Following combustion the counting solution contains a considerable amount of oxygen which, on standing in the counter, may slowly equilibrate with air (rapidly if opened for introduction of an internal standard). This increase of efficiency with time may lead to erratic results. A few minutes air purge avoids the problem. A nitrogen purge should not be used because of a similar effect but in the opposite direction. The use of the external standard method to obtain the counting efficiency of a sample saturated with oxygen is not recommended because it determines counting efficiency at a time different from the sample count and may lead to errors. Furthermore, a drifting count rate may conceal an instrumental or sample instability problem. Stable counting rates should be present, and an air purge avoids the variable oxygen effect. The advent of air-tight counting vials at low cost would, of course, change this.

Combustion train

Combustion tube analysis of organic samples for Carbon and Hydrogen is an old and highly developed procedure. Very high accuracy for Carbon and Hydrogen is achieved with such systems which employ hot (~800°C) CuO to complete the combustion of samples heated in a stream of oxygen. Various additional packings in the train are sometimes used to eliminate nitrogen oxides, halogens and sulphur. The procedure is well suited to the preparation of samples for liquid scintillation counting since it simply entails placing a trap on the exit end of the combustion tube to collect the $^{14}CO_2$, $^{35}SO_2$ or $^{3}H_2O$. It is not appropriate for non-volatile radioisotopes, and has been most often employed on samples in the milligramme range.

However, Peets et al.[46] used a triple combustion tube set-up to burn three samples at a time, up to 2 g each, and could process 24—30 samples a day, collecting the tritiated water in cold traps. Freeland[47] burned up to 6 g of carbon, collecting the CO_2 in caustic and liberating the CO_2 with acid to prepare methyl benzoate via a Grignard reaction. This method is largely of interest in age dating where activities are very low. Knoche[48] burned samples for tritium analysis in an argon/O_2 mixture and trapped the water for subsequent analysis. Griffiths and Mallinson[49] used 100—300 mg samples to run up to 40 samples a day on a conventional C/H apparatus. They trapped the water in dry ice and the CO_2 in a succeeding trap containing phenylethylamine plus methanol (15 ml of 2:1). Smith[50] used a Leco apparatus based on rapid induction heating of a sample mixed with iron chips, CuO, H_2SiO_3 and MoO_3. Combustion required 3 min for a 120 mg sample and an O_2 flow rate of 300 ml/min. Absorption was in an ethanolamine-based scintillator, two traps in series. The second trap absorbed 1—3% of the $^{14}CO_2$ at ice temperature. Our experience with an automatic combustion train, i.e. where the combustion rate is controlled by pressure sensors, is that 20—40 min per sample are required for samples from 100—500 mg. Trapping is via a small tower with a sintered disc disperser and either a dioxane or a phenylethylamine cocktail to trap water or CO_2 at ice temperatures. Recovery is quantitative and the 1—2% tritium memory effect is eliminated by following every tritium sample with 150 mg of water. All samples are automatically air purged at the end of the combustion.

The major disadvantages of the combustion train approach have been the high cost of the equipment (compared to oxygen flasks), the long analysis time and the rather careful attention required to ensure a smooth combustion. This is not the case with the controlled combustion system mentioned above. However, the proven quantitative aspects of the combustion train indicated that it was ripe for some major improvements to raise its appeal to radiochemists. This appeared in the Peterson[51,52] apparatus. This unit combines combustion, trapping and delivery into a liquid scintillation counting vial ready for capping and counting within 3 min. Samples up to 500 mg for Hydrogen-3 and slightly less for Carbon-14 are routinely handled. Samples are placed in Lexan or other capsules which are dropped into the combustion tube. Some studies made of the Petersen combustion train[53,54] showed it to be a fast, convenient unit requiring some operator skill and yielding absolute recoveries of Carbon-14 and Hydrogen-3 of 93—96% and 97% respectively. Memory effects are in the 0.2—0.4% region. The apparatus is certainly expensive but its speed and convenience make it an interesting possibility for the heavily loaded laboratory. According to Petersen et al.[51] the samples increase in counting efficiency with time. This is most likely due to slow equilibration with air and would be improved by including an air purge step. The consistently low recovery of Carbon-14 may be due to the dissociation of the carbamate. If so, keeping the absorbing cocktail cold will reduce this.

A final comment on methods of incorporating products of combustion into homogeneous counting systems. These involve chemical reactions to convert water or carbon dioxide into a low quenching organic liquid. Thus Moghissi and Kogube[55] reacted water with a benzyl Grignard to make toluene which could be distilled and counted. Only half of the tritium is recovered in this reaction but very large samples can be handled. Polack and Stipp[56] and Kearns[57] reacted CO_2 with lithium to make the carbide which was converted to ethylene and finally to benzene. This technique is of interest for large samples of low activity where the highest figure of merit — (sample size) x (efficiency) — is required, i.e. age dating, natural tritium concentrations, very high dilution experiments.

CONCLUSIONS

In conclusion, then, it seems that gases, inorganic samples and organic samples can be successfully analysed in homogeneous systems. One must be aware, however, that an homogeneous appearance may be desirable, but is not a sufficient, condition for accurate radioassay. It also may not be necessary if the β-energy is large compared to the particle size. Adsorption and deposition of the radionuclide must be ruled out by careful testing, such as the double ratio method of Bush,[7] by transferring the counting solution to a fresh vial, or by comparison of the sample channel ratio with an internal standard channel ratio. One should avoid extremely low concentrations of a labelled sample because the adsorption capacity of the vial wall is high enough to cause significant errors. The situation is far more hazardous with high molecular weight polar molecules which may be irreversibly adsorbed. Since the physical behaviour of these molecules is usually unpredictable it is safest to burn such samples, analysing the products of combustion in one of several proven scintillation cocktails. The partial digestion of such macromolecules yields a spectrum of unknown fragments whose physical properties cannot be predicted. The virtue of sample combustion lies not so much in the opportunity to count these homogeneously as in the fact that the radionuclides are in a known chemical form for which a suitable counting system is available.

Two major advances in combustion instrumentation, the flow oxygen flask of

Kaartinen and the high speed combustion train of Petersen, go a long way towards overcoming objections to this method of homogeneous sample preparation.

REFERENCES

1 D.L. Horrocks and M.H. Studier, Analyt. Chem. **36**, 2077 (1964).
2 D.L. Horrocks, Intern. J. Applied Radiation and Isotopes **19**, 859 (1968).
3 B.E. Gordon, H.R. Lukens and W. ten Hove, Intern. J. Applied Radiation and Isotopes 12, 145 (1961).
4 M.L. Curtis, S.L. Ness and L.L. Bentz, Analyt. Chem. 38, 636 (1966).
5 T.E.F. Carr and B.J. Parson, Intern. J. Applied Radiation and Isotopes 13, 57 (1962).
6 W.D. Fairman and J. Sedlet, Analyt. Chem. **40**, 2004 (1968).
7 E. Bush, Intern. J. Applied Radiation and Isotopes **19**, 447 (1968)
8 D.S. Glass, in Organic Scintillators and Liquid Scintillation Counting, (Ed. D.L. Horrocks and C.T. Peng), Academic Press, New York, 1971, pp.803–821.
9 H.J. Laurencot and J.L. Hempstead, in Organic Scintillators and Liquid Scintillation Counting (Ed. D.L. Horrocks and C.T. Peng), Academic Press, New York, 1971, pp.635–657.
10 J.D. Ludwick, Analyt. Chem. **36**, 1104 (1964).
11 J.D. Ludwick, Analyt. Chem. 32, 607 (1960).
12 D.L. Horrocks and M.H. Studier, Analyt. Chem. 30, 1747 (1958).
13 A.R. Ronzio, Intern. J. Applied Radiation and Isotopes **4**, 196 (1959).
14 C.E. Gleit and J. Dumont, Intern. J. Applied Radiation and Isotopes 12, 66 (1961).
15 S.W. Perry and G.T. Warner, Intern. J. Applied Radiation and Isotopes **14**, 397 (1963).
16 F.A. Blanchard and I.T. Takahashi, Analyt. Chem. 33, 975 (1961).
17 F.N. Hayes, USAEC Document, LA-1639, 1953.
18 C.G. Eicholtz, A.E. Nagel and R.B. Hughes, Analyt. Chem. 37, 863 (1965).
19 J.S. Binford and A.M. Gessler, J. Phys. Chem. **63**, 1376 (1959).
20 M.L. Smith, B.E. Gordon and R.C. Nelson, J. Phys. Chem. **69**, 3833 (1965).
21 G.J. Litt and H. Carter, in The Current Status of Liquid Scintillation Counting (Ed. E.D. Bransome, Jr.), Grune and Stratton, New York, 1970, pp.156–161.
22 M.F. Grower and E.D. Bransome, in The Current Status of Liquid Scintillation Counting (Ed. E.D. Bransome, Jr.), Grune and Stratton, New York, 1970, pp.263–269.
23 G.A. Bray, in The Current Status of Liquid Scintillation Counting (Ed. E.D. Bransome, Jr.), Grune and Stratton, New York, 1970, pp. 170–180.
24 D.A. Kahlben and A. Resvanie, in Organic Scintillators and Liquid Scintillation Counting, (Ed. D.L. Horrocks and C.T. Peng), Academic Press, New York, 1971, pp.149–166.
25 R.L. Little and M.P. Neary, In Organic Scintillators and Liquid Scintillation Counting, (Ed. D.L. Horrocks and C.T. Peng), Academic Press, New York, 1971, pp.431–439.
26 D.T. Mahin and R.T. Lofberg, Anal. Biochem. **16**, 500 (1966).
27 D.T. Mahin and R.T. Lofberg, in The Current Status of Liquid Scintillation Counting, (Ed. E.D. Bransome Jr.), Grune and Stratton, New York, 1970, pp.212–221.
28 L. Wheeler and A. Strother, Anal. Biochem. **53**, 42 (1973).
29 J.C. Bartley and S. Abraham, Atomlight **49** (1965).
30 G.N. Smith, Anal. Biochem. **17**, 24 (1966).

31 H. Sheppard and W. Rodegher, Anal. Biochem. 4, 246 (1962).
32 B.F. Scott and J.R. Kennally, Analyt. Chem. 38, 326 (1966).
33 V.T. Olivierio, C. Denham and J.D. Davidson, Anal. Biochem. 4, 188 (1962).
34 H.E. Dobbs, Analyt. Chem. 36, 687 (1964).
35 W.D. Conway, A.J. Grace and J.E. Rogers, Anal. Biochem. 14, 491 (1966).
36 G.N. Gupta, Analyt. Chem. 33, 1356 (1966).
37 T.H. Vickers, Analyt. Chem. 40, 2219 (1966).
38 R. Ronucci, G. Lambelin, M.J. Simon and W. Saudigm, Anal. Biochem. 26, 146 (1968).
39 H.E. Dobbs, in Organic Scintillators and Liquid Scintillation Counting, (Ed. D.L. Horrocks and C.T. Peng), Academic Press, New York, 1971, pp. 669–681.
40 D.W. Sher, N. Kaartinen, L.J. Everett and V. Justes, in Organic Scintillators and Liquid Scintillation Counting, (Ed. D.L. Horrocks and C.T. Peng), Academic Press, New York, 1971, pp. 849–869.
41 J.D. Davidson, V.T. Olivierio and J.I. Peterson, in The Current Status of Liquid Scintillation Counting, (Ed. E.D. Bransome, Jr.), Grune and Stratton, 1971, pp. 222–235.
42 L.M. Hunt and B.N. Gilbert, Intern. J. Applied Radiation and Isotopes 23, 246 (1972).
43 R.L. Hancock and D.L. Coleman, Anal. Biochem, 9, 487 (1964).
44 H.E. Dobbs, Analyt. Chem. 35, 783 (1963).
45 R.J. Carr and B.E. Gordon, J. Polymer Science Part C, No. 8, 71 (1965).
46 E.A. Peets, J.A. Florini and D.A. Buyske, Analyt. Chem. 32, 1465 (1960).
47 L.T. Freeland, Analyt. Chem. 36, 2055 (1964).
48 H.W. Knoche and R.M. Bell, Anal. Biochem, 12, 49 (1965).
49 M.H. Griffiths and A. Mallinson, Anal. Biochem. 22, 465 (1968).
50 L.W. Smith, Anal. Biochem. 29, 223 (1969).
51 J.I. Peterson, F. Wagner, S. Siegel and W. Nixon, Anal. Biochem. 31, 189 (1969).
52 J.I. Peterson, Anal. Biochem. 31, 204 (1969).
53 T.R. Tyler, L.E. Chapin, R.F. Alvaro and J.L. Cox, J. Radioanalytic. Chem. 12, 475 (1972).
54 T.R. Tyler, A.R. Reich and C. Rosenblum, in Organic Scintillators and Liquid Scintillation Counting, (Ed. D.L. Horrocks and C.T. Peng), Academic Press, New York, 1971, pp. 869–877.
55 A. Moghissi and K. Kogube, Intern. J. Applied Radiation and Isotopes 15, 165 (1964).
56 H.A. Polack and J.J. Stipp, Intern. J. Applied Radiation and Isotopes 18, 359 (1967).
57 D.S. Kearns, Intern. J. Applied Radiation and Isotopes 20, 821 (1969).

DISCUSSION

L.A. Wegner: You have presented a few results concerning the memory effects of commercially available combustion systems. What are the corresponding figures for the 'Pregl-chain' combustion system installed in your lab? What is the procedure involved to determine these values in respect to your choice?

B.E. Gordon: We simply run a blank after a sample and count the blank. For tritium we pass through about 0.25 ml water, for Carbon-14 we burn about 0.25 g hexadecane. Residual memory effect is in the region of a few-tenths of a percent. When combusting Carbon-14 samples the memory effect is of the order of 1/100 of a percent.

N.G.L. Harding: In viewing matters from this side of the crematorium it is apparent that different tissues combust at different rates (per unit mass). Tissue factors apparently determining the burning rate include water and salt content, tissue structure and fat content. Hence different samples require different combustion temperatures and oxygen concentrations. Do you know whether anyone has explored the possibility of optimising combustion temperatures using low-thermal-inertia systems such as a laser?

B.E. Gordon: Little work seems to have been done on the optimisation of combustion temperatures. Control of the combustion atmosphere is important.

B.W. Fox: I must stress that there are basic differences between true solubilisers and those mixtures containing surfactants and they should not be confused. A number of these mixtures described this morning contain surfactants and they belong to this afternoon's session.

M. Cosandey: Have you any information about cold combustion with Fenton's reagent, i.e. concentrated hydrogen peroxide with a trace of Fe^{2+} ion? In a recent report by IAEA, Sansom reported that all organic materials such as cationic exchange resins could be mineralised at low temperatures.

B.E. Gordon: I have not heard about this report or about this technique.

J.D. Eakins: With reference to the problem of controlling the rate of combustion in the furnace tube method of oxidation, we have overcome this at AERE, Harwell by the use of a nitrogen blanket. We use a single furnace tube on which two furnaces are wound. The first of these is for heating the sample and the second contains a platinum/aluminium catalyst maintained at 700 ^{o}C. Oxygen is fed in between the two furnaces. Up to 10 g of organic sample, contained in a silica boat, is inserted into the cold sample furnace. Nitrogen is passed over the sample and the temperature gradually raised to 500 ^{o}C. The volatile components of the sample are mixed with oxygen and flushed over the preheated catalyst where they are decomposed. The nitrogen blanket prevents ignition of the sample and a subsequent uncontrolled rise in temperature. When all the volatile components are evolved, oxygen is passed over the carbonised residue to complete the oxidation.

Chapter 11

Solubilisation of Mammalian Tissue for Scintillation Counting

J.G. Dent[a]

Wellcome Research Laboratories, Beckenham, Kent, England

P. Johnson[b]

Hoechst Pharmaceutical Research Limited, Milton Keynes, Buckinghamshire, England

INTRODUCTION

The problems facing the biologist in the preparation of biological samples for liquid scintillation counting are echoed in Hamlet's words: 'Oh that this too, too solid flesh would melt'.

There exists a plethora of commercially available solubilising agents, the main characteristic of many of them being their high cost. The need arose in our department to analyse large numbers of samples for Carbon-14-labelled material. The use of any of the commercial solubiliser systems would have led to an economic disaster in the department and the use of combustion techniques was impractical due to the large numbers of samples involved. We therefore decided to re-investigate some of the older published methods, taking as a basis sodium hydroxide digestion and the perchloric acid digestion method of Mahin and Lofberg.[1]

PRELIMINARY STUDIES

It was intended to develop a method whereby large pieces of tissue could be digested and from the final digestion mixture portions could be taken for counting. If aqueous sodium hydroxide was used for this purpose and the tissue involved contained large amounts of fat, sodium soaps resulted and these were, as a rule, insoluble in the final digest. This problem was overcome by the addition of Triton and methanol to the digestion mixture resulting in a solubiliser (SHT) with the composition shown in Table 1.

---

a Full-time postgraduate research student registered at the University of Surrey. This work was undertaken whilst in receipt of a Wellcome Studentship grant.

b Formerly of the Wellcome Research Laboratories, Beckenham, Kent.

Table 1. SHT solubiliser.

Sodium hydroxide	80 g
Distilled water	600 ml
Methanol	300 ml
Triton X405 (75% aqueous solution)	100 ml

SHT may be counted, after preliminary neutralisation, in a toluene/triton scintillator of the type described by Patterson and Greene.[2] The composition of this scintillator is shown at the top of Table 2 and is referred to as tT21. We also devised a scintillator (S.20),[a] the composition of which is shown at the bottom of Table 2, 10 ml of which would accept and neutralise 1 ml of a 2M aqueous alkali solution. Nonanoic and lauric acids are present in quantities just sufficient to neutralise the alkali. Dimethyl POPOP and PPO are used as the fluors rather than butyl PBD which shows a tendency to decompose in the presence of alkali. It should be noted that when SHT is used with scintillator S.20 the resulting solution does tend to chemiluminesce for 2 to 4 h.

Table 2. Composition of scintillators S.20 and tT21.

Scintillator tT21	
Toluene	1000 ml
Triton X100	500 ml
Butyl PBD	10.25 g
Scintillator S.20	
Toluene	600 ml
Triton X100	300 ml
Oxitol	30 ml
Nonanoic acid	20 ml
Lauric acid	12 g
Dimethyl POPOP	0.5 g
PPO	6.0 g

As mentioned previously, SHT when used with tT21 needs preliminary neutralisation with an acid. We tried several acids for this purpose and found that nitric acid gave the least quenching and had less tendency to cause precipitation. In the normal laboratory procedure we neutralised 1 ml of an SHT digest with 0.5 ml of 4.4M nitric acid. For the purposes of investigating the effect of excess acid on the counting efficiency of the SHT/tT21 system, we used 8M nitric acid.

It may be observed in Table 3 that the counting efficiency of SHT/tT21 is little affected by excess acid. The same is true of SHT tissue digests. Even if a sample received two additions of acid in error, the resulting counting efficiency would not be seriously affected. Table 4 shows the solubilising ability of SHT, in conjunction with the two scintillators already described. It also shows the stability of the tissue digest/scintillator cocktail combination. tT21 will accept up to 500 mg of homogenate, i.e. 250 mg wet weight of tissue, and give a stable counting solution. Scintillator S.20, however, has a lower capacity, 300 mg of homogenate, i.e. 150 mg of tissue. As would be expected, efficiencies decrease with increasing tissue concentration. However, 500 mg of

---

[a] Composition based on unpublished results of Dr. C.G. Raison.

Table 3. Effects of excess nitric acid on counting efficiency of SHT/tT21.

Acid (ml)	Excess acid (%)	Water (ml)	Efficiency (%) ^{3}H	^{14}C
0.25	0	0.25	19.5	85.0
0.3	20	0.20	19.5	85.0
0.4	60	0.10	15.5	82.0
0.5	100	0.00	12.5	78.0

1 ml SHT + 8M nitric acid, water and 10 ml tT21. Spiked with ^{14}C hexadecane or ^{3}H hexadecane.

Table 4. Solubilising ability of SHT

Weight homogenate (mg)	Stability 3 days tT21	S.20	Counting efficiency (%) ^{3}H tT21	^{3}H S.20	^{14}C tT21	^{14}C S.20
100	yes	yes	14	23	80	85
200	yes	yes	10	19	74	84
300	yes	yes	9	16	71	82
350	yes	no	9	14	71	81
400	yes	no	7	13	68	80
500	yes	no	4	11	61	75

Aqueous rat liver homogenate (1:1) was solubilised in 1 ml SHT and counted in 10 ml of scintillator.

liver may be counted in 20 ml of tT21 at a Carbon-14 efficiency of 60%. A more extensive description of the practical problems with and limitation of SHT, when used as a routine laboratory method, will be described elsewhere.[3]

The general performances of SHT solubilisation and perchloric acid digestion followed by treatment with hydrogen peroxide were compared with four commercial products: NCS (Searle), Protosol (New England Nuclear), Soluene 100 (Packard) and Biosolve (Beckman). At the time this study was undertaken Soluene 350 (Packard) and TS1 and TS2 solubilisers (Koch Light) were not available.

METHODS

The initial comparisons were carried out as described below.

Tissues were removed from rats and prepared for solubilisation in triplicate both as tissue snips of 100 ± 5 mg and homogenates (1:1 aqueous) of 200 ± 10 mg. Where commercial solubilisers were used the manufacturer's recommended methods were followed without modification. Solubilisation was brought about in an ultrasonic cleansing bath (Kerry Pulsatronic K901), the vials being held in a wire cradle in strong detergent solution (5% Lab Brite). The samples were visually assessed at regular intervals and when they had been completely solubilised, scintillator was added. The samples were spiked and counted in an Intertechnique SL40 ABAC 4K scintillation counter.

Details of the methods and of the solubiliser/scintillator combinations used are shown in Table 5. In all cases except SHT/S.20 a digestion temperature of 50°C was used. With SHT/S.20 it was found that a digestion temperature of 70°C was necessary to ensure that the sample did not precipitate out in the scintillator

Table 5.

Solubiliser	Volume of solubiliser	Scintillator	Extra steps	Time for satisfactory solubilisation (h)
NCS	1 ml	10 ml toluene butyl PBD (7.5 g/litre)	Add 30 μlitres glacial acetic acid after addition of scintillator	2
Soluene	2 ml	10 ml toluene, Me_2 POPOP 100 mg/litre PPO 6.0 g/litre		7
BBS	1 ml 2M NaOH	12 ml toluene/BBS3 (5:1) butyl PBD 6.25 g/litre	Add 2.5 ml BBS2 to NaOH digest prior to addition of scintillator	0.5
Protosol	1 ml	10 ml toluene Me_2 POPOP 100 mg/litre PPO 6 g/litre		22
Perchloric acid	0.4 ml $HClO_4$ 0.8 ml 60% H_2O_2	16 ml tT21		0.5
SHT	1 ml	10 ml tT21	Neutralise with 0.5 ml 4M HNO_3	0.5
SHT	1 ml	10 ml S.20		0.5

Table. 6. Mean counting efficiency (%) of Carbon-14 for solubilised rat tissues.

Solubilised tissue	NCS	Soluene 100	BBS	$HClO_4$	Protosol	SHT	SHT/S.20
Brain	86.9 ± 0.6	70.7 ± 18.7	67.4 ± 1.3	78.3 ± 1.0	87.4 ± 1.5	79.7 ± 4.3	82.4 ± 2.3
Skin	84.1 ± 1.7	87.7 ± 2.4	72.4 ± 6.4	80.1 ± 6.1	77.3 ± 2.1	76.1 ± 1.6	82.3 ± 2.2
Lung	61.9 ± 5.6	45.9 ± 14.0	63.1 ± 2.4	82.5 ± 3.7	51.6 ± 13.2	57.3 ± 4.1	76.1 ± 3.3
Heart	81.5 ± 3.9	66.7 ± 9.7	66.2 ± 4.0	77.7 ± 1.4	66.1 ± 12.8	70.5 ± 1.7	79.9 ± 2.8
Liver	70.1 ± 3.7	47.0 ± 14.9	60.6 ± 1.6	74.9 ± 0.7	41.0 ± 20.5	65.5 ± 3.6	80.3 ± 3.8
Spleen	17.7 ± 1.3	15.8 ± 5.0	46.6 ± 2.2	75.6 ± 2.2	15.7 ± 6.6	29.7 ± 11.5	63.9 ± 2.6
Kidney	72.2 ± 0.9	46.3 ± 10.6	53.8 ± 16.3	73.9 ± 1.2	58.9 ± 12.1	60.7 ± 2.9	77.8 ± 2.0
Intestine	87.0 ± 1.5	74.9 ± 11.3	67.2 ± 7.4	78.5 ± 2.0	80.2 ± 1.5	79.4 ± 3.6	81.5 ± 1.3
Muscle	87.0 ± 3.0	67.1 ± 14.6	65.9 ± 3.9	78.1 ± 1.3	79.1 ± 4.2	78.8 ± 2.0	81.5 ± 0.8
Adipose	90.3 ± 1.8	77.2 ± 19.6	78.4 ± 1.3	81.3 ± 1.5	68.7 ± 22.2	82.1 ± 2.9	83.7 ± 2.8
Blood	8.0 ± 0.4	5.1 ± 4.4	4.7 ± 1.6	77.2 ± 1.1	4.7 ± 2.7	2.5 ± 0.3	12.6 ± 1.0

Results are the means of six determinations ± S.D.

Table 7. Mean counting efficiency (%) of Hydrogen-3 for solubilised rat tissues.

Solubilised tissue	NCS	Soluene 100	BBS	$HClO_4$	Protosol	SHT/tT21	SHT/S.20
Brain	32.1 ± 2.9	6.1 ± 0.8	13.9 ± 0.8	11.4 ± 0.6	23.2 ± 0.8	15.9 ± 0.9	18.5 ± 0.9
Lung	14.6 ± 3.4	9.6 ± 6.8	6.3 ± 2.6	6.4 ± 0.8	13.7 ± 1.6	7.8 ± 1.2	9.1 ± 1.7
Liver	13.2 ± 5.7	4.8 ± 0.4	10.3 ± 0.9	8.1 ± 0.4	15.9 ± 1.4	6.4 ± 0.3	11.1 ± 0.2
Spleen	2.4 ± 0.5	2.0 ± 0.3	2.2 ± 0.9	7.6 ± 0.9	4.2 ± 0.1	3.1 ± 1.9	4.5 ± 1.4
Blood	0.4 ± 0.1	1.6 ± 0.1	0.1 ± 0.1	9.1 ± 0.4	0.5 ± 0.1	0.1 ± 0.1	0.2 ± 0.1
Stomach and contents	1.1 ± 0.1	2.6 ± 0.5	6.1 ± 0.9	8.8 ± 0.9	8.7 ± 1.4	2.3 ± 0.1	7.3 ± 0.9
Caecum and contents	1.2 ± 0.1	2.4 ± 0.1	4.6 ± 0.6	9.6 ± 0.5	8.4 ± 0.1	2.8 ± 0.4	8.3 ± 1.2

Results are the means of six determinations ± S.D.

cocktail. It can be seen that those methods involving NaOH, namely SHT and BBS, needed only 30 min for complete solubilisation as did perchloric acid. Whereas NCS needed 2 h, Soluene required 7 h and Protosol 22 h. The Protosol digestions were not carried out entirely in the ultrasonic bath. The first 7 h were in the bath and the rest of the time in an oven at 50°C.

CARBON-14

Table 6 shows the mean Carbon-14 counting efficiencies for solubilised rat tissues. In each case the results are the means of six determinations, three for homogenates and three for tissue snips. We did not find any significant difference between the counting efficiencies of tissue snips as compared with homogenates. However, in those cases where long digestion periods were necessary, the homogenates tended to solubilise more rapidly than the tissue snips. No appreciable differences were observed in the efficiencies with the various methods. In general, as would be expected, white tissues such as brain gave higher counting efficiencies than red tissues. The range for white tissues was 67-87% and the range for red tissues, such as liver, was 41-80%. BBS and perchloric acid gave the most consistent counting efficiencies and Soluene 100 exhibited the highest variations in counting efficiencies of individual tissues, as shown by the standard deviation.

HYDROGEN-3

Table 7 shows the mean Hydrogen-3 counting efficiencies of tissue digests. The results are the means of six determinations, three for homogenates and three for tissue snips. None of the solubilisers completely digested the fibrous material which is found in the rat stomach and caecum, as a result of digested food. As with Carbon-14, red tissues quenched most, white tissues quenched least but there were no general trends in the results. NCS and Protosol, however, gave the highest counting efficiencies, but perchloric acid gave the most consistent counting efficiency over the range of tissues.

IN VIVO STUDIES

The use of some of the solubilisers with labelled tissues obtained after in vivo dosing was investigated. The methods chosen for this investigation were SHT/tT21, SHT/S.20 and perchloric acid, the three cheapest methods, as may be seen from Table 10. NCS was also included as the cheapest commercial solubiliser and the one with which we had the most experience.

To obtain labelled tissues intraperitoneal injections of labelled glucose were administered to four rats, using Carbon-14 or Hydrogen-3. The rats were killed after one hour and tissues were removed and pooled. Homogenates of the tissues (1:1 aqueous) were prepared and then 200 mg homogenate were solubilised in triplicate by each method. We also combusted portions of each tissue. For Carbon-14-labelled samples a tube furnace was used and in the case of Hydrogen-3 a Model 305 Packard oxidiser was used. The radioactivity content of each sample in d.p.m. was computed by the external standard channels ratio method using quench curves prepared from biological samples.

The levels of Carbon-14 in tissues from rats dosed with D-(U- ^{14}C) glucose as measured by the different methods are shown in Table 8. The levels of Hydrogen-3 from rats dosed with D-(6- ^{3}H) glucose are shown in Table 9.

The solubilisation methods generally give higher results than combustion for Carbon-14 and lower results than combustion for Hydrogen-3. It would appear that the more quenched the samples, i.e. the more coloured the tissue, the greater

Table 8. Assay of Carbon-14 in tissues from rats dosed with D-(U-Carbon-14) glucose.

	Solubiliser				
Tissue	SHT/tT21	SHT/S.20	$HClO_4$	NCS	Furnace combustion
Liver	27600	26300	26900	25300	24500
	±310	±230	±550	±700	±650
Heart	2900	3000	2600	2700	2700
	±50	±30	±40	±40	±130
Brain	5000	4900	4700	4800	4900
	±190	±80	±230	±10	±210
Lung	2700	2800	2600	2600	2000
	±50	±50	±10	±30	±10
Muscle	5700	5800	4700	6100	5500
	±160	±140	±260	±290	±510
Blood	2400	2300	2700	2000	2800
0.1 ml	±100	±160	±160	±90	±170
Plasma	13500	13500	12700	13400	13300
0.5 ml	±30	±90	±10	±240	±520

d.p.m./100 mg tissue ± S.D. of mean.

Table 9. Assay of Hydrogen-3 in tissues from rats dosed with D-(6-Hydrogen-3) glucose.

	Solubiliser				
Tissue	SHT/tT21	SHT/S.20	$HClO_4$	NCS	Packard oxidiser
Liver	129700	128900	141800	134900	152900
	1400	2300	2800	8500	2600
Heart	51300	45500	59200	34000	55600
	500	900	2600	1500	900
Brain	70600	66300	72600	61300	68800
	420	1000	1500	3100	2800
Lung	50100	44800	57200	39600	55600
	1800	1500	1600	900	900
Muscle	58000	55000	58700	50000	59000
	2000	350	1500	1600	900
Blood	55300	58300	69200	—	70700
0.1 ml	3600	1600	3000		1400
Plasma	392000	364300	405000	375300	399200
0.5 ml	7700	5600	4100	6700	27300

d.p.m./100 mg wet weight tissue ± S.D. of mean.

the difference between the results obtained by solubilisation as compared with combustion. This was initially taken to indicate a possible inaccuracy in the quench curve.

However, we digested samples of the in vivo-labelled liver with all the solubilisers and counted them, using the internal standard technique to calculate d.p.m. The results obtained in this way were identical to those obtained with a quench curve.

We also incubated a liver homogenate with D-(1-^{14}C) glucose and subsequently assayed portions of this for Carbon-14 activity by each of the five methods; no significant differences were observed in the results obtained by each method. The present observations do not allow a satisfactory explanation of the discrepancies observed with in vivo-labelled tissues.

COST OF METHODS

Perchloric acid and SHT in conjunction with either tT21 or scintillator S.20 are much cheaper than the other methods. The most expensive of these three methods, perchloric acid, costs one-third of the price of NCS, the cheapest commercial product.

Table 10. Cost comparison of solubilisers (costs in £ for 100 samples (100 mg tissue)).

Solubiliser	Cost		
	Solubiliser	Scintillator	Total
BBS	7.5	6.93	14.43
Protosol	9.60	1.15	10.75
Soluene 100	5.16	1.15	6.31
NCS	4.80	0.93	5.73
$HClO_4$	0.20	1.66	1.86
SHT/S.20	0.20	1.54	1.74
SHT/tT21	0.20	1.04	1.24

SUMMARY

Experiments using different solubilisers indicated that different methods gave different counting efficiencies. It is apparent that none of the methods investigated are generally superior to any other except in terms of time and cost. The assays of in vivo-labelled tissue indicate that the results obtained by solubilisation differ from those obtained by combustion, in some cases by up to 12% for Carbon-14 and by up to 40% but more generally about 15% for Hydrogen-3. These discrepancies may be due to different interactions between the various solubilisers and metabolites of glucose.

Table 11 shows the ideal qualities of a solubiliser. We believe the three inexpensive systems described conform to these requirements as well as any of the commercial systems and a good deal better than some. They are certainly cheaper than all the commercially available solubilisers. Dr. Gordon has commented that only combustion is really reliable but he also agrees that with large

Table 11. Requirement for solubiliser.

1.	Inexpensive in large quantities
2.	May be used with an inexpensive scintillator
3.	Will digest all small animal tissues
4.	Gives a stable solution with scintillator cocktail
5.	Gives 'good' and reproducible counting efficiency
6.	Ideally should be suitable for Carbon-14 and Hydrogen-3
7.	Solubilises rapidly
8.	Minimal number of steps involved in sample preparation
9.	Sample preparation should not require great precision or accuracy

numbers of samples combustion is impractical. Certainly we have shown that with some types of samples and some solubilisers agreement with combustion methods is very good but with other samples there can be large discrepancies, for reasons as yet unknown.

ACKNOWLEDGEMENTS

The authors are indebted to Dr. C.G. Raison for helpful discussions and early work on the composition of Scintillator 20. They are also most grateful to Miss Elaine Butler for her excellent technical assistance and to Mr. Mike Allen for invaluable help in preparing the quench curves and counting the samples.

REFERENCES

1 D.T. Mahin and R.T. Lofberg, Anal. Biochem. **16**, 500–509 (1966).
2 M.S. Patterson and R.C. Greene, Analyt. Chem. **37**, 854–857 (1965).
3 J.G. Dent and P. Johnson, in preparation.

DISCUSSION

T.J. Rising: I noticed that your d.p.m. values for Carbon-14 for $HClO_4$ digestions were low when compared to other solubilisation and combustion methods. Could this possibly be due to the evolution of $^{14}CO_2$, since the tritium values from $HClO_4$ digests were comparable with those from other techniques?

J.G. Dent: Yes it could be loss of CO_2. We investigated this possibility by digesting liver homogenates (after dosing ^{14}C-glucose) under conditions where we were able to collect any $^{14}CO_2$ in phenylethylamine. With perchloric acid digestion, 2–4% of the estimated total activity in the sample appeared in the phenylethylamine. This could explain the lower Carbon-14 results obtained with perchloric acid.

P. Johnson: I would add that we do not claim to have solved other peoples' chemical problems. The onus is on the worker concerned to prove that with his given material in a given solubiliser system he is not getting loss of carbon dioxide or other volatile components. Also the hazard exists not only with perchloric acid, and we have experienced one compound which underwent a small loss of labelled carbon dioxide when treated with sodium hydroxide. If volatility prolems are experienced, they can be overcome by use of a 'sealed' vial as described at the last Brighton Symposium {P. Johnson, in Liquid Scintillation Counting, Vol. 2, (Eds. M.A. Crook, P. Johnson and B. Scales) Heyden & Son, London, 1972, p.241}.

K. Painter: Neutralisation with glacial acetic acid is preferable to either nitric acid or hydrochloric acid as it buffers the cocktail; titration of each sample is not necessary. Workers should be cautioned against perchloric acid wet digestion techniques as there are many reports of serious explosions in the literature.

J.G. Dent: Concerning Dr. Painter's point about the choice of acid for neutralisation, we have tried a whole range of acids including glacial acetic and in our experience nitric acid is preferable. It does not tend to cause precipitation and gives the most stable samples in the tT21 system.

B.W. Fox: If the Mahin Hofberg procedure is carried out on wet or dry samples, i.e. 20 mg sample + 0.2 ml perchloric acid (70%) and 0.4 ml H_2O_2 heated to not more than 80°C (usually 60°C), then we have not obtained any explosions in many such measurements. If one exceeds 100°C, chlorine peroxide is formed and explosions occur, usually on handling the vial.

P. Johnson: Our experience agrees with that of Dr Fox, although one should always treat perchloric acid with respect.

B. E. Gordon: Wet oxidation with $HClO_4$ is safe if preceded by complete pre-oxidation with H_2SO_4/HNO_3 until a clear solution is obtained. Then 0.1 ml 72% $HClO_4$ added to the cooled solution followed by heating to boiling H_2SO_4 will complete the oxidation.

Chapter 12

Automated Tube Furnace Combustion for Liquid Scintillation Sample Preparation

E. Rapkin

180 White Oak Ridge Road, Short Hills, New Jersey 07078, U.S.A.

INTRODUCTION

Almost since the first use of the liquid scintillation counting technique as a routine method for the measurement of Hydrogen-3 and Carbon-14 activities in biological materials, the potential value of combustion as a part of the sample preparation process has been evident. Combustion converts all such test substances to just two species — HTO and $^{14}CO_2$. If they can be solubilised in suitable counting solutions — and, of course, they can — then the problems of the investigator are obviously simplified. The combustion products are colourless and colour quenching is eliminated. Only one chemical species of each isotope need be counted; even if the counting mixture does exhibit chemical quenching, counting efficiencies should be near constant. The requirement for various counting mixtures and complex (and expensive) solubilisation techniques is eliminated; only two well defined and easily soluble combustion products need be handled. Materials that contain both isotopes give combustion products that are easily separated; dual label-counting, with its complications of instrument settings, statistics and interpretation of results, is essentially eliminated.

But, sometimes reality does not parallel theory. While all of these advantages of combustion have long been appreciated, and have in fact been attained on a small scale, with few exceptions it has never been possible to achieve them with the ease required when many hundreds of samples must be processed. Until recently, the combustion method of choice has been that of Schoniger[1] as practised on samples of limited size[2] or even on a micro scale.[3] With much manual effort, huge amounts of glassware, and dedication, a few busy laboratories have managed to keep up with their sample processing requirements via Schoniger combustion, but these are the exceptions. For the most part, conventional solubilisation methods are practised because combustion techniques have not offered the throughput necessary to satisfy normal needs. Until recently, none of the several combustion methods of the past,[4] almost all of which give excellent results for a few samples, have been converted to the automatic or semi-automatic basis which might make such methods attractive to the typical laboratory.

Recognising that the problem was not fundamental, but only one of selecting the most suitable existing means and then automating that, a group at the National Institutes of Health, Bethesda, Md. under J.I. Peterson evaluated existing combustion methods with a view to choosing that most suitable for liquid scintillation sample preparation. They concluded that vertical tube furnace combustion[5,6] offered the most promise and devised an improved system[7,8] in which substances containing Hydrogen-3 and/or Carbon-14 can be combusted and the combustion products, trapped in appropriate counting fluids, automatically collected in counting vials which need only be manually capped before being counted. An independent assessment of the Peterson system has been provided.[9] Because it forms the basis for the Intertechnique Oxymat combustion system,[10] discussed in detail in this chapter, the Peterson method will now be briefly described.

THE PETERSON SYSTEM

The two essential elements of the Peterson apparatus are illustrated in Fig. 1. On the left is the sample introduction mechanism, the furnace and the HTO trapping system; the $^{14}CO_2$ collection unit is pictured on the right. The test substance, dry, moist or even in aqueous solution or suspension, is weighed out in a combustible capsule. Gelatin capsules were originally employed, but more recently polycarbonate resin, less affected by water and not giving large amounts of undesirable nitrogen oxides when burned, has been substituted. The capsule is

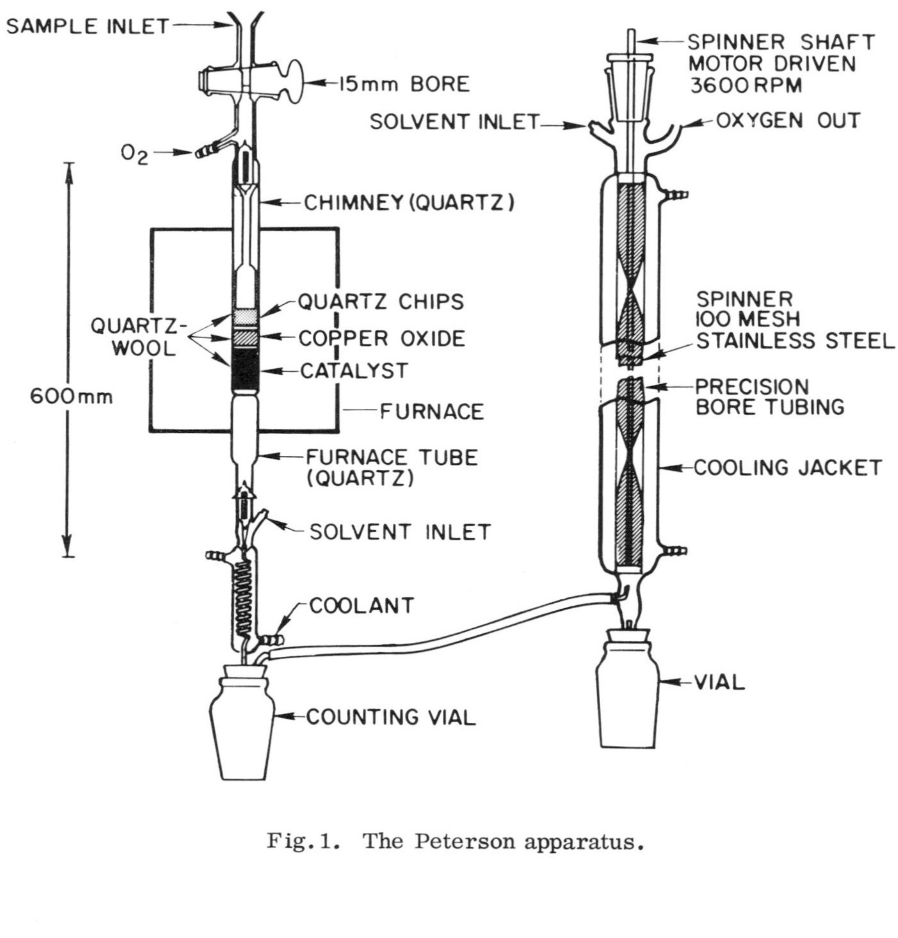

Fig. 1. The Peterson apparatus.

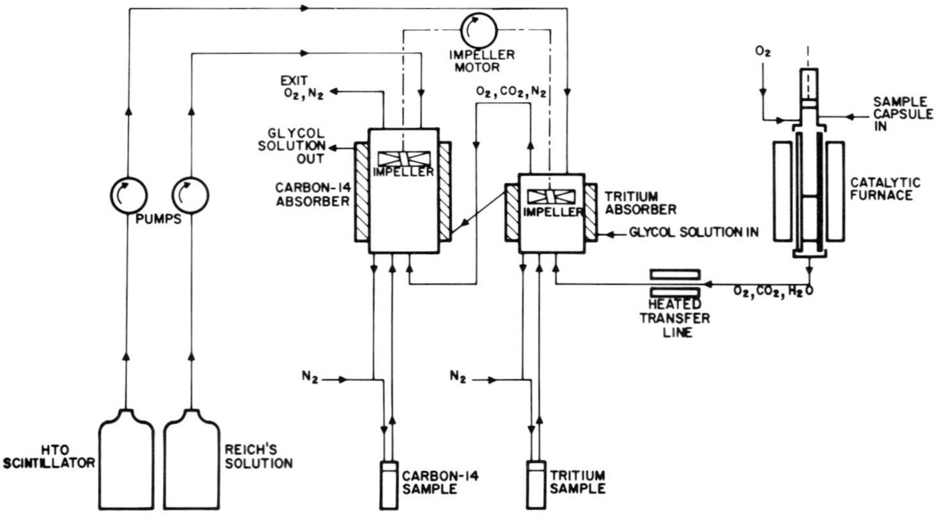

Fig. 2. Oxymat flow diagram.

Fig. 3. The Oxymat. Product collection vials are in place; $^{14}CO_2$ is collected on the left, HTO on the right.

placed in the large diameter (15 mm) blind-bore stopcock on top of the quartz combustion furnace. Upon rotating the stopcock the capsule drops into the furnace on top of a bed of quartz chips where, at 600—650°C, and in a stream of oxygen flowing at 1 litre/min, the capsule bursts into flame. Volatiles are flashed off, surely some pyrolysis occurs, and finally the test substance burns. With sample size held to recommended limits — 1 g dry weight for Hydrogen-3, 0.5 g for Carbon-14 and no more than a total volume of 1 ml for aqueous solutions — combustion is over and there is no visible flame after 2 min. Combustion products, pyrolysis products and volatiles are swept downwards by the oxygen flow through a packed bed of copper oxide needles and hopcalite, a copper—manganese oxide catalyst. At the elevated temperature, and with the large oxygen reserve provided by both the flowing gas and the copper oxide bed, total oxidation is assured; only water vapour and CO_2 (together with nitrogen, rather than nitrogen oxides — possibly due to the catalyst) and, of course, excess oxygen, leave at the bottom of the furnace.

HTO, if present, is almost immediately trapped in a stream of liquid scintillator (methanol/toluene or dioxane/toluene with naphthalene, PPO and POPOP) which is introduced just below the bottom of the furnace. The scintillator stream, the combustion products and excess oxygen flow through an externally cooled coil and then into a counting vial. Scintillator flow rate is 10 ml/min for 2 min; HTO trapping is near quantitative. Once the scintillator flow stops and a few seconds are allowed for drainage, the vial may be removed for capping and counting. It has been shown, however, that counting efficiency is enhanced if the solution is bubbled for a few seconds with dry air or dry nitrogen; dissolved oxygen is removed; perhaps trace quantities of nitrogen oxides and more significant amounts of dissolved CO_2 are also.

If $^{14}CO_2$ is to be trapped, whether or not HTO is present, the HTO counting vial must make a tight seal with the system so that no gases are lost. Non-condensed gases passing the HTO trap are directed to the $^{14}CO_2$ trap where an alkaline scintillator solution (toluene/methanol/β-phenylethylamine with PPO and dimethyl-POPOP) sequesters CO_2. The CO_2 trap is a centrifugal wiped film absorber in which a high speed rotor (~3600 r.p.m.) assures maximum gas turbulence and creates a thin film of trapping solution with a view to presenting maximum surface area. External cooling takes up the heat of reaction between CO_2 and β-phenylethylamine and also, by maintaining reduced temperatures until the counting solution is in the vial, favours the reversible equilibrium towards increased carbamate formation. Once scintillator flow has ended — it usually begins a few seconds after the flow of the HTO scintillator would normally start and lasts about 30 s longer (but with about the same total fluid volume) — the vial is removed, capped and counted. For $^{14}CO_2$, air or nitrogen purging of dissolved oxygen is less important than for HTO but, nevertheless, is still useful.

THE INTERTECHNIQUE OXYMAT

A pictorial representation of the Oxymat is given in Fig. 2; photographs of the system and some of its key elements are shown in Figs. 3 and 4. The Oxymat is a direct descendent of the Peterson apparatus. As such, it retains its advantages and overcomes the disadvantages. It does, of course, include those refinements which are normally expected when laboratory apparatus is engineered into a commercial product.

Sample introduction is no longer by stopcock. Instead the self-standing 1.2 ml polycarbonate resin capsule is thrust into the vertical furnace through a port which opens when a heavy duty, spring-loaded solenoid lifts its stainless steel plunger

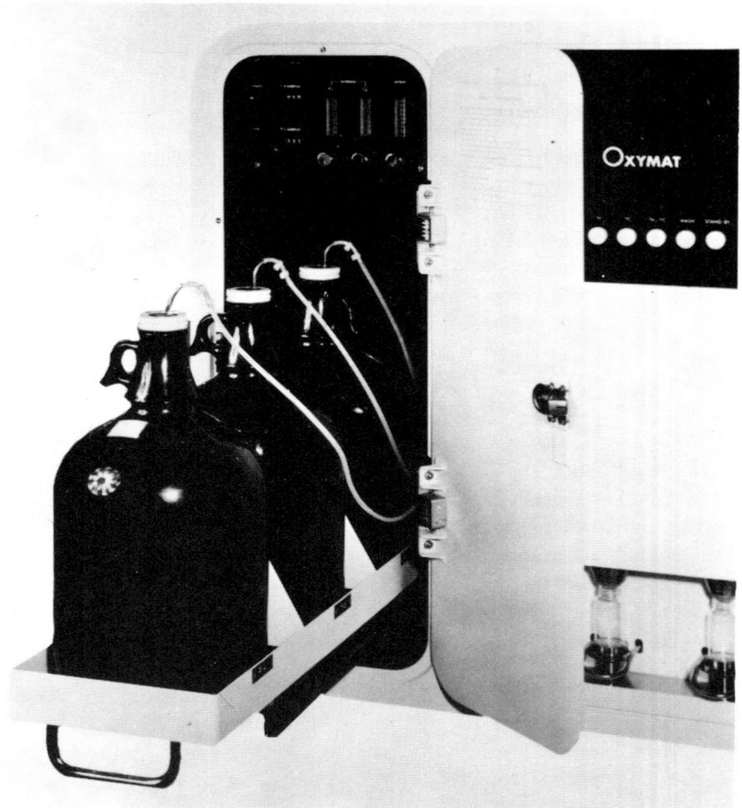

Fig.4. With the Oxymat's right front panel open, the furnace compartment is at right. The heated transfer line, a part of which can be seen below the furnace, leads combustion products to the HTO absorber. To the left is the $^{14}CO_2$ absorber and then one of the two metering pumps.

from a Viton seat. The port is opened, capsule introduction takes place and the port is closed again in less than 3 s. The quartz tube furnace (operating at 650—700°C) remains vertically oriented but its packing (Fig. 5) has undergone considerable change from that of Peterson. Combustion of the capsule and its contents takes place on top of a bed of coarse silicon carbide chips. Below that, most of the packed zone is filled with 99% cupric oxide pellets (E. Merck, Darmstadt # 2768) which have an appreciably higher oxygen content and are less prone to sintering and clogging than the 80% Cu_2O/20% CuO needles formerly used. Finally, the small remaining lowermost zone is filled with a catalyst (W.R. Grace, Grade 908, Type SMR-7 35 35) of the type currently being evaluated for control of automobile exhaust pollution. The catalyst presumably eliminates nitrogen oxides in the product stream leaving the furnace. That nitrogen oxides are no problem is fact; whether or not the catalyst is really the cause of their elimination has not been adequately tested.

Combustion products leaving the furnace pass through a heated (~200°C) transfer line to the HTO absorber. Use of the transfer line allows the HTO absorber to be placed adjacent to the furnace rather than below it as in the Peterson apparatus. This makes for a better dimensioned system which allows the user easier access

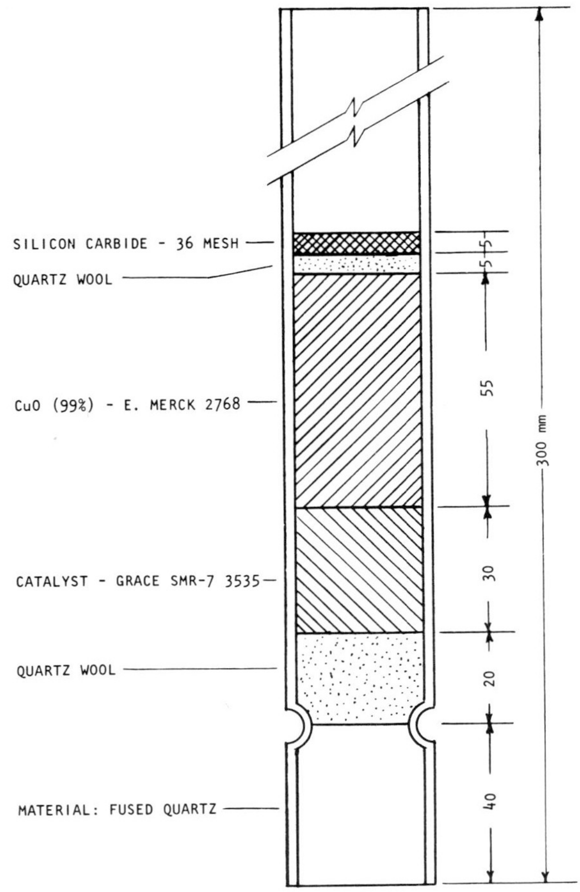

Fig. 5. Oxymat combustion catalyst tube.

since it does not have excessive height. It also removes the absorber with its operating content of inflammable solvent from close proximity to the furnace and its flame.

Both the HTO and $^{14}CO_2$ absorbers use the centrifugal wiped film principle found so effective for $^{14}CO_2$ by Peterson. In this case, the $^{14}CO_2$ absorber is about twice as long as that used for HTO, and the $^{14}CO_2$ rotor spins at twice the speed, both in recognition of the greater difficulty of collecting $^{14}CO_2$. External cooling (~4°C) is employed for both absorbers and is provided by a circulating refrigerant system which is an integral part of the apparatus. Solutions leaving either of the absorbers are directed to counting vials which are held on spring-loaded platforms and make gas-tight seals to the unit via soft silicone rubber cones firmly fastened to the fluid outlets. But, before the scintillator solutions actually enter the counting vials they pass through nebulisers, one directly at the outlet of each absorber, where they meet a nitrogen purge. Deoxygenation takes place, as does CO_2 removal from the HTO solution. Without the latter, instances have been observed in which almost 5% of the $^{14}CO_2$ content from a dual-labelled sample came into the HTO vial. With the nitrogen purge, the maximum $^{14}CO_2$ content of the HTO vial does not exceed 0.5% of the total $^{14}CO_2$ and is usually less.

The Oxymat incorporates those features and interlocks required of a commercial device. Adjustable timing cams allow precise turn-on and turn-off of the two stainless steel/Teflon controllable flow metering pumps (Fluid Metering Inc., Oyster Bay, New York) at the times most suitable for optimum product collection. Pressure sensors check the rates of oxygen and nitrogen flow and do not allow sample introduction if the rates are incorrect. Similarly, the temperatures of the furnace, transfer line and circulating refrigerant are monitored and operation is not possible if they are outside acceptable limits. Assuming adequate gas pressure, gas flows are automatically established; the operator is not required to manipulate valves other than to make certain that they are open. Temperature control is also automatic; the operator need not make any adjustments.

The Oxymat's combustion/collection cycle has a duration of 3 min; the next sample can be burned immediately after a cycle has been completed. Control is restricted to only a few pushbuttons. The operator denotes the nature of his sample — Hydrogen-3, Carbon-14 or Hydrogen-3/Carbon-14 — and after placing the combustion capsule with its test material on the introduction slide, depresses a single pushbutton. From there on, except for capping the vial, everything is automatic. The capsule is thrust into the furnace, combustion occurs, the pumps operate and trapping takes place, the nitrogen purge is carried out, and the counting vials are filled, all without operator assistance. In fact, during the 3 min combustion/collection cycle, the instrument may be left unattended, providing a good opportunity for the operator to weigh out or otherwise prepare the next sample. At the end of the day, or if the unit is not to be used for some hours, depression of the Wash pushbutton causes a methanol/toluene rinse to clean the system thereby preventing possible clogging of the fluid lines as solvent evaporates and scintillator is left behind.

OXYMAT PERFORMANCE

For the most part, the parameters that must be checked in evaluating the performance of a device such as the Oxymat are comparable to those for almost any analytical method. They are:
(a) Precision.
(b) Recovery.
(c) Linearity with sample type.
(d) Memory.
(e) Linearity with sample size.
Other parameters particular to radioisotope measurement which must be examined include:
(f) Counting efficiencies.
(g) Dual isotope separation.
Each of these factors will be considered separately.

Precision

Table 1 provides results obtained on combustion of diluted whole blood labelled with ^{14}C-dihydroxyphenylalanine (DOPA). Twelve identical 200 μl samples were combusted sequentially; there was essentially no delay between the end of one combustion and the beginning of the next. Samples were counted twice; the average of the two counts of each sample was taken as the activity. The standard deviation of the twelve averages is seen to be of the same order as would be expected had the same quantity of activity been directly pipetted into a counting solution and measured. Clearly, this suggests that Oxymat reproducibility is of a high order. Data presented below for other purposes supports this contention.

Table 1. Combustion of diluted whole blood.

Sample	1	2	Avg.
1	12313	12405	12359
2	12175	12206	12100
3	12345	12067	12206
4	12006	12125	12065
5	11737	11793	11765
6	12179	12191	12185
7	11903	12067	11985
8	11968	11866	11917
9	12095	12278	12186
10	12075	11874	11974
11	12112	11973	12042
12	12030	12139	12084
			Mean = 12072 ± 154.7

Recovery

Recovery measurements are based upon combustion of known amounts of activity. Tables 2 and 3 indicate better than 97% radioisotope recovery for combustion of bovine muscle homogenates to which known amounts of ^{3}H- or ^{14}C-cholesterol had been added. Such performance is typical and has been observed both by ourselves and our users over more than 20,000 combustions covering test substances of all types.

Table 2. Combustion of bovine muscle homogenate (^{3}H-cholesterol).

	Sample 1	2	3	4	5	Mean
c.p.m.	103186	111607	115337	116869	115475	112494 ± 4.94%
d.p.m.	275898	289137	279944	276940	276255	279634 ± 1.98%
			Average recovery — 97.9%			

Table 3. Combustion of bovine muscle homogenate (^{14}C-cholesterol).

	Sample 1	2	3	4	5	Mean
c.p.m.	401.2	405.9	408.3	399.2	396.8	402.3 ± 1.18%
d.p.m.	658.8	665.9	669.5	654.4	651.7	661.7 ± 0.90%
			Average recovery — 96.0%			

Linearity with sample type

That performance does not change with differing substrates is suggested by Table 4. Here, identical quantities of ^{14}C-cambendazole were added to 300 μl of various tissue homogenates or sucrose solution. Again, recoveries of activity were above 97%. There were no apparent differences related to the material being burned. With a mean count rate of 484.5 c.p.m. and a standard deviation of 12.05 for a total of 19 samples, it surely must be concluded that combustion considerably reduces the variability between sample types normally observed when conventional solubilisation procedures are employed.

Table 4. Combustion of ^{14}C-cambendazole — c.p.m.

Liver	Fat	Kidney	Lung	Muscle	Sucrose
488	481	487	482	479	497
484	491	489	472	510	496
471	498	485	454	480	
486					
476					
Mean of 19 samples — 484.5 ± 12.05					

Memory

In collecting the data for Tables 5 and 6, blanks were burned between samples and the amount of activity from one sample that might have found its way into the next had there been no blank ('memory') was determined. The average of the 10 blanks for each isotope, expressed as a percentage of the preceding activity, is given together with the standard deviation.

Unfortunately, memory decrease does not follow a straight line function. A second blank burned immediately after the first typically exhibits 20% of the activity of the first or about 0.1% of the activity of the substance combusted.

Memory decreases as the non-use time interval after a combustion is lengthened, probably due to the slow release of activity from binding sites as a result of temperature, continued oxygen flow and possibly further drainage of the last few drops of scintillator solution from the absorbers. Within perhaps 30 min essentially all memory has been eliminated. Preliminary experiments with a brief steam purge immediately after combustion suggest that memory can be reduced by better than a factor 5, but it is presently premature to state whether or not the effect is long term or if such a purge will prove practical in actual use.

Linearity with sample size

The effect of increasing sample weight on Oxymat performance, while still keeping within the suggested limits of 0.5 g for Carbon-14 and 1.0 g for Hydrogen-3, with the volume of aqueous solution not to exceed 1.0 ml under any circumstance, is less obvious than one might imagine. HTO counting efficiencies (see below) are dependent upon the quantity of water that must be trapped. Observed count rates, of themselves, do not have great significance and must be corrected for counting efficiency. When that is done, it is apparent (Table 5) that recovery is unaffected by sample size.

Table 5. Recoveries versus sample weight (Hydrogen-3).

	Butter mg		Fat mg		Liver mg		Testes mg		Whole blood mg	
	100	200	250	500	250	500	250	500	250	500
Recovery %	98.6	96.3	95.0	96.6	100.8	96.6	101.2	97.0	100.7	98.5
Memory %	0.37	0.37	0.26	0.19	0.21	0.93	0.45	0.05	0.29	0.18

Mean of 10 measurements: Recovery — 98.13 ± 2.18%
Memory — 0.33 ± 0.24%

The problem of collecting $^{14}CO_2$ is more difficult than collection of HTO where a physical process is so largely responsible. For that reason, maximum quantities of Carbon-14-labelled substances are limited. Again, observed count rates vary with the quantity of product collected, but here, because the β-phenylethylamine is more of a quencher than is its reaction product with CO_2, the count rate per unit sample weight increases with increasing sample size. However, as shown in Table 6, when corrections are made for counting efficiency, there is no significant variation in recovery as sample size is altered.

Table 6. Recoveries versus sample weight (Carbon-14).

	Butter mg		Liver mg		Muscle mg		Testes mg		Whole blood mg	
	50	100	250	500	250	500	250	500	250	500
Recovery %	98.7	99.0	105.2	98.4	100.1	101.0	100.5	97.8	101.0	101.9
Memory %	0.35	0.35	0.40	0.25	0.25	0.38	0.34	0.38	0.61	0.25

Mean of 10 measurements: Recovery — 100.36 ± 2.15%
Memory — 0.36 ± 0.11%

Counting efficiencies

Counting efficiencies observed for scintillator solutions leaving the Oxymat are functions of solution composition, the quantity of combustion products trapped and, of course, the counter in which they are measured. Formulations have been chosen to give the best compromise between trapping capacity, physical characteristics (especially viscosity), stability, absolute counting efficiency and constancy of counting efficiency with change in the concentration of trapped combustion products. The HTO solution — 700 ml dioxane, 300 ml toluene, 20 g naphthalene, 7 g butyl-PBD — up to its maximum useful water content of about 6% gives relatively good efficiencies when compared to more usual water counting preparations at the same water concentrations. In a modern counter, and with a non-quenching internal Hydrogen-3 standard, 40% counting efficiency should be expected. Data for increasing quantities of water produced by combustion of increasing quantities of sucrose are given in Table 7.

Table 7. Combustion of increasing sample weights (Hydrogen-3).

	Sucrose (mg)			
	248	350	450	553
Hydrogen-3 eff. %	37.6	35.8	35.4	33.3

While the HTO trapping solution is, for a solution of its type, relatively free from quenching, the same is not true of the $^{14}CO_2$ solution — 330 ml β-phenylethylamine, 220 ml methanol, 400 ml toluene, 50 ml water, 7 g butyl-PBD. With its high contents of β-phenylethylamine and methanol, and its rather low toluene content, it is highly quenched. However, since it is used only for Carbon-14, this is not a serious problem and, in fact, may be an advantage since the small amount of Hydrogen-3 contamination (see below) does not count. Counting efficiency for an unquenched Carbon-14 internal standard is typically about 75% and varies little with increasing quantities of $^{14}CO_2$ (Table 8).

Table 8. Combustion of increasing sample weights (Carbon-14).

	Salicylic acid (mg)			
	70	140	280	420
Carbon-14 eff. %	73	75	75	71

Dual isotope separation

Dual isotope separation is best measured by combusting singly labelled materials under dual isotope conditions. The two pairs of solutions so produced are counted, first in a wide window to ascertain the maximum amount of product collected in the absorber intended for the other isotope and then in the narrower window that is normally used to eliminate unwanted cross-contribution. In the first instance about 0.5% of each isotope comes into the vial for the other. When Carbon-14 contaminates the HTO vial, proper window setting (rejection of all counts above the Hydrogen-3 spectrum) will eliminate at least a third, and probably more, of the undesirable Carbon-14 counts. For HTO in the $^{14}CO_2$ vial, the situation is even better; the counting mixture itself serves to suppress the Hydrogen-3 counts making careful spectrometry largely unnecessary.

USER EXPERIENCE

It is difficult for the instrument manufacturer to anticipate all of the test substances likely to be encountered in the biomedical laboratory which various investigators might wish to subject to combustion. Therefore, it was of interest to us to inquire of our users as to their experiences with the Oxymat. Some representative examples are cited.

Combustion of Rat Faeces

In an experiment extending over a 30-day period, the daily excretion of ^{14}C-cholesterol in faeces was measured for 48 rats. A single collection from each animal was made daily and a representative sample of the pooled faecal material was counted after suitable processing. Initially, serious problems were encountered in obtaining uniformity and duplicates processed by combustion and by other means did not exhibit sufficient agreement. The difficulty was overcome by digestion of the pooled material with alcoholic KOH in capped, cone-bottomed tubes at about 85°C for 24 h. The digest was then neutralised (otherwise the alkalinity was found to attack to Lexan combustion capsule), agitated by hand to disperse remaining solids, and then duplicate aliquots were pipetted directly into the Lexan capsules. Activity after digestion was thought to be only in the liquid phase; solid particles were pipetted into the capsule only with the view of obtaining an average sample representative of the entire day's collection.

Open capsules were allowed to air dry overnight. The dried residue was burned without difficulty; over the 30-day period an average of 120 samples was combusted each working day. The entire experiment involved almost 3000 samples including standards and blanks. After each 12 combustions the Oxymat was put through a wash cycle to remove water which accumulated in the HTO trap and which, if not removed, led to the appearance of colour within the system. Ash was removed from the combustion furnace after each 200 samples.

Counting was performed with a Packard 3380/544. Inherent counting efficiencies were high but, by virtue of the 'electronic quenching' system of the unit,[11] it was not possible to know the maximum. Overall recovery for Carbon-14, as indicated by the combustion of commercially available plastic combustion standards (New England Nuclear Corp.), was 97% or better. Some of the experimental results are given in Table 9.

Table 9. Combustion of faecal samples — day 10.

Rat	ESR*	Counts/5 min	d.p.m.	Rat	ESR*	Counts/5 min	d.p.m.
1	0.2962	38974	9904	7	0.3025	60856	15493
1	0.3006	39565	10055	7	0.3004	59191	15068
2	0.3035	42620	10835	8	0.3054	19787	5003
2	0.3065	41436	10533	8	0.2935	21440	5425
3	0.3021	29648	7522	9	0.2980	30014	7615
3	0.2994	32681	8297	9	0.2953	31335	7953
4	0.3032	33238	8439	10	0.3020	18781	4746
4	0.3056	35963	9128	10	0.3100	19655	4969
5	0.2923	21392	5413	11	0.3048	25606	6489
5	0.3041	24284	6152	11	0.3051	30378	7708
6	0.3027	48201	12261	12	0.2969	33674	8550
6	0.3014	48041	12220	12	0.1982	31148	8932

* External Standard Ratio

Table 10. Activity distribution in a whole animal (Carbon-14).

Organ	ESR*[a]	Wet wt.[b]	d.p.m.	d.p.m./g	Total d.p.m.
Liver	0.3034	0.223	3084	13831	94355
"	0.3029	0.235	3158	13441	91693
Pancreas	0.3032	0.249	1041	4181	2989
Kidney	0.3084	0.188	936	4981	6994
"	0.1983	0.189	863	4567	6412
Adrenal	0.1977	0.033	163	4958	317
Spleen	0.2031	0.265	744	2810	1284
Ovary	0.2995	0.065	300	4618	591
Uterus	0.2034	0.187	872	4665	2286
Bladder	0.3030	0.050	113	2266	227
Lung	0.1959	0.174	476	2737	2893
"	0.3028	0.163	430	2639	2790
Adipose	0.2969	0.089	824	9260	1519
Muscle	0.2010	0.224	122	546	338
Brain	0.1957	0.246	77	312	535
"	0.2006	0.257	74	286	489
Heart	0.1995	0.288	435	1512	1034
Pituitary	0.3042	0.005	15	2920	15
Stomach	0.3003		10	1425	2251
"	0.3043		13	1324	2093
S. intestine	0.3002		582	2746	8923
"	0.2991		644	2709	8801
L. intestine	0.3007		482	47781	61446
"	0.2960		502	54398	69956
S. int. wash	0.2998		1117	364	10198
"	0.3037		1019	374	10478
L. int. wash	0.3046		1678	79131	1820008
"	0.3096		1487	81298	1869845
Stomach wash	0.3032		28	760	2910
"	0.3002		22	561	2577
Faeces					5431088
Urine					6050000
Serum					22866
Injected dose					15147000 = 89.5% Recovery

* External Standard Ratio.

[a] 0.3 = 79% eff.
0.2 = 70% eff.

[b] Dry faeces was combusted.

Total activity distribution in an animal

A 1,5-^{14}C six carbon carbohydrate was administered to female Charles River rats by oral intubation; the dose was approximately 1.5 x 10^7 d.p.m. per animal. At intervals after administration animals were dispatched; blood was collected, urine and faeces were recovered, stomach contents were segregated and organs were excised from the carcass. Representative samples of all materials were combusted and the total radioactivity content of each part of the body was determined. All tissue samples and other materials, with the exception of faeces, were burned wet; faeces were lyophilised and pulverised prior to combustion. The quantity of material combusted in each case was approximately 0.2 g or 0.2 ml; the pituitary, adrenals and ovaries were oxidised in entirety.

Counting of the products of combustion with an Oxymat was performed in a Packard 3380/544. Because of the nature of the instrument operation, it was not possible to know the inherent counting efficiencies of the solutions that were counted. However, only two external standard ratios were observed, one corresponding to approximately 79% efficiency and the other to 70%. As the particular instrument in question is known to frequently 'drop through' a high external standard ratio to a lower one, it is presumed that all samples had a value greater than 79%. Results for a single animal are given in Table 10.

Combustion of plant leaf extracts

Leaves taken from plants grown in the presence of Hydrogen-3/Carbon-14 doubly labelled compounds were homogenised with acetonitrile. Solids were discarded and the intensely coloured solution was collected for counting. Combustion of the neat acetonitrile solution proved unsuccessful due to extreme vigour and rapidity. However, when combustion was retarded with the aid of cellulose powder, quite satisfactory results were obtained. In each of the experiments cited in Table 11, 100 mg of cellulose powder was added to the combustion capsule before the acetonitrile solution was introduced. Blanks, used to check memory after each five samples contained only cellulose.

Experimental data is provided for five replicates of each of three different acetonitrile volumes. Within each group, precision is quite acceptable. Unfortunately, though external standard ratios indicate slight differences in counting efficiency as more combustion product is collected, corrections for quenching were not made. Therefore it is not surprising that the relationship between count rate and volume is not quite linear being 1.944:1 and 1.952:1 for Hydrogen-3 and Carbon-14 respectively rather than the expected 2:1 and 3.806:1 and 3.613:1 instead of the anticipated 4:1. However, when the Hydrogen-3:Carbon-14 ratios are examined, they are seen to be almost identical in all three groups, an indication that the degree of change in quenching with product concentration is about the same for the two isotopes in the weight range employed.

Combustion of polyacrylamide gel slices

A bacterial protein into which ^{14}C-isoleucine had been incorporated was subjected to gel electrophoresis. After staining and measurement of the bands, the gel cylinder was cut into 1 mm segments which were placed in combustion capsules and burned in an Oxymat. Counting was performed in an Intertechnique SL30; channels ratio was used as an index of counting efficiency since the counting data was further processed by computer using a program based on this method of quench correction. Counting data from a typical experiment is presented in Table 12. Counting efficiency is seen to be essentially constant; on examination of those

Table 11. Combustion of plant leaf extracts.

Sample	ml	Hydrogen-3 c.p.m.	Hydrogen-3 ESR*	Carbon-14 c.p.m.	Carbon-14 ESR*	$^{3}H/^{14}C$
1	0.05	4288	6.98	2043	2.63	2.10
2	0.05	4295	7.17	2155	2.65	1.99
3	0.05	4012	6.57	2057	2.61	1.95
4	0.05	3992	6.73	2032	2.46	1.96
5	0.05	4227	7.12	2128	2.65	1.99
Mean		4163	6.91	2083	2.60	2.00
Std. dev. %		3.58	3.72	2.64	3.08	2.93
Blank		6(0.14%)[a]	7.17	10(0.48%)[a]	2.81	
6	0.1	8072	6.90	4078	2.54	1.98
7	0.1	8119	6.86	4118	2.52	1.97
8	0.1	8030	6.52	4083	2.51	1.97
9	0.1	8019	6.64	3932	2.45	2.04
10	0.1	8230	6.90	4131	2.55	1.99
Mean		8094	6.76	4068	2.51	1.99
Std. dev. %		1.06	2.57	1.95	1.56	1.46
Blank		28(0.34%)[a]	7.26	18(0.44%)[a]	2.80	
11	0.2	15335	6.35	7304	2.26	2.10
12	0.2	15976	6.59	7045	2.47	2.27
13	0.2	16110	6.54	7754	2.46	2.08
14	0.2	15806	6.51	7898	2.44	2.00
15	0.2	16002	6.64	7634	2.39	2.10
Mean		15846	6.53	7527	2.40	2.11
Std. dev. %		1.93	1.69	4.61	3.59	4.63
Blank		59(0.37%)[a]	7.26	55(0.58%)[a]	2.81	

* External Standard Ratio

[a] Memory %

channel ratio values that seem to be aberrant, it is evident that the number of counts above background totally accumulated were too low for the ratios to be meaningful.

Table 12. Combustion of polyacrylamide gel slices (Carbon-14).

Slice	Gross count	Time	Net c.p.m.	Ch. ratio
1	20038	3.89	5128	1.344
2	20006	8.75	2263	1.364
3	20000	17.38	1127	1.495
4	20008	18.13	1080	1.489
5	585	20	6	2.038
6	20007	10.10	1958	1.453
7	783	20	16	1.710
8	20010	9.65	2050	1.510
9	20014	6.01	3307	1.490
10	20007	7.29	2721	1.651
11	9772	20	465	1.526
12	20033	6.73	2953	1.511
13	20002	5.46	3640	1.542
14	20020	4.56	4367	1.545
15	20036	4.53	4400	1.598
16	20033	5.93	3355	1.724
17	20016	9.01	2198	1.491
18	20008	18.34	1068	1.661

Miscellaneous

Several potentially interesting applications have been reported to us, but with insufficient experimental data for them to be described in detail. They are cited here, however, to suggest that the scope of the combustion method is broader than treatment of biological material:

(1) In a health physics screening programme, the possible content of Hydrogen-3 and Carbon-14 in urine is separated from natural Potassium-40 by combustion in an Oxymat.

(2) In column chromatography where recoveries are not quantitative, samples of ion-exchange resin from along the column length have been burned to locate bands of activity.

(3) Labelled soil samples are being burned. Due to the high ash content, the catalyst tube must be cleaned after each 20 combustions. Though such frequent cleaning is rather time consuming, no other method previously examined has allowed these same samples to be counted.

REFERENCES

1 W. Schoniger, Mikrochim. Acta 23, (1955).

2 F. Kalberer and J. Rutschmann, Helv. Chim. Acta **44**, 1956 (1961).

3 L.A. Wegner and H. Winkelmann, Atompraxis **16**, 19 (1970).

4 E. Rapkin, Digitechniques No. 6, Intertechnique S.A., 1973.

5 L.T. Freeland, Analyt. Chem. **36**, 2055 (1964).

6 M.A. Tamers and M. Diez, Intern. J. Applied Radiation and Isotopes **15**, 697 (1964).

7 J.I. Peterson, F.Wagner, S. Siegel and W. Nixon, Anal. Biochem. **31**, 189 (1969).

8 J.I. Peterson, Anal. Biochem, **31**, 204 (1969).

9 T.R. Tyler, A.Reich and C. Rosenblum in Organic Scintillators and Liquid Scintillation Counting (Ed. D.L. Horrocks and C.T. Peng), Academic Press, New York, 1971, p.869.

10 E. Rapkin and A. Reich, American Lab. **4** (10), 35 (1972).

11 R.J. Herberg, in Organic Scintillators and Liquid Scintillation Counting (Ed. D.L. Horrocks and C.T. Peng), Academic Press, New York, 1971, p.783.

DISCUSSION

B.E. Gordon: How important is the Hopcalite or other CO oxidiser? Most train combustions use only CuO to get complete oxidation to CO_2.

E. Rapkin: We have never made use of Hopcalite due to our inability to obtain it on a commercial scale. We do use an automobile exhaust emission control catalyst because it seems logical to ensure complete conversion of CO to CO_2 and nitrogen oxides back to N_2. As we have gained experience with the Oxymat, our tendency has been to reduce the volume of catalyst and increase the bed of copper oxide with small, but noticeable, improvements in recovery and especially in reduction of memory. I suppose that inertia, together with the thousand or more samples that must be burned for a complete test, has kept us from evaluating the performance with copper oxide alone.

B. Scales: You rejected the use of gelatine capsules as containers for your radioactive samples because of unwanted combustion products. Does this mean that the combustion of certain raw biological samples can also give rise to undesirable combustion products, and can therefore cause difficulties?

E. Rapkin: We avoid gelatine because it has been reported that combustion of gelatine capsules leads to the formation of nitrogen oxides which, in Peterson's original apparatus, apparently found its way to the counting vials where it acted as a quencher. Since we could specify almost any plastic which we thought best as we were having capsules made to our requirements, we chose Lexan, a polycarbonate resin which contains no nitrogen and which avoids the problem if, in fact, there is one. With biological samples, we have never observed any problem from nitrogen oxides and we have never sought for the reasons as to why we don't have any problems. It may be that the nitrogen purge which we use, and which Peterson didn't, is enough to overcome any difficulties. It may also be that our automobile emission control catalyst breaks down nitrogen oxides.

A.R. Ware: In suggesting that combustion is a good replacement for conventional solubilisation techniques for biological samples, you are adding to the problem of measuring counting efficiency accurately that of measuring the percent of Carbon-14 and Hydrogen-3 from the original sample. I accept that the counting efficiency determination is much improved by this technique as near constant quenched samples are obtained. How often is calibration necessary in order to obtain an accurate recovery figure?

E. Rapkin: We advise our users to combust standard samples at least once each day the Oxymat is in use. If it is being used continuously throughout the day, we would like our users to burn standards morning and afternoon. A system with mechanical seals and a catalyst with a finite life (2–3000 samples) ought to be checked routinely as a matter of good laboratory practice. Combustion of a standard requires only three minutes.

A.R. Ware: Does the recovery figure vary at all for different materials or is a single standard material adequate?

E. Rapkin: As long as the suggested weight restrictions are adhered to, recovery is near constant for all substances which we have tested so long as they do not burn explosively. Explosive combustion has never damaged any of the equipment but the rush of gaseous products produced thereby is too great for the trapping system. Combustion of standards is not suggested as a measure of the recovery of any particular substance but rather as a measure of the integrity of the apparatus and the efficacy of the copper oxide/catalyst section.

Chapter 13

Semi-automatic 'In-Vial' Combustion with the Micro-Mat System

L. A. Wegner and H. Winkelmann

Bayer AG, Isotopen-Institut, Friedrich-Ebert-Str. 217,
56 Wuppertal 1, W. Germany

INTRODUCTION

Since 1956 when Schöniger[2] first introduced the micro-determination of halogens and sulphur in organic substances, many different systems of oxygen flask combustion have been proposed in the literature,[3-7] some of which have become routine analytical methods. The most important advantage of combustion methods is the provision of uniform and favourable measuring conditions for a variety of specimens in biological and chemical studies involving Carbon-14 or Hydrogen-3. However, there are some obvious disadvantages of oxygen flask combustion methods.

For each sample preparation, a combustion flask, a pipette (to take an aliquote of solution for counting) and a combustion head are needed. Several sets of these devices must be on hand for practical work. The equipment involved is therefore elaborate, sample treatment as well as cleaning is time-consuming and throughput is not adequate for preparing large numbers of samples. To overcome these drawbacks we modified Gupta's 'in vial' combustion method[7] with respect to the manual handling of large numbers of samples. The result was the development of the 'Micro-Mat' system, a device for the semi-automatic combustion of organic material. This paper will attempt to show that the method yields good results under a variety of conditions and can be applied to many different investigations involving Carbon-14 or tritium.

SAMPLE PREPARATION

The samples to be examined are weighed unwrapped into small, one-way, wire-gauze baskets lined with a layer of easily combustible, thin, black paper. Up to 20 baskets can be stored and handled on a special sample rack (Fig. 1) having numbered positions. With the aid of a supporter (Fig. 1), sample baskets can be removed from the rack and returned without the supporter coming into contact with the radioactive material. A Sartorius 'Electronic Micro-Balance Type 4135' is used to record the weights of the samples. Such weighings are performed in the range up to 99.99 mg with a sensitivity of 10 μg. To permit easy handling of the one-way baskets, the balance pans were modified.

By means of special sample holders (available from the manufacturer), the baskets together with weighed samples are inserted into the vials but are not sealed

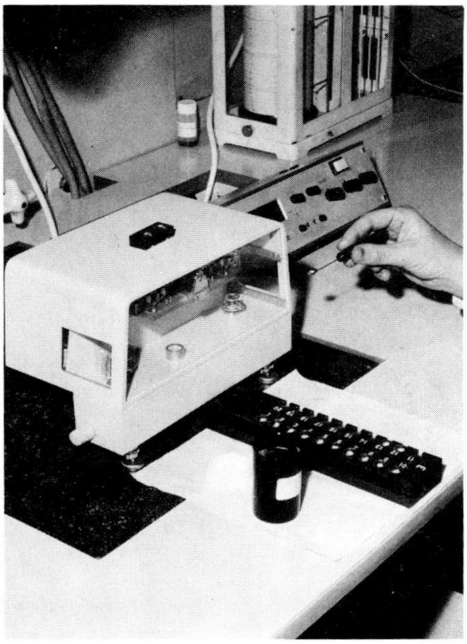

Fig. 1. Using a supporter to handle samples. Sartorius Electronic Micro-Balance, supporter and sample rack.

Fig. 2. Vial ready for evacuating and oxygen filling, containing the absorber, inserted sample holder fastened to the screwcap, sample-basket and unwrapped sample.

to be gas-tight (Fig. 2). For the absorption of CO_2 gases, as well as for tritium water collection, 0.5 ml of phenylethylamine is initially placed on the bottom of each vial. Up to 40 vials (two turntables) are evacuated and filled with oxygen (Fig. 3). When filling is complete the screw caps are gas-tight sealed, the turntable is placed on the 'Micro-Mat' (Fig. 4) and automatic combustion is started.

The samples are ignited stepwise by focussing a light beam from a low-voltage halogen lamp onto the samples. The black paper starts burning within about 1.5 s. In the case of biological specimens the samples burn smoothly with a visible flame. When combustion of the last sample is complete (for one turntable — 20 samples — it takes about 5 min from oxygen filling to complete combustion) the turntable is allowed to stand at room temperature for about 10 min while the carbon dioxide which is formed is absorbed. For complete collection of tritium water at room temperature about 45 min are required. After the sample holders have been removed 10 ml of a mixture of 85% toluene, 15% methanol and 8 g/litre butyl-PBD is added to each vial which may contain Carbon-14 or tritium samples. After gentle shaking so that the carbamate so formed is solubilised in the cocktail the samples are ready for counting.

THE WEIGHING SYSTEM

It must be emphasised that sample weighing is the most time consuming step of the preparation and weighing is one of the most susceptible sources of error. This is of particular importance when large numbers of samples have to be examined (as in the case in pharmacokinetic studies). We therefore use a system of three balances at

Fig. 3. Device for evacuating and oxygen-filling.

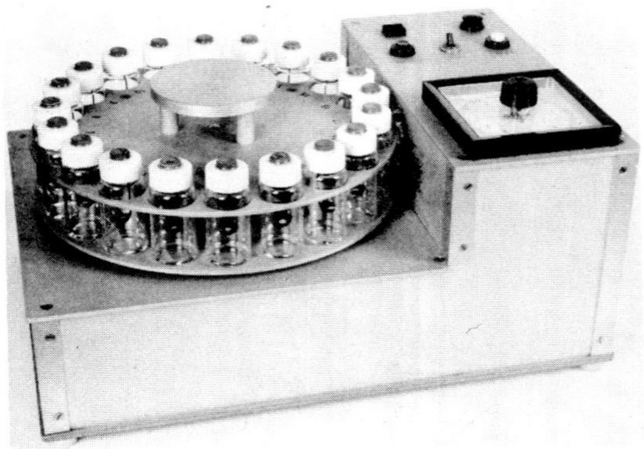

Fig. 4. Micro-Mat with 20 samples ready for combustion.

a time and a central Teletype output which enables us to punch out the weight as well as the sample identification on paper tape.

Keyboards in connection with the balances enable the operator to add a five-digit sample identification number ('I' and 'K' in Table 1) to each sample weight The number must be entered twice. Conformity of the first and second entries is automatically checked and when found to conform number and weight are printed out. If there is no conformity between the first and second entries a 'wrong' signal appears on the data-transfer-unit and the operator has to enter identification again. In this way it is ensured that the sample weight and identification number belong to each other. When sample identification has been checked, weighing can be continued using the same sample identification by simply pushing a 'continue' button. In this way the number of weighings is automatically increased by one ('J' in Table 1).

Table 1. Example of computer-processed net weight (mg), activity (d.p.m.) and specific activity (d.p.m./mg) with the mean value and standard deviation.

Organ	Animal (I)	(K)	(J)	N-NR	Weight (mg)	Act. (d.p.m.)	Spec. Act. (d.p.m./mg)	Rfl. conc. (P*)
Lung	19	583	1	183	4.12	94661.	2.298E+04	
			2	184	3.27	74574.	2.281E+04	
			3	185	3.71	84581.	2.280E+04	
			4	186	3.24	73559.	2.270E+04	
			5	187	3.74	85387.	2.283E+04	
			6	188	4.11	93939.	2.285E+04	
			7	189	3.65	82293.	2.255E+04	
			8	190	4.11	93472.	2.274E+04	
			9	191	3.73	85241.	2.285E+04	
			0	192	4.86	111340.	2.291E+04	
			M.	V. :			2.280E+04	
			S	(%) :			0.6	0.0

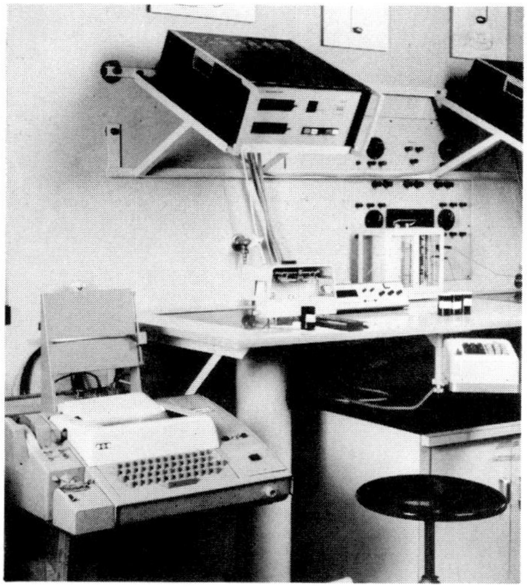

Fig. 5. Weighing system with keyboard for sample identification, balance, data-transfer-unit and Teletype.

EXAMPLE OF COMPUTER-PROCESSED DATA

The raw data output from the balances as well as from the liquid scintillation spectrometer are automatically coded in ASCII-Code on paper tape and are computer processed by an IBM 370/145. A modified Nuclear Chicago Fortran-program was used for fitting polynomials for quench correction curves by the least squares procedure with up to 20 standards with increasing amounts of quenching agent. These investigations (Table 1) are made to test the performance of a combustion system taking into consideration the different specimens to be examined (Table 3).

INVESTIGATIONS OF THE MICRO-MAT SYSTEM

For the determination of the recovery of the activity we took pains to have exactly the same cocktail composition for the combusted samples and the solubilised samples. We therefore prepared these samples in the following way (See Fig. 6).

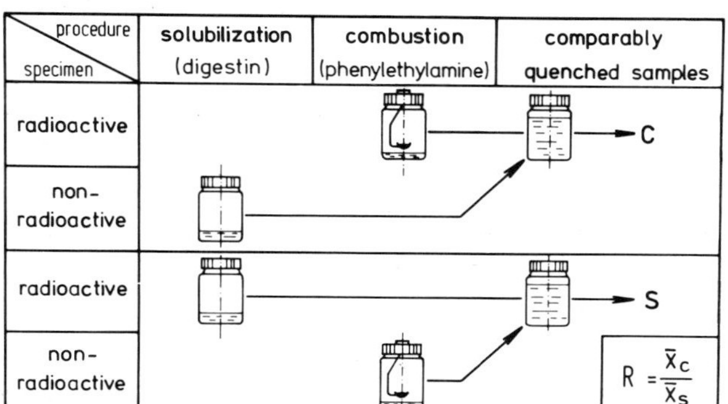

Fig. 6. Schematic representation of sample preparation for recovery determination.

The non-radioactive material is solubilised with Digestin® whilst the radioactive material is combusted as mentioned above. The solubilisation vial is filled with 10 ml of cocktail, shaken and poured into the radioactive combustion vial.

The sample ready for counting is the combusted sample C with a specific activity X_c (d.p.m./mg). For the preparation of the solubilised sample S the procedure is reversed, thus ensuring that the counting samples C and S exhibit exactly the same degree of quenching. Provided the activity found by solubilised samples was 100% of sample activity (assuming no loss of activity during solubilisation), we had recoveries of over 98% for Carbon-14 (on average 99%) — Table 2 — and more than 97% for tritium (on average 98%).

Table 2. Recoveries for different Carbon-14 labelled specimens.

Specimen	R	S_R (%)	Specimen	R	S_R (%)
	0.99_8	0.5	lung	1.00_8	1.1
liver	0.98_5	0.3	kidney	0.99_0	0.7
	0.99_7	0.5	guanacline	0.99_5	0.3
brain	0.99_0	0.4		0.98_8	0.5
bloo d	1.01_4	1.2	succinic acid	0.99_2	0.2
		$\overline{R} = 0.99_6$			

Generally, it was found that the samples exhibited a minimum variable degree of quenching. We found an average efficiency of

$$E = 85\% \text{ with } s < 1\% \text{ for Carbon-14}$$
$$E = 40\% \text{ with } s < 2\% \text{ for Hydrogen-3}$$

The limit of sample weight which is combustible without an ash residue and decreased recovery depends on the nature of the material. Fuchs and De Vries[8] reported that with the Micro-Mat system they combusted amounts of approximately 10 mg of freeze-dried plant material with good recoveries and leaving only very small ash residues. We have also combusted filter papers and cellulose powders of up to 15 mg with only small residues. For freeze-dried biological materials, e.g. liver containing Hydrogen-3 compounds, we found amounts of up to 6 mg combusted without ash residues, but with increasing sample weights ash residue increased (Fig. 7). Nevertheless, results were found to be entirely reproducible up to 9 mg.

The criterion used to assess the reproducibility of the Micro-Mat combustion system was the standard deviation of the assay of the specific sample activity. Seven different rabbit tissues were examined for both isotopes in groups of 10 samples per specimen and day on three different days. In addition, Carbon-14 labelled succinic acid was tested, but with 30 consecutive samples (results have therefore not been taken into consideration in the calculation of Σn and $\bar{s}_x$). The standard deviations, $\bar{s}_x$, of the different Carbon-14 groups (Table 3) ranged from 0.6 to 1.2% with an average, $\bar{\bar{s}}_x$, of 0.9%, while the ^3H-values ranged from 1.1 to 1.6% with an average of 1.4%. In earlier investigations of five different tritium-labelled rat tissues we found $\bar{\bar{s}}_x = 1.1\%$ for 300 activity determinations.

For the time being we do not recommend that one can conclude from these results that there exists a definite distinction in standard deviations when preparing Carbon-14 or Hydrogen-3 materials by Micro-Mat combustion. Additional investigations will have to be performed concerning the influences of time and persons involved in the sample handling.

specimen	isotope: ^{14}C n/day 1.	2.	3.	$\bar{s}_x$ [%]	isotope: ^{3}H n/day 1.	2.	3.	$\bar{s}_x$ [%]
brain	10	10	10	1.2	10	9*	9*	1.6
testicles	10	10	10	0.7	10	10	10	1.6
kidney	10	10	10	1.0	10	10	10	1.2
liver	10	10	9*	1.2	10	10	10	1.4
lung	10	9*	10	0.6	10	10	10	1.2
blood	10	10	10	0.9	10	10	10	1.1
muscle	10	10	10	0.8	9*	9*	10	1.6
succinic acid			(29)*	(0.7)	—	—	—	—
	Σn = 208			$\bar{\bar{s}}_x$ = 0.9	Σn = 206			$\bar{\bar{s}}_x$ = 1.4

s_x = standard deviation; $\bar{s}_x$ = average of s_x; $\bar{\bar{s}}_x$ = average of $\bar{s}_x$

n/day = number of samples per specimen and day
* = number of non-aberrant observations

Table 3. Standard deviations of Carbon-14 and Hydrogen-3-labelled specimens combusted using the Micro-Mat system.

$$S_x = \frac{100}{\bar{x}} \sqrt{\frac{\sum_{i=n}^{n} (x_i - \bar{x})^2}{n-1}} \quad ; \quad \bar{S}_x = \frac{1}{n} \sum_{i=1}^{n} S_{xi}^{\ 2}$$

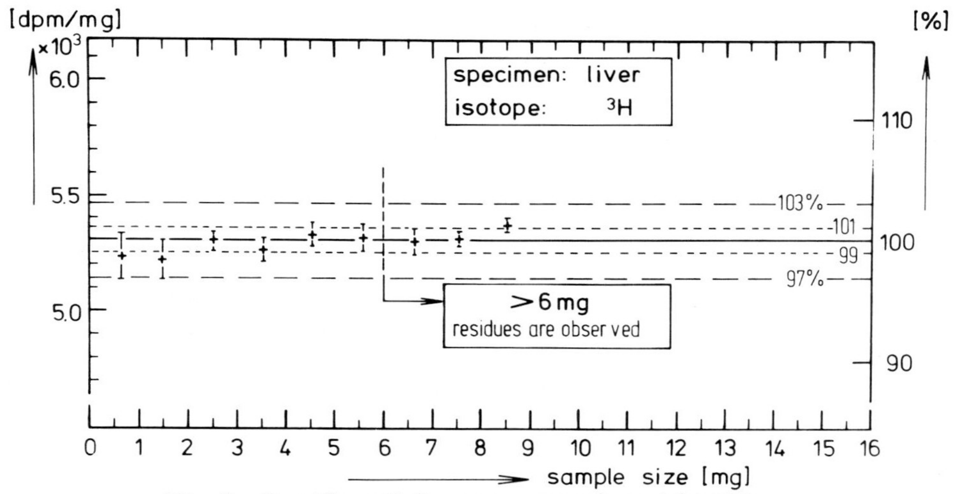

Fig. 7. Specific activity versus sample weight (H).

Samples were counted predominantly in a Nuclear Chicago Mark I liquid scintillation spectrometer equipped with automatic external standardisation and Teletype-output. A few samples were counted with a Philips LSA and a Packard Analyser, both equipped with paper tape punch output and automatic external standardisation. Counting was started not earlier than 30 min after pre-cooling. Samples were counted at 2°C using the external channels ratio technique.

ADDITIONAL APPLICATIONS OF THE MICRO-MAT SYSTEM

In autoradiography the determination of Carbon-14 and Hydrogen-3 distribution in whole-body sections of small animals has been quantified. By means of special

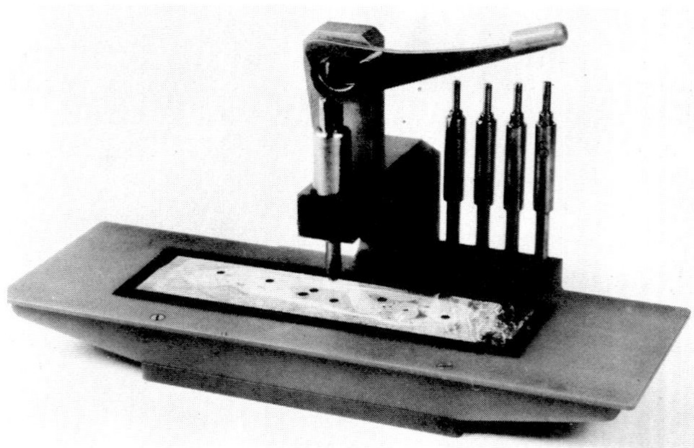

Fig. 8. A whole-body rat-section on special punch deck with different interchangeable punching tools.

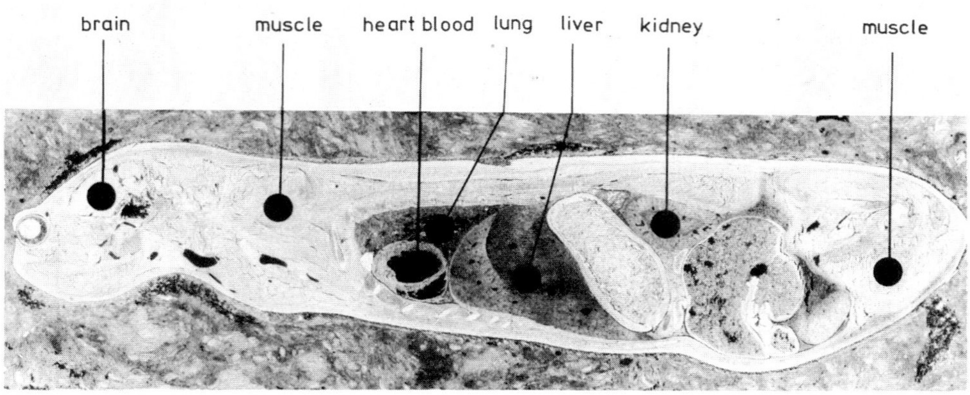

Fig. 9. A whole-body rat-section after punching out disks of tissues to be examined.

punching tools (Fig. 8), tissue disks of whole-body sections ranging from 2 to 6 mm diameter with a thickness of 50 μm are punched out (Fig. 9). Volumes of approximately 1 mm^3 of biological material are obtained by this technique when punching disks of 5 mm diameter. This method therefore requires no weighing of the samples and offers a quick and reliable means of determining activity concentrations in sections with a satisfactory degree of accuracy (as has been demonstrated in an extended investigation to be published).

Another interesting application is the repeated determination of the time-dependent activity concentration in the blood of the same rat; the tail of the animal is cleaned with water (it is not necessary to shave the tail to find the veins). A vein is pierced using a skin needle and spontaneous extravasating drops of blood are collected by means of pre-weighed filter-paper-disks of 8 mm diameter. Amounts of 0.5 to 5 mg of dry blood, i.e. about 4 to 40 μl blood, are collected. The samples are combusted as already described. Small amounts of blood can be obtained repeatedly (10-20 times per day) using this procedure. The activity concentrations are calculated as described earlier (Table 1).

REFERENCES

1 L.A. Wegner and H. Winkelmann, Atompraxis, **16**, 19 (1970).
2 W. Schöninger, Mikrochim. Acta 869 (1956).
3 R.G. Kelly, E.A. Peets, S. Gordon and D.A. Buyske, Anal. Biochem. 2 267 (1961).
4 F. Kalberer and J. Rutschmann, Helv. Chim. Acta **44**, 1956 (1961).
5 V.T. Oliverio, C. Denham and J.D. Davidson, Anal. Biochem. **4**, 188
6 H.E. Dobbs, Analyt. Chem. **35**, 783 (1963).
7 G.N. Gupta, Analyt. Chem. **38**, 1356 (1966).
8 A. Fuchs and F.W. De Vries, Intern. J. Applied Radiation and Isotopes **23**, 361 (1972).

DISCUSSION

J.F.Stoutjesdijk: During the absorption of the CO_2 a film of solid material is formed on the surface of the absorption liquid. How can a complete absorption of the CO_2 be possible?

H.Winkelmann: The carbamate precipitates and fresh liquid is exposed to the gas phase.

B.E.Gordon: How much of the oxygen is consumed by the sample + paper assuming a 6 mg sample?

H.Winkelmann: A freeze-dried 6 mg tissue sample and the layer of the thin paper of the basket consume approximately 12 ml of oxygen.

J.G.Dent: What is the minimum detectable level of Carbon-14 in tissue samples allowing a reasonable counting time?

H.Winkelmann: The question concerning the reasonable counting time can only be answered with respect to the desirable counting statistics. In our laboratory the minimum detectable activity concentration in dry tissue samples is approximately 1 nCi/g. That means about 10 d.p.m. per sample when 5 mg samples are examined. Assuming 80% counting efficiency the net counts per 100 min are 800.

P.Johnson: Have you any results with fatty tissues — for instance, to take the worst example, have you tried combusting adipose tissue?

H.Winkelmann: It is reported that in flask combustion methods only very small amounts of fatty tissue can be combusted. The samples must be wrapped in relatively large amounts of wrapping material, e.g. filter paper. In our case one of the principles of the Micro-Mat technique is the handling of unwrapped samples. Therefore the combustion of small amounts of adipose tissue using the Micro-Mat system is only possible when the sample is mixed with an approximately three to five times surplus of cellulose powder, sawdust or similar material.

B.Scales: One certain advantage of a self-contained combustion system is that even though the combustion may be rather incomplete, all the products of combustion are available to the scintillator. In contrast, pipe lines leading from the combustion vessel to the absorption trains often accumulate deposits of pyrolysis products, leading to poor total recovery of Hydrogen-3 and especially Carbon-14.

Chapter 14

Sample Preparation with an Automated Oxygen Flask Combustion Apparatus for Liquid Scintillation Counting of ^{3}H-, ^{14}C- and/or ^{35}S-labelled Material

P. Rauschenbach and H. Simon

Organisch-Chemisches Laboratorium, Technische Universität München, Germany

INTRODUCTION

Insoluble materials and quenching differences for soluble materials cause difficulties in sample preparation and thus diminish considerably the advantages of sensitive determination of weak β-emitters by means of liquid scintillation counting. In addition, measurement of insoluble materials in emulsions or suspensions rarely gives exact analyses, and the measurement of quenching from very different materials by one of the channel ratio methods, for example, easily leads to falsified results.

For organic or biochemical compounds, however, these difficulties can be overcome by combustion of the samples. From Hydrogen-3, Carbon-14 and Sulphur-35 labelled materials, one obtains uniformly HOT-, $^{14}CO_2$- and oxides of Sulphur-35 respectively — singly or in combination. All these oxidation products are soluble in appropriate liquid scintillation systems, with defined and known quenching characteristics. A disadvantage of these procedures lies in the considerable manual labour associated with the combustions, irrespective of the particular method selected.

The first part of this paper, therefore, describes an automated apparatus, in which an oxygen flask combustion[1,2] is carried out. This apparatus is commercially available with the designation 'Pyroszint'.

The second part of this paper attempts a comparison of advantages and disadvantages of the automated oxygen flask combustion and the oxygen stream technique.

APPARATUS

The automated apparatus consists of the combustion apparatus (Fig. 1), the shaking device (Fig. 2) and the racks. Each rack can hold eight normal Erlenmeyer flasks with a volume of 0.5 or 1 litre.

Fig. 1. Combustion apparatus.

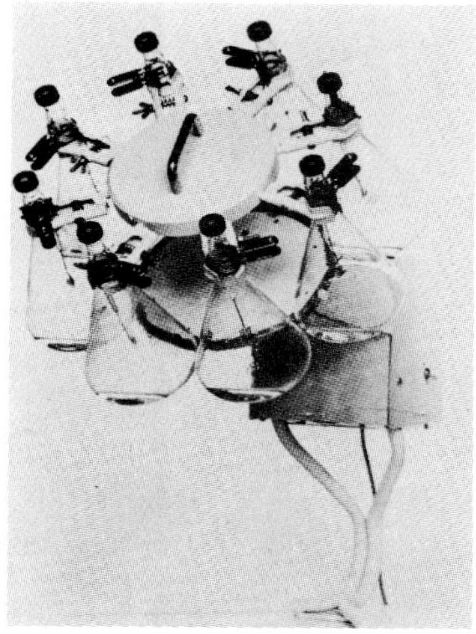

Fig. 2. Shaking device.

The combustion apparatus is equipped with a device for filling the flasks with oxygen, a projector lamp for the ignition and a dispenser for the exact delivery and injection of a scintillator-absorbent solution from a storage vessel. The combustion flask is closed manually by a sample holder (Fig. 3) which holds the filter paper-wrapped sample in a wire basket. The holder seals the flask by means of a screw cap, carrying a septum. Through this septum the scintillator-absorbent solution is automatically injected, after the combustion.

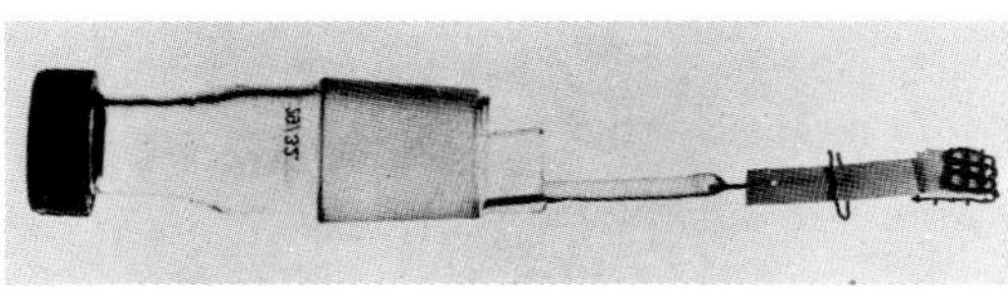

Fig. 3. Sample holder.

The actual procedure in detail is arranged as follows :

1 For solid samples, samples of up to 100 mg are weighed on ashless filter-paper cut-outs (Fig. 4). (These special cut-outs are commercially available.) For liquid samples, up to 100 µl of aqueous aliquots can be applied directly to a filter tablet without drying.

2 The filter paper-wrapped samples or the filter tablets are mounted on the sample holders, which are securely equipped with septum and screw cap. The holders can then be stored in a rack (Fig. 5), until they are inserted into the combustion flasks.

3 A rack with eight Erlenmeyer flasks is then set into the combustion apparatus. The oxygen flush is initiated by push-button. Two flasks are flushed at the same time for 1 min, approximately, at an adjustable flow rate of about 2 litres/min. Immediately after the flush, the two flasks are closed with the prepared sample holders.

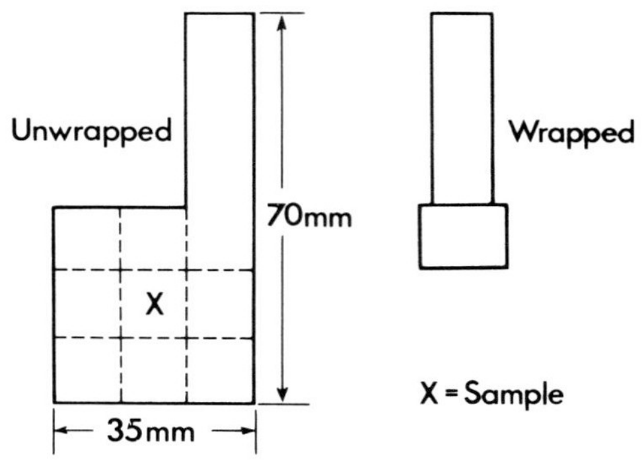

Fig. 4. Filter paper cut-out.

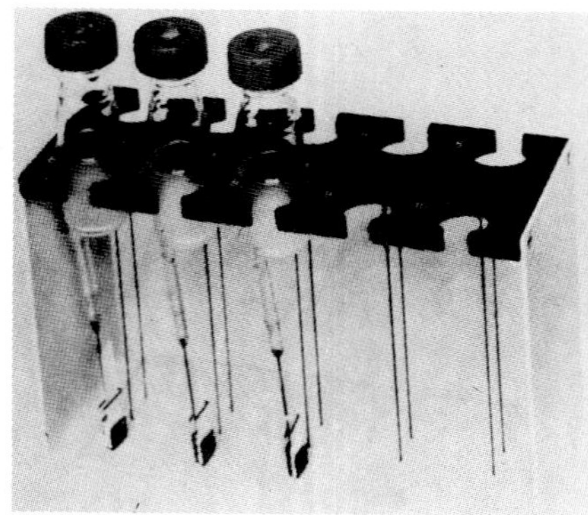

Fig. 5. Sample holder rack.

4 When all flasks are filled with oxygen and closed with sample holders, the push-button activates the ignition of the sample by means of a focussed beam from a projector lamp.

5 The apparatus then stops for 6 min for cooling after the last ignition. A defined volume of scintillator-absorbent solution is then automatically injected through the septum of the sample holder of each flask.

6 After the last injection, the rack is transferred to the shaking device — 10 min is needed for shaking.

7 An aliquot from each flask is then transferred to a counting vial. The sample is then ready for spectrometric measurement and evaluation.

A flow sheet procedure is shown in Fig. 6. It can be seen that approximately 5 min is needed for each sample. While the apparatus runs one batch of analyses, sufficient time is available to make weighings for the next group of samples.

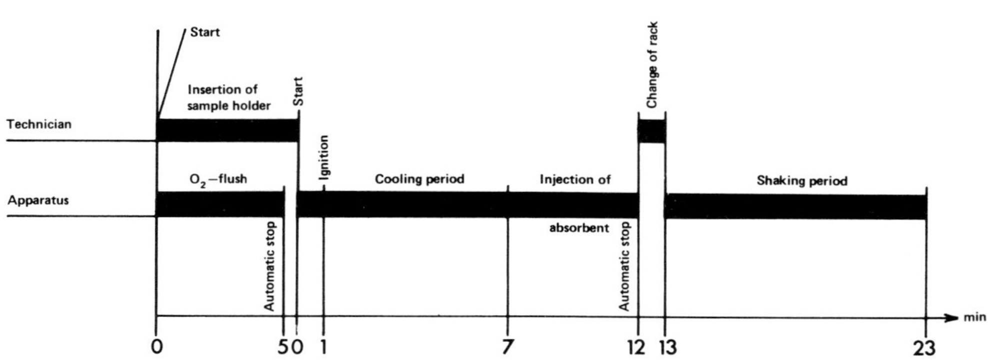

Fig. 6. Flow sheet of procedure.

RESULTS

The following scintillator-absorbent solution has proved to be sufficient:

180 ml	2-Phenylethylamine,
250 ml	Methanol,
570 ml	Toluene,
5 g	PPO (i.e. 2,5-Diphenyloxazole),
and 0.5 g	Dimethyl-POPOP (1,4-bis-2-(4-Methyl-5-phenyloxazolyl)-benzene).

This mixture is a modification of the one described by Woeller[3] and it permits the simultaneous measurement of Hydrogen-3 together with Carbon-14 and of Hydrogen-3 together with Sulphur-35. For Sulphur-35 analyses, 4% of water has to be added to the mixture. In modern spectrometers counting efficiencies for singly labelled materials are 20% for Hydrogen-3 and 85% for both Carbon-14 and Sulphur-35, or, for doubly labelled materials, 16% for Hydrogen-3 and 50% for Carbon-14 or Sulphur-35, with a spillover of 10% Carbon-14 or Sulphur-35, respectively, into the Tritium counting channel.

Tables 1 and 2 show some results that were obtained with the apparatus. All samples are strong quenchers and/or insoluble. Each sample in a weight range of 10–100 mg was analysed 8–10 times and standard deviations of singly labelled materials of 1.0–2.1% have been obtained. Doubly labelled materials show standard deviations dependent on the isotope ratio. For a ratio Hydrogen-3/Carbon-14 of about 0.4 the deviations are 3.3–5.7% for Hydrogen-3 and 1.5–1.6% for Carbon-14. Isotope ratios of 4.5–8.7 show deviations of 1.2 and 1.0–2.9%, respectively.

Before continuing further, some of the advantages of the described procedure in comparison to the 'classical' flask combustion should be pointed out. The oxygen flush of the flasks, the ignition and the delivery of defined aliquots of scintillator-absorbent solution are completely automated. No cooling materials are needed whatsoever; singly-labelled, and either Hydrogen-3/Carbon-14 or Hydrogen-3/Sulphur-35 doubly-labelled samples can be analysed by a single procedure.

The sample combustion in streaming oxygen has also been automated,[4-6] and some apparatuses are commercially available. Both procedures, flask and stream, have inherent advantages and disadvantages that are shown in Tables 3 and 4.

Table 1. Results for singly labelled materials.

Compound/nuclide	Number of samples	d.p.m./mg	% S.D.	% Recovery
Nicotinic acid / ^{3}H	8	10719	1.25	94.6
Nicotinic acid / ^{14}C	8	5303	0.99	95.2
Glucosazone / ^{3}H	9	3189	1.76	—
Glucosazone / ^{14}C	10	1513	1.24	—
p-Nitraniline / ^{3}H	9	10658	1.48	—
Liver / ^{3}H	8	1001	1.50	—
Faeces (rat) / ^{14}C	8	626	2.14	—
Thiourea / ^{35}S	10	1481	1.90	100.9

Sample weight range, 10–100 mg.

Table 2. Results for Hydrogen-3/Carbon-14 doubly labelled materials.

Compound	No. of Samples	d.p.m./mg ^3H/% S.D.	d.p.m./mg ^{14}C/% S.D.	Isotope ratio (measured)	% S.D.	Isotope ratio (theoretical)
Nicotinic acid I	11	1990/3.26	4394/1.51	0.453	3.60	0.460
Nicotinic acid II	10	7512/1.05	1657/1.01	4.53	1.45	4.61
Glucosazone I	8	505/5.68	1240/1.60	0.407	5.90	–
Glucosazone II	10	2437/1.22	281/2.86	8.67	3.11	–

Sample weight range, 10–100 mg.

Table 3. Oxygen flask combustion (Pyroszint).

Advantages	Disadvantages
Without memory effects	No chemical separation of ^3H from ^{14}C [a]
Accuracy of results not influenced by chemical composition	For each combustion a flask has to be cleaned
Accuracy of results not changed by differing sample weights	Decreased sample number per time unit, as compared to stream technique
Procedural errors visually observable	Sample weights up to 100 mg
Minimal technical experience is needed	

[a] Physical separation by the spectrometer is reliable up to $^3H/^{14}C$ ratios of $\geqq 0.5$.

Table 4. Oxygen stream combustion

Advantages	Disadvantages
Increased sample number per time unit, as compared to flask technique	Memory effects up to 1 % [a]
Cleaning of combustion vessels unnecessary	Empty combustion decreases the background only asymptotically
Chemical separation of ^3H from ^{14}C	Accuracy of results influenced by the chemical composition
Sample weights higher than 100 mg are possible	Accuracy of results influenced by differences in sample weight
	Procedural errors not directly observable and can only be found with test runs
	Apparatus requires experienced technician

[a] Following a sample with 10^5 c.p.m. a blank sample shows an activity of 10^3 c.p.m.

As was shown, the two analytical methods do not exclude or replace each other, but have special features which permit adjustment to the particular experimental problem that is at hand.

ACKNOWLEDGEMENT

Financial support by the Bundesministerium für Bildung und Wissenschaft of the FRG is gratefully acknowledged.

REFERENCES

1 F. Kalberer and J. Rutschmann, Helv. Chim. Acta **44**, 1956 (1961).
2 D.A. Buyske, R. Kelly, J. Florini, S. Gordon and E. Peets, Atomlight 20, 1 (1961).
3 F.H. Woeller, Anal. Biochem. 2, 508 (1961).
4 J.I. Peterson, F. Wagner, S. Siegel and W. Nixon, Anal. Biochem. 31, 189 (1969).
5 J.I. Peterson, Anal. Biochem. 31, 204 (1969)
6 N. Kaartinen, Technical Bulletin No. 18, April 1969, Packard Instrument Company, Downers, Grove, Ill., U.S.A.

DISCUSSION

G. Stoydin: It is stated that 6 min cooling be allowed following the combustion in an oxygen flask. Is the period of cooling critical from the safety point of view and are there other considerations also involved?

P. Rauschenbach: The cooling period should not be smaller than 2 min. Under this time period there is a possibility of explosions, but the time of 6 min excludes securely this hazard.

C. Palais: What is the total volume of scintillator-absorber added to the combustion flask and what is the volume of the aliquot used for counting?

P. Rauschenbach: The total volume is 17.5 ml from the dispenser and the aliquot volume 15.0 ml from a pipette.

B.E. Gordon: Has anyone taken into consideration that the reaction of CO_2 + amine $\rightleftharpoons$ carbamate is an equilibrium reaction which is temperature dependent? Thus quantitative recovery may not be achieved if the temperature is not low enough ($\simeq 0^{\circ}C$).

P. Rauschenbach: In the work of Woeller {Anal. Biochem. 2, 508 (1961)} there is no comment concerning a temperature dependence at room temperature, and we too could not observe such a dependence.

Chapter 15

A Comparison of Combustion Techniques Used in the Preparation of ^{14}C-labelled Samples for Liquid Scintillation Counting from Biological Materials

R.G. Cooper

Pharmaceuticals Division, Imperial Chemical Industries Limited, Macclesfield, Cheshire, U.K.

INTRODUCTION

The preparation of homogeneous samples for the radioassay of β-emitters in a wide range of biological materials is a problem which has almost as many answers as there are analysts and opinions. It is, however, generally accepted that, for labelled tissues, combustion is the most accurate and frequently the most convenient method. This chapter confines itself to the author's personal experience of combustion techniques and of their application to samples labelled with Carbon-14 only. The more important of the methods I have tried are outlined with brief descriptions of the apparatus and the results obtained. A comparison is then given of the methods in current use in our laboratories, particularly with respect to reliability in the multi-user situation.

THE PEETS AND FLORINI TUBE FURNACE

The method developed by Peets and Florini (Lederle Laboratories, American Cyanamid 1960)[1] is of current interest in that a low-cost biological oxidiser based on the same principles is now marketed by Beckman and the Harvey Instrument Corporation. The apparatus is an enlarged and modified version of the standard carbon/hydrogen microanalysis furnace. It can oxidise samples of up to 2 g dry weight. The filling of the silica tubes consists of a loose roll of acid-washed nickel foil about 20 cm long and a 2 cm thick pad of loosely matted silver wool. Three such tubes are mounted in an electrically heated furnace. The catalyst section is maintained at about 950°C, the sample section is gradually heated to 750°C. Current to the heating elements is controlled by a timer and switching device to give an adjustable rate of temperature increase. The Lederle group quote Pregl and Grant in claiming this type of filling to be as efficient as the conventional copper oxide filling when used as a combustion catalyst, and also claim that it has no tendency to pack and retard gas flow as does copper oxide. Water of combustion is collected in a borosilicate glass trap immersed in a slurry of dry ice and chloroform. Carbon dioxide is collected in a spiral containing 10 ml of 5N sodium hydroxide which has a maximum capacity equivalent to 300 mg carbon. This undesirable absorbent has since been replaced by a cocktail containing an organic amine.

Using the original apparatus we found the recommended cycle time of 45 min led to explosions with some samples; in fact, for some fatty materials, double the suggested cycle time was more realistic. The surface of the nickel catalyst was attacked and a build-up of material blocked the interstices of the nickel roll. Recovery was determined both chemically and radiochemically as a little over 90%. The losses, which were in the trapping of the carbon dioxide, could be accounted for when a second absorption spiral was introduced in the train. We used the technique exclusively for very low activity samples and 'memory effect' proved no problem. The Beckman/Harvey instrument, which is claimed to process eight samples an hour, is being used by our colleagues at I.C.I. America. Their reports have been good but further comment would be unwise before trying the instrument ourselves.

THE KALBERER AND RUTSCHMANN OXYGEN FLASK (Fig. 1)

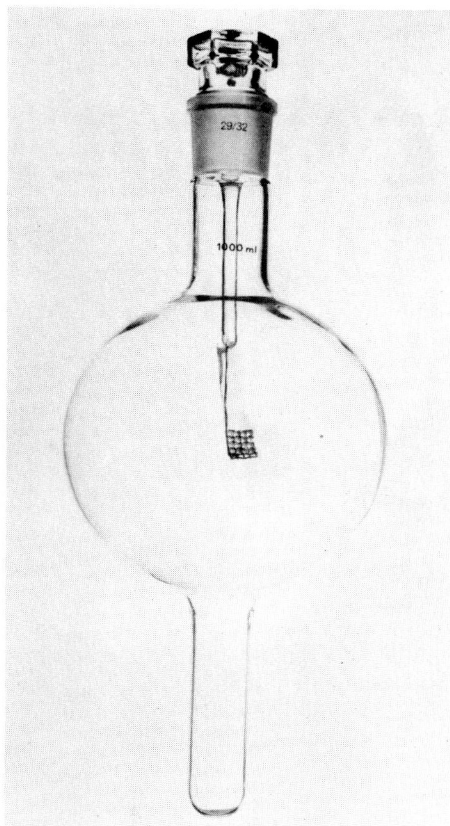

Fig. 1. Photograph of Kalberer–Rutschmann flask

As in the classical Schöniger technique for elemental analysis, organic material is burned in pure oxygen in a closed vessel using a platinum support. There are many excellent techniques based on this approach, including that of Dobbs.[3] The

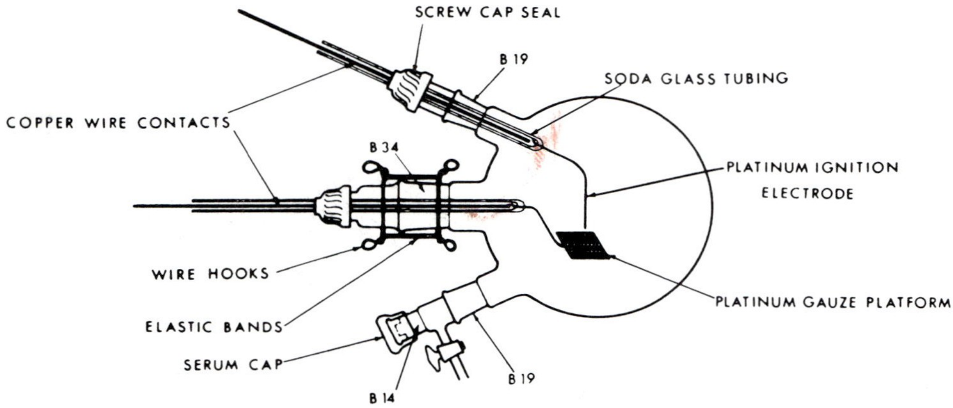

Fig.2. The Dobbs flask

Dobbs flask (Fig. 2) will cope with larger samples than the Kalberer–Rutschmann flask (Fig. 3). Impressive results were obtained in our laboratory. However, for various, including logistic, reasons we continued with the Kalberer–Rutschmann

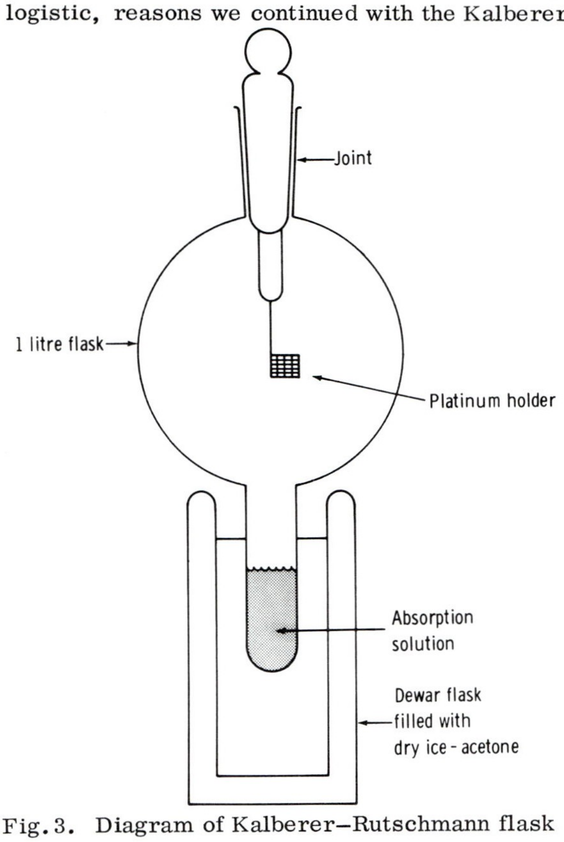

Fig.3. Diagram of Kalberer–Rutschmann flask

approach in which the absorption medium (12% ethanolamine in methanol) is contained in a finger-like extension to the combustion flask which is cooled in dry ice—acetone contained in a Dewar vessel. This lowers the vapour pressure of the methanol and prevents the formation of an explosive solvent—oxygen mixture in the flask. This process is normally carried out in a shielded stand which can be rotated electrically to ensure efficient absorption of combustion products. 0.5 ml blood or up to 400 mg tissue is taken and the sample is air dried. The absorption solution is introduced into the cold finger and the flask is then flushed with oxygen. The paper tail is ignited. Combustion is rapid and safe. The flask, normally in the special stand, is rotated for 20 min to facilitate absorption of the combustion products. An aliquot of the absorption solution is added to butyl PBD in toluene for scintillation counting. Recoveries are good but, as in all such techniques, should be checked using samples spiked with the radio-labelled compound of interest at a level relevant to the problem in hand. Table 1 shows such a series at low levels of activity. Memory effect is non-existent with a well organised system for cleaning the apparatus.

Table 1. Recovery of radioactivity added to control muscle tissues.

Added ^{14}C (d.p.m.)	100-min count	Δc.p.m.[a]	Corrected[a] for efficiency and sampling	% Recovery
4.9	3,003	3.03	4.9	100
4.9	3,031	3.3	5.4	110
9.8	3,357	6.6	10.7	110
49.1	5,006	23.1	37.7	76.7
49.1	5,767	30.7	50.0	102
98.2	8,688	59.9	97.6	99.5
98.2	8,398	56.4	92.0	94.0
147.3	10,988	82.9	135.0	91.5
147.3	11,117	84.2	137	93.0
196.4	13,851	111.5	182	92.5
196.4	13,822	111.2	181	92.0
245.5	16,965	142.5	232	94.5
245.5	16,713	140.1	229	93.0
				$\bar{x}$ = 96.1
				s = 8.7(9.0%)

Packard Tricarb Series 3003 used.
[a] Average background 27.0 c.p.m. and average counting efficiency of 68.0% used in these calculations.

THE PACKARD 305 SAMPLE OXIDISER (Fig. 4)

The sample is placed in the Platinum—Rhodium ignition basket which is held in place by a pneumatic piston. Combustion is controlled by operator regulation of the oxygen flow and basket heater current. Water from the combustion is trapped in a condenser at 2°C and caught in the counting vial under the condenser. Carbon dioxide and other gases pass through the condenser and vial to a CO_2 reaction column filled with ethanolamine. This column is automatically washed with scintillator and the products recovered directly in a counting vial. The system is flushed with nitrogen after the combustion.

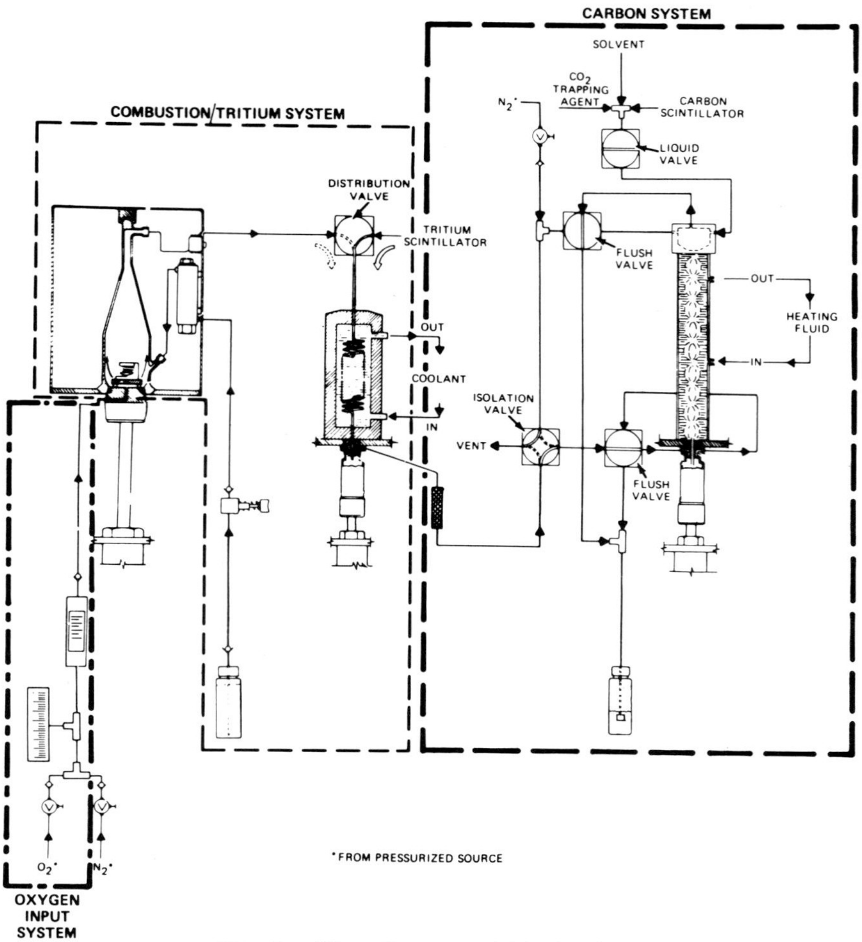

Fig.4. Flow diagram of Packard 305

A Packard 305 has been in our laboratory for $2\frac{1}{2}$ years and used routinely by our metabolic groups for a little over one year. A considerable number of mechanical difficulties were experienced and enthusiasm waned as it became apparent that both time and care had to be lavished on the instrument to maintain a good level of performance. Usage of the instrument dwindled and stopped completely before the end of the second year. At best, recovery and memory were up to the manufacturer's specification, but as Dr. Scales indicates elsewhere (p.211), an accumulation of errors can lead to poor recoveries before the relevant faults in the instrument are obvious to the casual user. The major difficulties were caused by blockage and leakage in various lines and valves. A retrospective examination of routine results (Table 2) showed that when the Packard 305 was working well the assays were as reproducible as those produced by the flask combustion technique.

THE INTERTECHNIQUE OXYMAT SAMPLE OXIDISER (Fig. 5)

In this instrument, which is based on a system devised by Peterson (National Institute of Health), combustion takes place in a stream of pure oxygen. The sample, in a small polycarbonate sachet, is introduced by an electromechanical

Table 2. Retrospective comparison of Packard 305 and Kalberer—Rutschmann flask.

Percentage standard deviation for d.p.m./ml blood		
Range (d.p.m. ^{14}C/ml)	100—1,000	1,000—10,000
K.R. flask	7.9%	5.6%
Packard 305	5.9%	3.7%

Each value represents 15 samples taken at random from the results of about 10 operators.

system into a furnace heated to 700°C. A catalyst completes the oxidation. The combustion products pass through a heated transfer line and water and carbon dioxide are separated and trapped by two cooled, wiped film centrifugal absorbers. The descending film of trapping fluid is met by the countercurrent stream of combustion products. The combination of a thin liquid film and the turbulence impressed on the gas ensures rapid and efficient trapping.

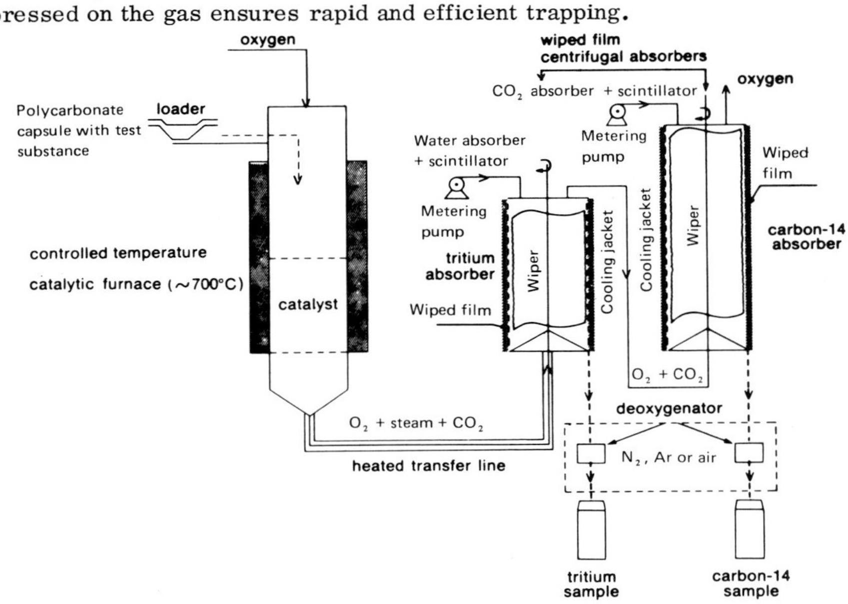

Fig.5. Flow diagram of the Intertechnique Oxymat

In the first absorber, water vapour is condensed and absorbed in a dioxane—methanol—water phosphor system. Carbon dioxide is trapped in the second absorber using a toluene—phenylethylamine—methanol—water phosphor system. As the liquid flows into its respective counting vial it is de-oxygenated by purging with a suitable gas such as nitrogen.

The first of three attempts to evaluate the instrument was in early 1972. After less than 100 samples a large amount of carbon was deposited throughout the system and massive temperature fluctuations were noted in the furnace. The manufacturers were informed and later modified the combustion tube design, holding back the solid particles with a bed of silica sand covered by coarse carborundum chippings. They also uprated the furnace.

In the second attempt at evaluation (June 1972), good recoveries (>98%) and low memory (<0.1%) were obtained with small dry samples. However, after 70 samples were burned in rapid succession, recovery dropped rapidly to 60-70%. On the second day only 30 samples were combusted successfully before large losses were noted. This was later attributed by the manufacturers to leakage at the seal of the drive shaft to the first absorber. At our request a series of 100 small dry cellulose samples were combusted by the manufacturers on yet another instrument. The recoveries were good but, since this was only a fraction of the required testing indicated by our previous experience, and took six months to accomplish, we were not yet entirely convinced.

Because of our requirement for a rapid instrumental solution to our problem we purchased an Oxymat in June 1973 subject to certain stipulations regarding performance. There were assembly faults which we were able to adjust ourselves. Only a modest number of results can be reported since, unfortunately, at the most interesting point in our investigations the solenoid controlling the sample entry port self-destructed. Seven of our staff each performed a series of combustions as indicated in Table 3. In addition they combusted an equal number of non-radioactive samples which indicated that memory was less than 0.4%. Recovery was excellent and similar to that obtained by the manufacturers in the experiment

Table 3. Operation of the Oxymat. Recovery of added ^{14}C from blood, muscle (minced and part dried) and cellulose.

Added ^{14}C	Percentage recovery			
	Blood 0.5 ml + 500 mg cellulose	Muscle ~400 mg	Cellulose ~500 mg	Mean
10,000 d.p.m.	94.86	98.87	92.51	95.41
500 d.p.m.	97.56	102.77	98.67	99.67
Mean	96.21	100.82	95.59	97.54

Average standard deviation = 2.5%.
Each value is the average result obtained by seven operators.

mentioned previously. We decided to investigate losses we observed when trying to combust wetter tissue and moistened cellulose samples. Using the simplest of traps, to avoid introducing back pressures, the effluent gases were shown to contain carbon dioxide including a high proportion of the Carbon-14 added to the samples. I was not surprised to note that the luminous stage of some of these combustions was not completed until most of the absorbent was already in the vial. We did not, at this stage, wish to take a hacksaw to the timer cams in order to prolong the absorption step but we did improve recoveries by introducing 60 Hz mains cogs into the timer drive (Fig. 6). This prolonged all stages controlled by the timer, and of course necessitated adjustment of the absorbent pump flow rates. We have only one series of results, for the reasons mentioned earlier, but these do show a trend worthy of further investigation. It appears in fact to bring the instrument up to the manufacturers specification, for these more awkward samples.

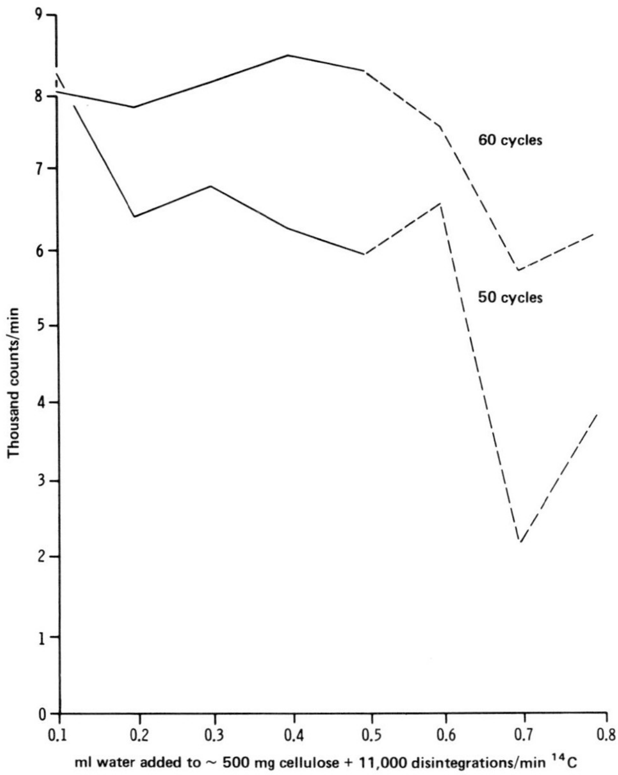

Fig. 6. Effect of 60Hz drive wheels on the performance of the Oxymat

THE PACKARD 306 SAMPLE OXIDISER

The appearance of the model 306 confirms its description as a 'second generation' model. Although the basic layout is similar to the model 305, there has been considerable streamlining to eliminate the less reliable features of the earlier model. As the flow diagram (Fig. 7) indicates, the principle remains the same as the model 305, a single push button initiates the positioning of the vials and combustion basket, combustion, injection of liquid scintillation cocktails, unloading and a cleaning cycle. After the first 20 samples we noted a reduction in the delivered volume. This was traced to the metering syringes, which were cleaned and re-assembled by the Packard engineers, but a recurrence of the fault necessitated replacement of the parts. After a few more samples, losses in volume were noted which were traced to a blockage at the top of the column obstructing the flow of the absorbent. Packard were aware of this fault which they believe they have now eliminated. The column was, however, easy to strip and clean.

Recovery of added Carbon-14 from cellulose, plasma and fat was quite good but losses were noted when moist muscle was burned (Table 4). In fact, recovery appeared, to some extent, inversely proportional to the required combustion time. Packard recommend the use of combustion aids with some samples and it is perhaps unfair not to have followed their advice, but this was intended as a preliminary investigation to acquire the 'feel' of the instrument. Memory was impressively low at less than 0.03%.

An additional series of cellulose samples with 50 d.p.m. Carbon-14 added and part dried muscle with 500 d.p.m. Carbon-14 added gave recoveries around 95%. At this point we thought that we could settle down to a more serious evaluation of

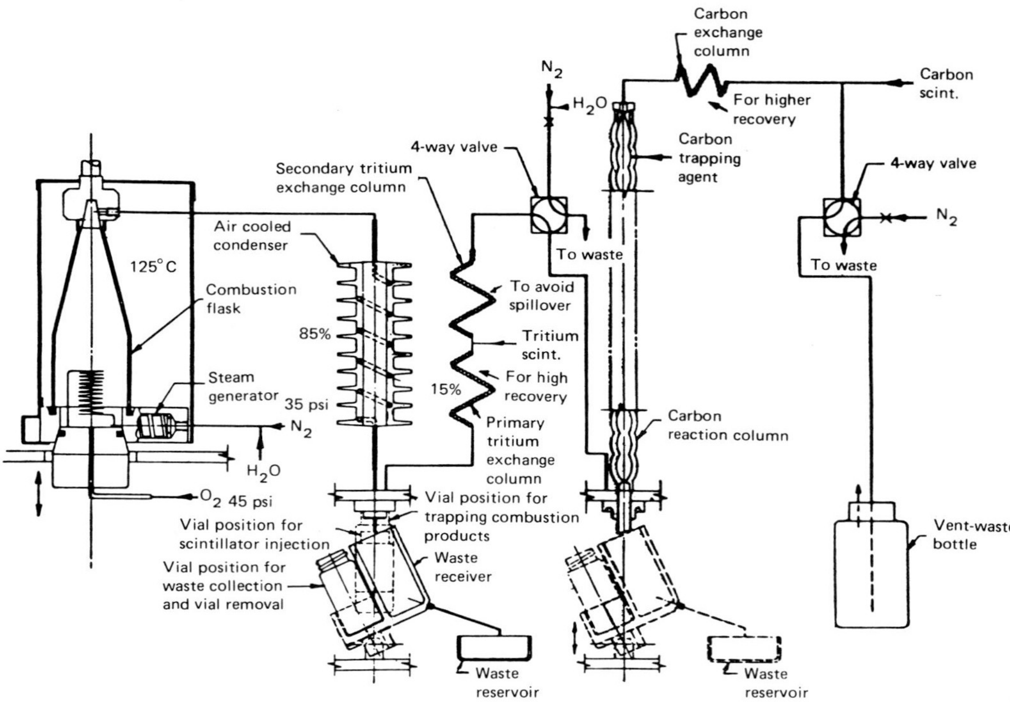

Fig. 7. Flow diagram of Packard 306

the instrument. Unfortunately, we had further recurrences of the liquid dispensing problem. We were informed by Packard that this was due to the high ambient temperatures in our laboratories during a recent spell of very hot weather. This

Table 4. Operation of the Packard 306. Recovery of added ^{14}C from cellulose, plasma, fat and muscle (minced and part dried); 10, 000 d.p.m. ^{14}C added to each sample.

Sample		Percentage recovery
Cellulose	500 mg	100.1
Plasma	0.5 ml + 350 mg cellulose	92.1
Fat	250 mg + 250 mg cellulose	90.3
Muscle	500 mg	72.7

Average standard deviation = 5.4%.
Each value is the average of 10 analyses.

fault had not occurred in the instrument prototype which was developed in air-conditioned laboratories. We were told that modified seals were now being fitted which had overcome this problem. We can only look forward to evaluating a modified instrument at the earliest opportunity.

CONCLUSIONS

The oxygen flask technique stands out for its reliability. Bearing in mind the time wasted on breakdowns and maintenance of commercially available equipment, manual techniques can often be quicker and cheaper in the long run. In fact, if an equal financial investment were made in support equipment for a flask combustion set-up it would require a very good instrument to better it. The Packard 305 can be regarded as a first venture into instrumental methods and we have learned a great deal from it. The Packard 306 appears promising if the one major fault so far detected has now been overcome. The Oxymat impressed us as an elegant concept but the manufacturers were a little slow to make the final adjustments to design which would bring its performance up to specification. It is, however, an instrument which the user can adapt to suit his own requirements.

ACKNOWLEDGEMENTS

I wish to thank both Packard Instruments Ltd. and Intertechnique Ltd. for their help and interest, and also my colleagues at I.C.I. Pharmaceuticals Division for assistance and encouragement.

REFERENCES

1 E.A. Peets, J.R. Florini and D.A. Buyske, Analyt. Chem. 32, (11), 1465–7 (1960).
2 H.E. Dobbs, Intern. J. Applied Radiation and Isotopes **17**, 363–4 (1966).
3 F. Kalberer and J. Rutschmann, Helv. Chim. Acta **44**, 1956 (1961).

DISCUSSION

P. Johnson: You mentioned early in your talk that it is desirable — in fact essential — to check combustion recoveries by spiking with the compound itself which is under investigation. In other words, the standard should be closely related structurally to the test sample. This in my opinion is one of the problems with combustion methods. With a solubiliser method, as I pointed out earlier, one has to prove that there is no loss of volatile materials but this does not require any knowledge of the labelled components — it either happens or it doesn't — and in any case can be prevented by using vials sealed with septa. In combusting biological samples, however, we are often dealing with mixtures of unknown labelled components; therefore, how can one choose a structurally related compound for spiking? If recoveries were not dependent on structure, then one would of course be free to standardise with anything; in fact, many people do so, and risk the probably small number of times that it will cause trouble.

R.G. Cooper: It is probably not essential to choose a standard which is structurally similar to the compound under investigation, and as you say it is often impossible, but we do compromise by using the parent compound in a metabolic investigation.

P. Johnson: But you don't identify the metabolites and use them as standards.

R.G. Cooper: We have on occasions, but not routinely.

M. Stiasni: You showed that you combusted 0.5 ml blood + 500 mg cellulose. This is a remarkably big amount and disagrees with Dr Gordon's paper in which the recovery drops down with sample weights more than 200 mg.

R.G. Cooper: Please ask Dr. Gordon. I don't remember the exact composition of his absorption solution, which may, of course, be highly relevant.

B.E. Gordon: Our absorption solution contains about 27% v/v of phenethylamine in a cocktail of methanol plus toluene plus the usual fluorescers. The volume of absorbing solution is sufficient to neutralise about 18 mmoles of CO_2 or 216 mg of carbon. This value is not reached because before all the phenethylamine is quantitatively converted to the carbamate, precipitation occurs (the trap is chilled with ice) and CO_2 is no longer quantitatively absorbed. Use of a cocktail with a better solvent power for the carbamate, use of more absorbing solution plus a counting vial of greater capacity than those currently in vogue would increase the amount of CO_2 which could be trapped. The bottleneck is not in the burning but in the trapping.

N. Kaartinen: What is the capacity of the Packard 306 CO_2 trapping agent?

R.G. Cooper: 40 mmoles CO_2/13 ml Carbosorb, including the margin in the degree of saturation. A remark concerning the low blood recovery. When wet samples are combusted without Combustaid, and thus without a flame, pyrolysation products escape through the exhaust. A sufficient amount of Combustaid should be used.

SECTION IV

SAMPLE PREPARATION : PART II

Chapter 16

Heterogeneous Counting Systems

V. Tarkkanen

Vondellaan 86, 8100 Haren, The Netherlands

In liquid scintillation counting (LSC), the accuracy of the instrumentation and subsequent measurement is reproducible and optimal only if the sample is correctly prepared. The general requirements are intimate contact between the liquid scintillator and the sample throughout, and absence of colour and substances inhibiting the scintillation process. A major difficulty in the use of LSC is the final form of the sample which often differs greatly from the form in which it was originally collected. I will here describe the preparation of the most frequently occurring samples produced by recent techniques and discuss some typical sources of error and their prevention. At present, more than threequarters of the samples measured by LSC contain tritium or Carbon-14. These nuclides are widely used in biomedical research for measurement of their distribution in tissues and body fluids to determine dynamic effects at the molecular level of the cell, or in routine analysis or diagnostic work. These samples are predominantly aqueous and as such are insoluble in aromatic scintillation systems. This represents the main problem, especially in routine LSC measurements. Therefore these aqueous biological fluids containing Carbon-14 and Hydrogen-3 nuclides will be the main theme discussed here.

A sample in its final form consists of three main components: the material being counted, the solvent and the scintillator. The main sample types can be listed as follows: (1) aqueous solutions or colloids such as body fluids and water soluble substances like buffer solutions and electrolytes; (2) lipophilic compounds, directly soluble in aromatic scintillation solvents, such as lipids, steroids and drug metabolites; (3) solid compounds like protein precipitates, animal tissues and their homogenates, which can be modified into a soluble form and equally distributed into scintillation solvents; (4) insoluble or only partial soluble samples, such as connective tissue, bone, cartilage, certain proteins and polysaccharides.

In an ideal case, the sample is completely soluble in the scintillation solvent. The choice of scintillation solvent is, however, ultimately based on the primary criterion, namely the energy transfer of the tracer in the sample to the scintillator. Unfortunately, the second group includes only a few percent of samples measured by LSC and the incompatibility of the aqueous or solid samples with solvent having adequate energy transfer properties is the chief limitation in sample preparation. Various approaches have, therefore, been used in order to obtain a homogeneous system of the three main components in order to secure the necessary 4π counting geometry.

The first group of samples, toluene based systems containing organic solvents such as dioxan, alcohols, alkoxyalcohols and naphthalene have been used to count pure water up to 1/5 of the final sample volume. Figures 1 and 2 illustrate the performance of such systems. Furst and Kallmann,[1] Okita et al.,[2] Kinard,[3] Bray,[4] Butler,[5] Bruno and Christian[6] used more polar solvents to count water, together with toluene or naphthalene. The disadvantages of these systems are

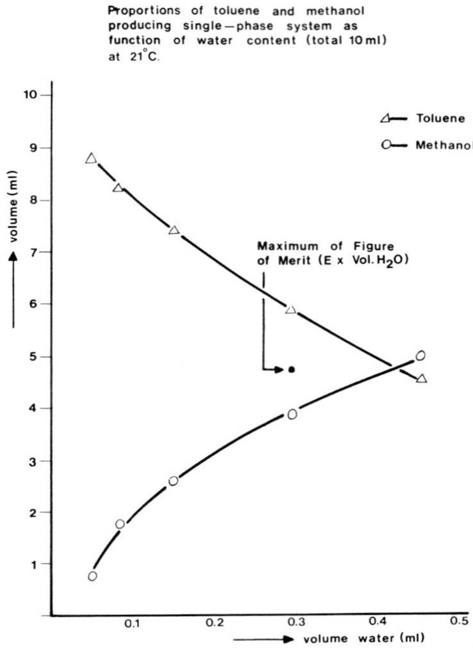

Fig. 1. Behaviour of a multi-component solvent system based on methanol and toluene. ^{3}H-toluene as an internal standard.

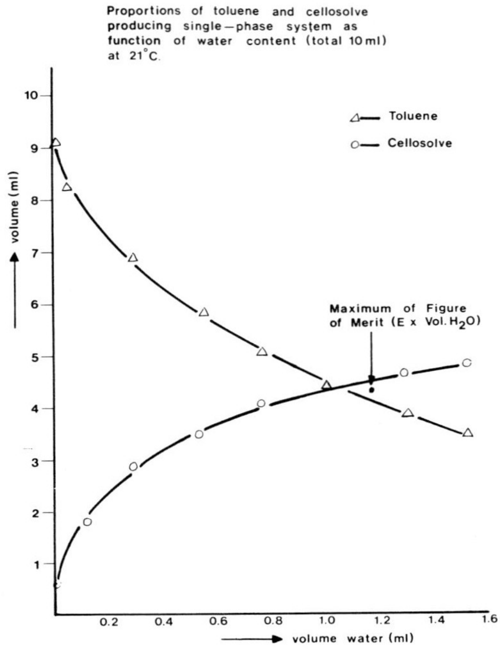

Fig. 2 Behaviour of a multi-component solvent system based on ethylcellosolve and toluene. ^{3}H-toluene as an internal standard.

low counting efficiencies at water concentrations above 5 percent and inability to count samples containing dissolved solids. If the suitability of these systems had been tested for the counting of fluids, for instance for competitive protein binding techniques, they would not have been accepted as standard solutions for such a long period of time. Precipitation, like turbidity and even phase separation, have been accepted in the final sample form. Therefore, LSC has not been as precise a method as it should and can be, if the first step in the measurement, the sample preparation, had been treated correctly instead of as described in statements such as 'the radioactivity was measured after addition of a standard scintillator'. Figure 3 illustrates the typical deviation in the count rate observed when multi-component systems based on organic solvents of various types are used for the counting of proteinaceous systems. For comparison a modern heterogeneous (colloidal) counting system is given.

From the foregoing and from the general inability of organic solvents to dissolve aqueous biological systems it was obvious that other types of approach were needed for ultimate solutions. Thus, a fourth component, other than an additional organic solvent, scintillation solvent and scintillator is needed in order to meet the objective for reproducible measurement of samples represented by groups 1, 3 and 4. Problems associated with the third and fourth group of samples have generated various efforts to obtain reagents to degrade intact samples like tissue into a form which is soluble in the scintillation solvents. These solubilising reagents are various types of quaternary ammonium hydroxides dissolved in alcohol or in a scintillation solvent. Thus when dissolved in toluene they can be used to produce completely homogeneous counting systems for both homogenates and dry or wet biological samples. The sample, up to a few hundred milligrams, would normally undergo a solubilisation procedure probably involving a heating period of one or more hours. The numerous attempts to find a 'fourth compound' for group 1 samples — the most common group — have frequently failed due to excess of quenching, limited sample size and complicated preparation.

Developments of heterogeneous counting systems, those in which the bulk of the sample is not in the same phase as the bulk of the scintillator, were not widely accepted when presented by Erdtmann[7] and Meade[8] in 1962. The following reasons for this non-acceptance have been cited: substantial changes in counting results caused by temperature and pH dependences of the systems, self-absorption phenomena, discontinuities in reproducibility and recovery, and chemiluminescence. However, following a later systematic study investigated by Patterson,[9] these heterogeneous systems have now been recognised as realistic possibilities in obtaining final samples with high load and efficiencies. They are also easy to prepare and many recent publications indicate that numerous efforts have been directed to the further development of previous ideas based on the utilisation of such heterogeneity. The first heterogeneous systems were based on sorbitol esters and castor oil derivatives as the fourth component to effect a process of solubilisation of the aqueous sample into toluene through formation of emulsions. Reasonable results were reported by Shapira[10] and Gordon,[11] using large surface area silicon oxide (Cab-O-Sil and Aerosil) with toluene or dioxane/naphthalene to obtain a suspension of samples like barium carbonate, body fluids and tissue homogenates. Apparent self-absorption, especially for tritium, caused by the original sample or the silica, and instability due to the use of these kinds of thixotropic systems, were not adequately overcome even after further investigations. A basis for further development was the discovery of the usability of ethylene oxide polycondensates of alkylphenols,[7,8,9] which were capable of dissolving aqueous samples in the presence of toluene or xylenes by forming colloidal systems which were seemingly clear in certain proportions, and to a given extent, thermodynamically stable. The general formula of these alkylphenoxy-polyethoxyethanols is

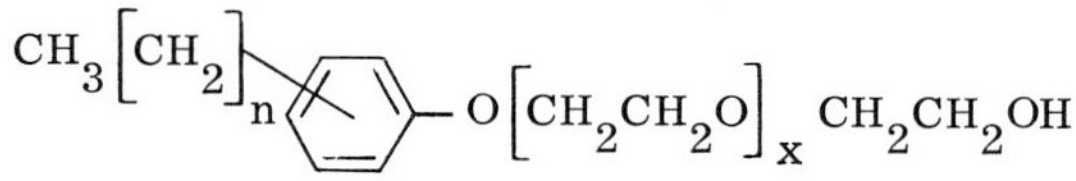

Patterson and Greene used a polyethoxyethanol of the above general formula, where n = 7 and x = 9 (Triton-X—100). They showed initially the useful limits of application of this emulsifier with toluene (in ratios 3:1 to 1:1) for various samples, frequently derived from biological experiments. Subsequent publications have confirmed these earlier studies or have reported refinements of the systems such as varying the length of the polyethoxy or the alkyl chain of the tenside molecule, the solvent or the scintillator.

For the described type of tenside wherein the hydrophobic portion contains polyoxyethylene, a quantitation of emulsifying properties is given by the so-called HLB value. This value is computed by dividing the oxyethylene weight percentage (EO) of the total molecular weight of the tenside by five. Thus the HLB = EO x 1/5. Tensides having an HLB value lower than 7 are suitable for water-in-oil (w/o) emulsions, where the water is finely dispersed in a hydrophobic solvent. Molecules with higher HLB value form emulsions of the type oil-in-water (o/w). For Triton-X—100 the calculated value is 13.6, but the practical value varies.

Special attention should be paid to the terminology relating to heterogeneous counting systems as many varying expressions appear in the LSC literature. The usual terms emulsion or suspension are widely used, but these terms are not quite correct because they refer to dispersions of one liquid in another whereby the dispersed liquid is in the form of droplets with a diameter of about 0.1 to 100 microns. Such systems are also thermodynamically unstable. The same applies even for so-called microemulsions[12,13] which are transparent, isotropic systems and should not be referred to as emulsions in terms of the expression above.

The alkylphenol derivatives (APPE) with the general formula described above contain a hydrophobic and a hydrophilic portion. Figure 4 shows the arrangement of the hydrophilic ethylene oxide part. On the left we have an oblique-angled and

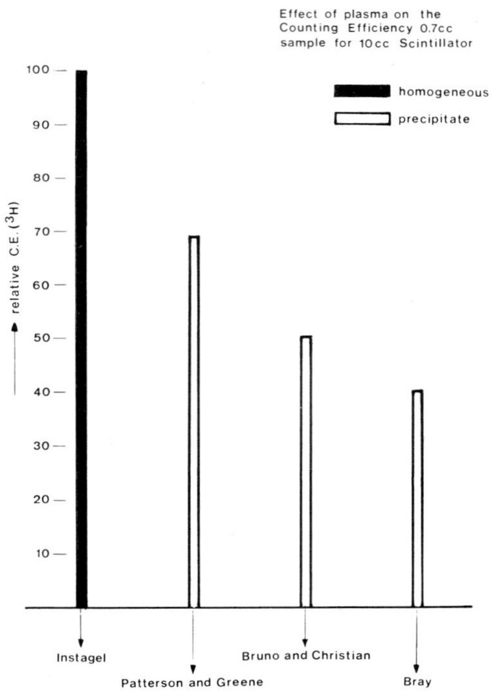

Fig. 3. Effect of phase separation.

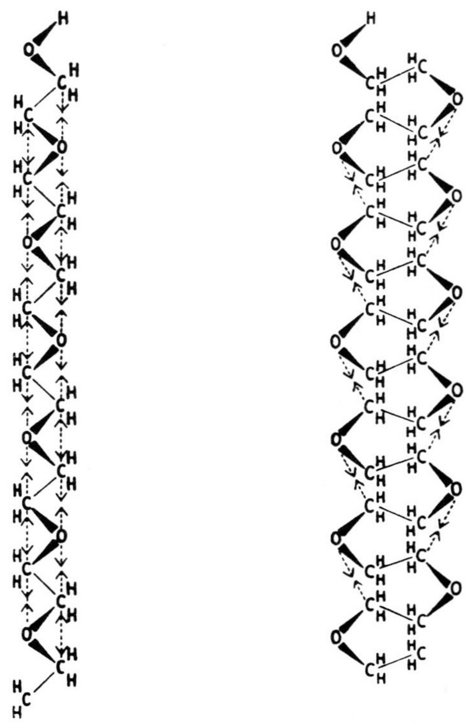

Fig. 4. Arrangement of the hydrophilic polyoxyethylene chain.

equilateral zig-zag line whereby the oxygen atoms are attracted by the adjacent CH_2—groups. A meander-form (on right) has been proposed as a result of the contraction of the chain in aqueous solutions. This chain can be hydrated by water by means of hydrogen bonding, thus making the products water soluble and therefore amphiphilic. The ratio of hydrophilic to hydrophobic portion (HLB value), the Van der Waals forces and the hydrogen bonds fundamentally affect the behaviour of these molecules, although they are chemically rather inert; but their major advantage is that they are able to form particles of colloidal size, so-called micelles, which can incorporate insoluble substances into a scintillation solvent. In this connection, Fig. 5 shows different shapes of micelles where the black end represents the hydrophilic part of the tenside molecule. The term

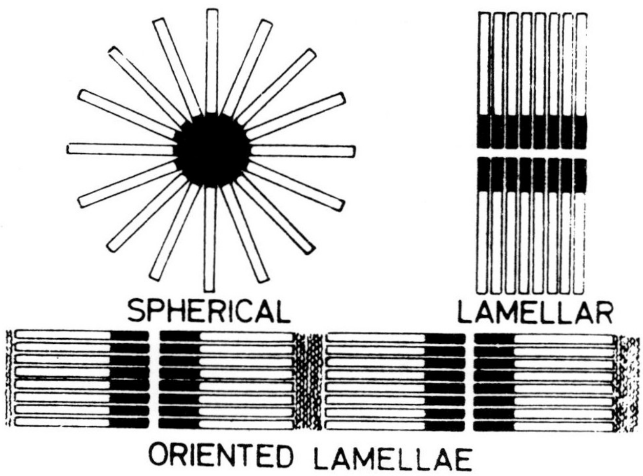

Fig. 5. Shapes of a micelle.

solubiliser is used with LSC for the mentioned quaternary hydroxides with digestive properties. Thus, it would not be correct to adopt the term solubiliser with regard to surface chemistry in order to avoid possible misinterpretations. Figure 6 shows count rates after direct incorporation of equivalent amounts of tritium activity of dry palmitic acid and sucrose. It should be noted that dry hydrophilic samples are not solubilised, while in aqueous solution they can be equally disitributed in a few seconds and counted with high figures of merit.

In colloid science, the solubilising molecules were named tensides about 30 years ago. These terms have not been used in LSC for heterogeneous systems, but rather the expressions gel and emulsion are widely accepted. Fox[14] first introduced the most correct term of 'colloid scintillation counting' in 1971. At lower concentrations of aqueous solutions these colloidal systems of APPE form sols and at higher concentrations so-called gels, which are more rigid in form. In contrast to the term liquid scintillator solution, which refers to a true solution, we will name the sols and gels as colloidal scintillator systems.

Conductivity measurements have shown that almost all of the colloidal systems utilised in LSC are of the type of water-in-oil, which means that, in the case of a spherical micelle, the droplet is water surrounded by the organic scintillator solution. It is of importance to understand the term 'heterogeneous counting system' correctly. The inclusion of emulsions and suspensions in this category has led to inaccuracy and non-reproducibility. A colloidal system consists of two discrete phases (bulks) and is, therefore, referred to as 'heterogeneous'. In an ideal colloidal scintillator, however, the contact of the two phases is

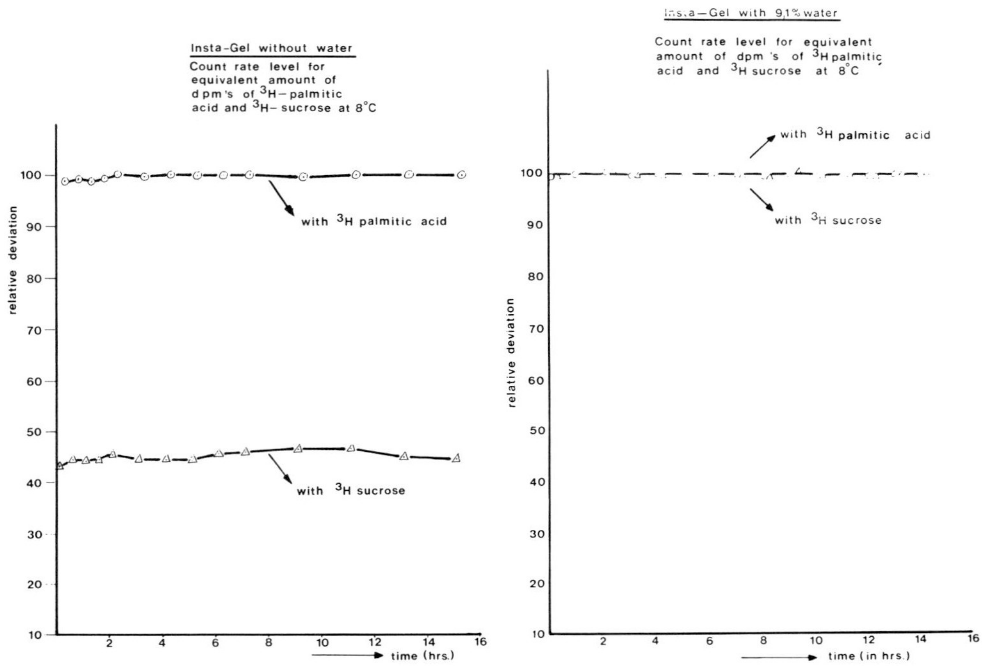

Fig. 6. Demonstration of the necessity of dissolving solid hydrophilic samples in water prior to counting in a colloid scintillator.

equivalent to true solution.

If we consider a tracer in the aqueous phase, little absorption of the beta particle as a function of the mass of the water should occur at low water concentrations. Neither should quenching occur for the scintillator as a function of the aqueous sample load as it is significant in the case of a true solution, i.e. by dioxan. If we compare Fig. 7, where counting efficiency of an aqueous tracer is plotted against its volume percentage of the total sample volume for both a colloidal scintillator and a liquid scintillator solution (i.e. dioxan based Bray solution), we can observe far less quenching for the colloidal scintillator system (i.e. Monophase—40). The rapidly decreasing counting efficiency for a true solution is caused by direct contact of the quencher (water) and the scintillator. In case of the colloidal scintillator the slowly decreasing counting efficiency is to be referred to as a function of the load of the aqueous phase: there are more layers or spherical micelles in the neighbourhood and even if the size of the micelle does not have any measurable effect on the counting efficiency by absorption of beta particles, the radiation must pass through a number of such aqueous layers between the layers of light-producing scintillator solution and result in a certain degree of quenching.

It is evident that the thickness or diameter of the aqueous micellar phase is of the order of less than 20 nm for an ideal colloidal scintillator. In the case of turbidity or opalescence the size approaches the visible regions or near 400 nm. This can be demonstrated by comparing the counting efficiencies of hydrophilic and lipophilic tracer solutions. In a good colloidal system both give identical results within the deviations of the accuracy of counting of the tracer available. Therefore, it is important that the size of the micelle is the smallest possible

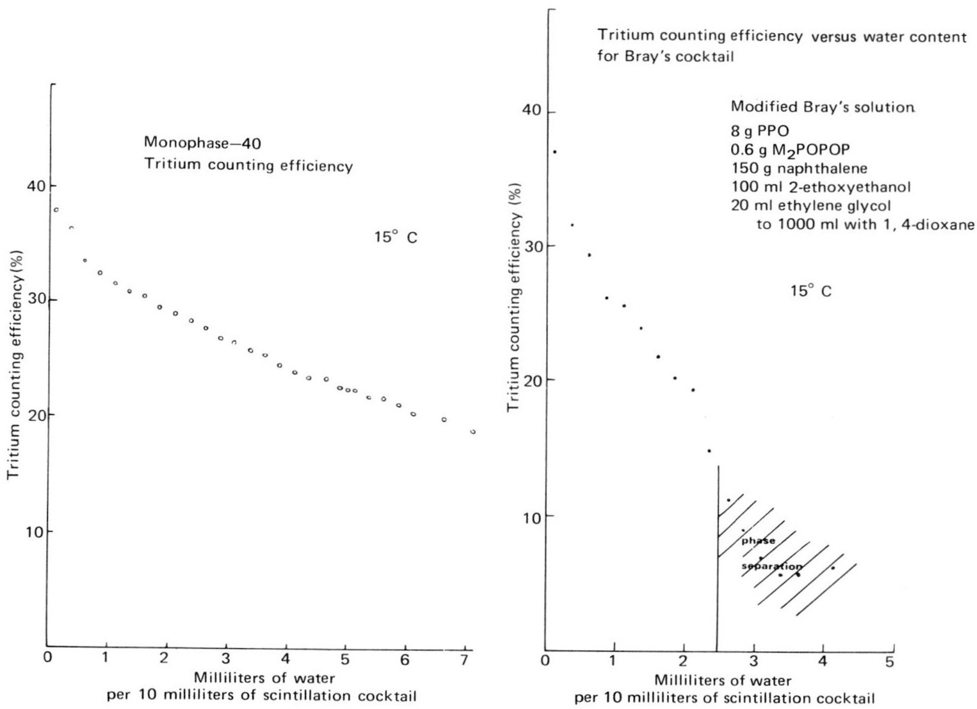

Fig. 7. Counting efficiency versus sample load of a modern colloidal counting system (left) and a dioxan-based scintillation fluid.

and subject to variations to the lowest possible extent. Fluctuations in the shape of the micelle can be caused by the ratios of the aqueous sample, scintillation solvent and the tenside. Furthermore, the electrolyte content of the sample and temperature changes have similar effects, resulting in thermodynamic instability of the systems.

Obviously, preference in LSC should be given to systems which are less affected by these instability factors. Therefore, knowledge of any remaining instability factors of the colloidal system are of importance in obtaining greater benefit from this most promising technique.

The most frequently occurring samples for LSC contain various combinations of inorganic salts and organic anions as buffer solutions, together with proteins. The property of organic scintillation fluids has been used to separate low energy beta emitters, soluble in organic solvents, from insoluble biological substances. In contrast, many assays of immunology and enzymes require the counting of an isotope bound to protein and any precipitation, even if invisible but detectable by the instrument, alters the bound/free substrate ratio and the validity of the radiotracer method. It is not in the scope of this paper to compare different organic scintillator systems but rather to give comparative data on the properties of available heterogeneous counting systems, capable of reproducible counting.

Experiments made by Van der Laarse[15] using phase diagrams of three components showed the necessary concentrations of tensides for various purposes. Fox[14] extended this work with studies on the appearance or stability of various systems of tensides and samples. It was not possible to predict the appearance, and even slight changes of the sample consistency or temperature variations unpredictably influenced the counting efficiencies; hence a universal system is obviously most preferable in routine experimental work.

As there is no exact instrumental method to compensate the faults caused by self-absorption (sedimentation, turbidity, phase separation) of scintillation fluids containing, for example, dioxan or Triton, and as the composition of biological material is not constant, these fluids are of limited use. The following nomograms (Figs. 8—11) illustrate comparative studies on the usability of formulations frequently cited in the literature such as toluene—Triton-X—100 (2:1) with 6 g PPO

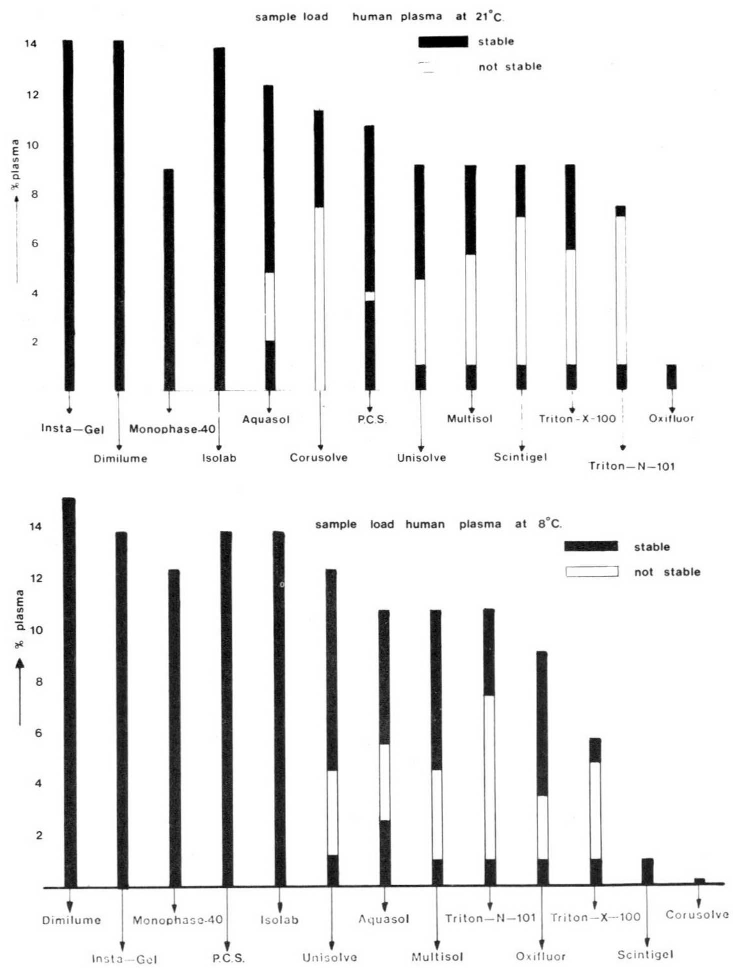

Fig. 8. Stability of commercial and experimental formulations for human plasma samples for counting temperatures of 8°C and 21°C.

and 0.5 g POPOP per litre and p-xylene—Triton-N—101 (2.5:1) with 7 g PPO and 1.5 g bis-MSB per litre and of the available commercial scintillation fluids. The usability of these systems is referred to as the reproducible stability of the count rate of the actual sample in the colloidal scintillator within a one percent

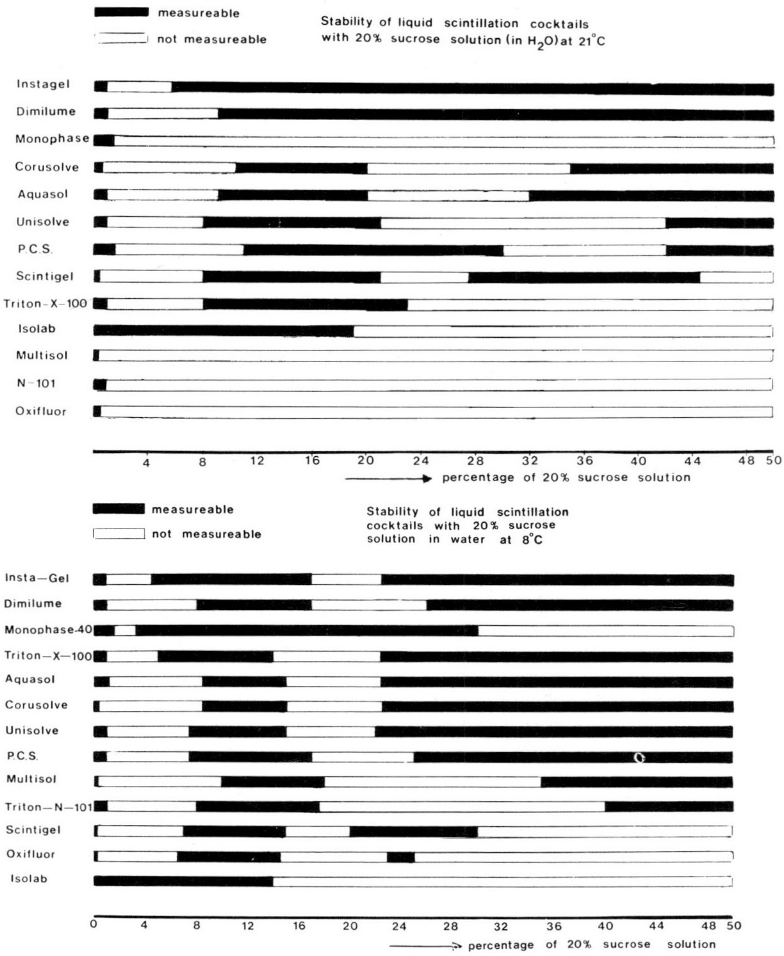

Fig. 9. Stability of various formulations with 20% aqueous sucrose solution at 8°C and 21°C.

error during a period of 170 hours following temperature equilibration for 2 hours in the sample compartment of a Packard Tri-Carb model 3380. When separation of the systems into two layers occurs (also referred to as two-phase range) the results are invalid and further counting statistics are not meaningful. The filled part of each nomogram indicates a reproducible counting system. First, Fig. 8 illustrates the ability to count human plasma after a mixing procedure with various colloidal counting systems and Fig. 9 similarly shows a sample of 20 precent sucrose in water. Figure 10 shows the influence of electrolytes, in this case 0.1 M sodium sulphate in water. Incubation of electrolytes lowers the repelling energy within the micelles, resulting finally in the two phases visibly separating into two layers. The influence of multivalent ions is greater than that of monovalent and therefore, it can be demonstrated that the colloidal systems can solubilise up to twenty times more concentrated sodium chloride than sulphate (Figs. 11 and 12).

Figure 13 shows the advantage of high counting efficiency and resistance against

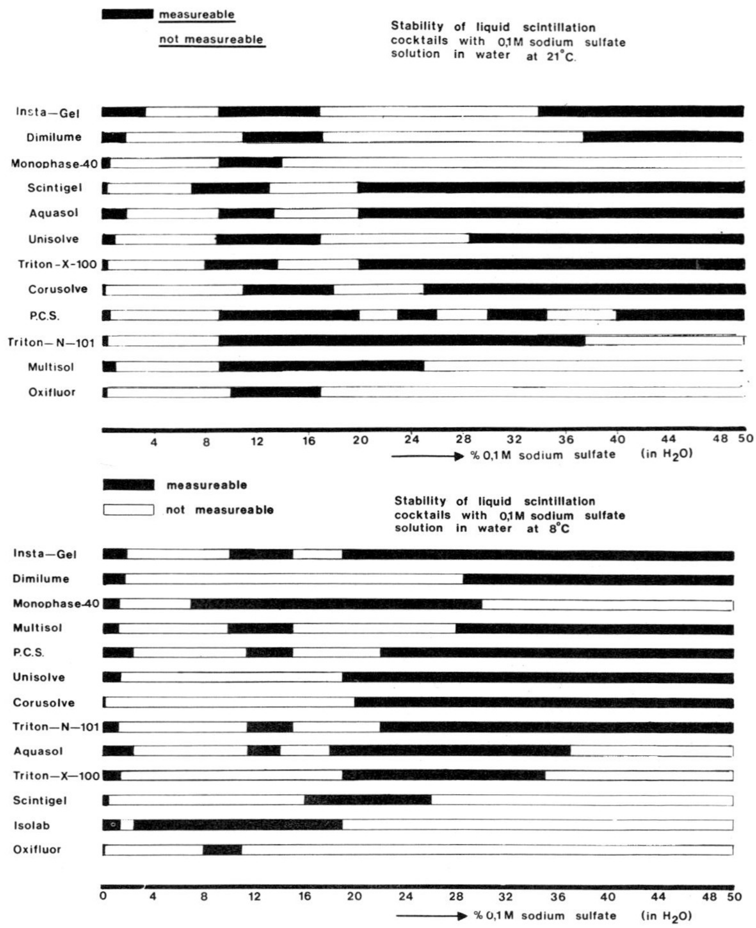

Fig. 10. Stability of various formulations with 0.1 M sodium sulphate.

quenching with equal loads of various inorganic and biological samples. This demonstrates the advantage of the presence of the sample in a phase other than the scintillator. Even here the phase contact is equal to that in a true solution for radioactivity measurement because no difference of recovery of tritiated water and tritiated toluene is detectable.

The earlier mentioned temperature dependence can be seen in Figs. 14, 15, 16, 17 and 18 where one can observe a discontinuity divided into one or more regions as the sample load or temperature changes. The sample measured contains only water and it is to be expected that even more peculiarities in the pattern of usable sample loads will occur for practical biological samples. At lower water percentages there is an apparent liquid form which results in the organic solvent and aqueous sample separating out into two distinct layers with a more rigid, gel-like, appearance. This phenomenon is influenced by changes in the shapes of the micelle. The blank area noted as 'two-phase' or 'frozen' gives different counting efficiencies for hydrophilic tritium (HTO) and for an equivalent

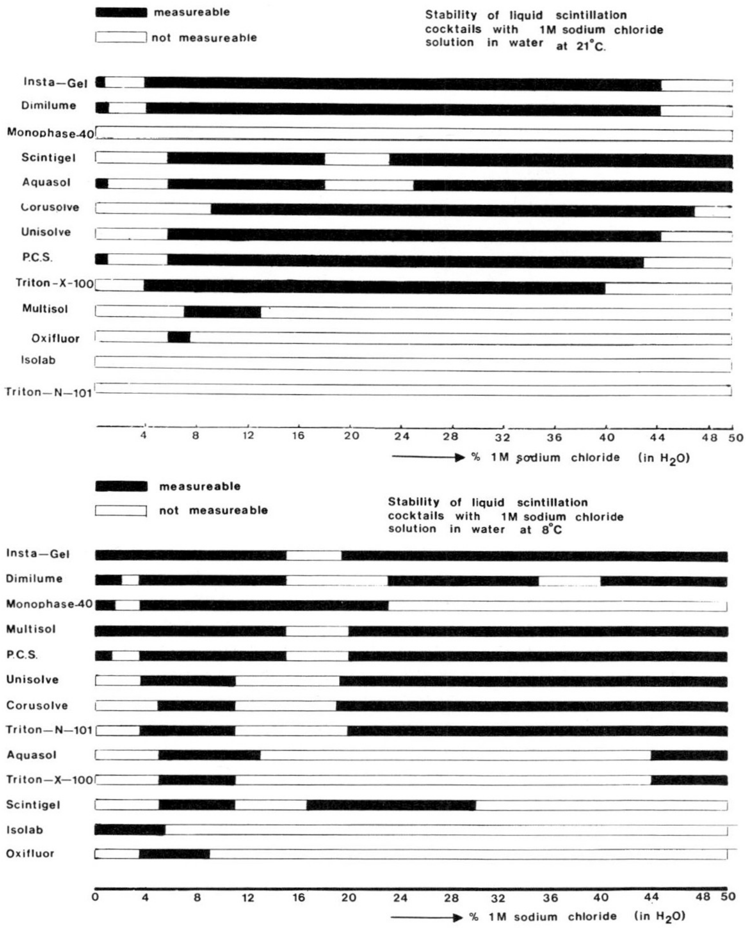

Fig. 11. Stability with 1 M sodium chloride solution.

amount of hydrophobic tritium (^{3}H-hexadecane) with count rates which also vary during the measurement. The micellar structure is found to be of the water-in-oil type at lower water concentration, changing at higher loads to the so-called reversed micellar structure of the oil-in-water type. This is demonstrated by conductivity measurement of water, with dissolved electrolytes: water-in-oil results in a marked conductivity increase compared with the reverse type.

Optical characteristics of these systems indicate a controversial structure of lamellar-type micelles. This would give a better classification to the usability of the colloidal system for reproducible counting at higher water load, where the appearance is more rigid and consists of the hydrophobic phase of scintillator dispersed in the aqueous sample.

Temperature dependence was observed following an extended study by Williams.[16] She recommended warming of the colloids to 40°C followed by several hours temperature stabilising in order to obtain the desired accuracy for low level tritium counting at high sample loads of water. By using a simple

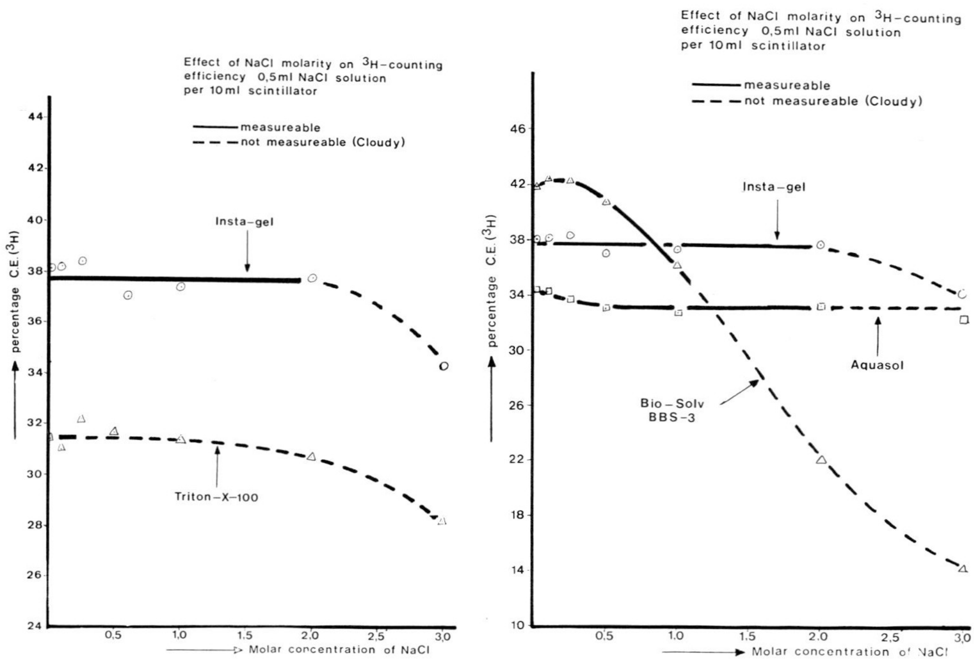

Fig. 12. Effect of sodium chloride concentration on the counting efficiency of tritium (as HTO).

mixing procedure at room temperature followed by a temperature equilibration of one hour maximum in the sample compartment of a temperature stabilised LSC spectrometer, we have observed adequate accuracies and stability over the range in which the system is generally used. It is desirable that the intimate phase contact is reached rapidly and this can be reached by the recommended sample preparation, i.e. for Insta-Gel: by the addition of the aqueous sample into the scintillator, followed by vigorous mixing.

The use of external standardisation for determination of counting efficiency for colloidal scintillators has generated many questions. The usable counting ranges versus temperature and sample concentration, as described here in various examples, are compatible with the automatic external standardisation (A.E.S.) method for quench correction. By using the recommended sample preparation method and the usable sample load ranges the behaviour of a modern colloidal scintillator is similar to any ideal homogeneous sample. This was not the case when Triton-X—100 was used, [17] probably due to the temperature sensitivity and to the already mentioned influences of slight alterations [14,15] in sample composition. Studies of Iwakura[18] are of interest in this connection. He used the ratio external standard to internal standard (Se/Si) as an indicator of the degree of phase contact for colloidal scintillators. The best phase contact yielded a factor Se/Si equal to 1. This fact was found to be dependent on the counting time, temperature and on the kind of tenside.

The temperature dependence and the stability of the count rate are influenced by the tenside system in a modern colloidal scintillator. Figure 17 shows absolute dependence of tritium counting efficiency as a function of temperature and similarly the AES ratio plotted against temperature. It can be observed that these curves

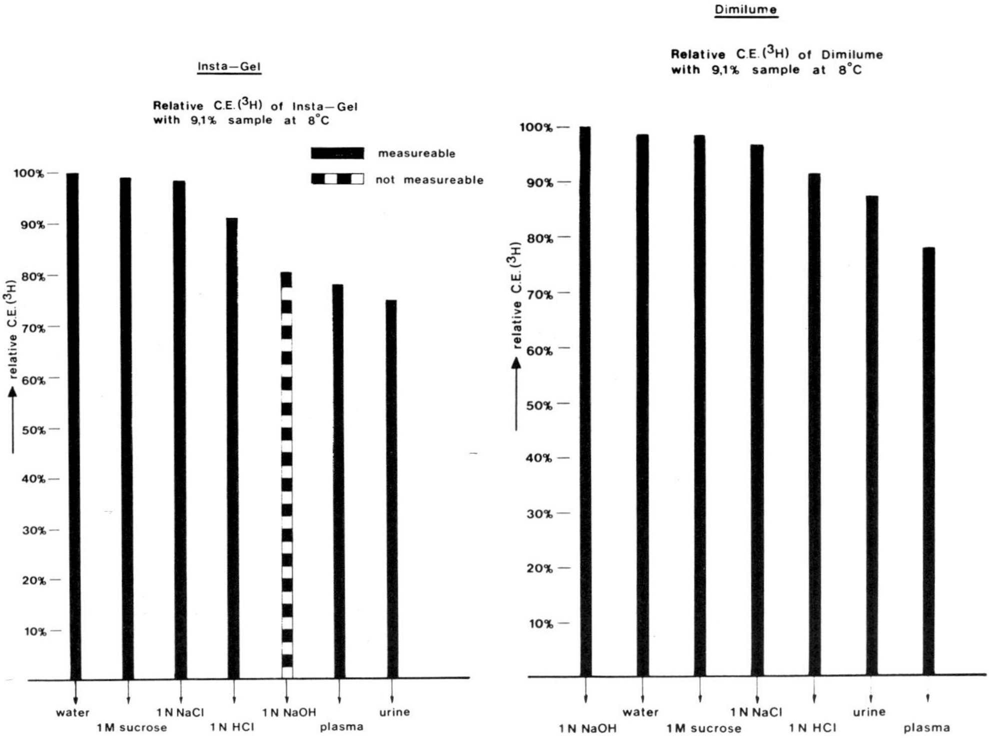

Fig. 13. Comparison of relative counting efficiencies of equivalent loads of various samples in Insta-Gel and Dimilume. 1 ml sample added to 10 ml scintillator.

match each other very well in shape and the external standard is usable, independent of the temperature, provided that the correct load and sample concentration are used as illustrated earlier. It can be seen that the external standard is applicable for Insta-Gel. However, there are some colloidal systems for which the physical changes are so temperature dependent that the correlation of changes in AES ratio with temperature does not match with the changes in efficiency, thus making the determination of counting efficiency inaccurate.

A practical standardisation curve for biological fluids shown in Fig. 20 illustrates the usable range of a typical colloidal scintillator for routine samples. The external standardisation method has been criticised by Rauschenbach[19] in terms of the type of standard used and the type of vials. Discrepancies due to these variables were reported when standards, other than Radium-226, were used. Also the use of polyethylene vials was not recommended. Figure 21 clearly shows the penetration of scintillator and/or activity through the wall of a plastic vial. For periods longer than a few hours there is little or no difference between colloidal scintillators and solutions of toluene and scintillator in terms of this penetration.

Stability of colloidal samples can be affected if the composition of the scintillator is incorrect and by variations in counting efficiency during the time of measurement or recounting after storage. No enhancement in count rate was detected by Handler[20] when counting mixtures containing up to 33 percent aqueous

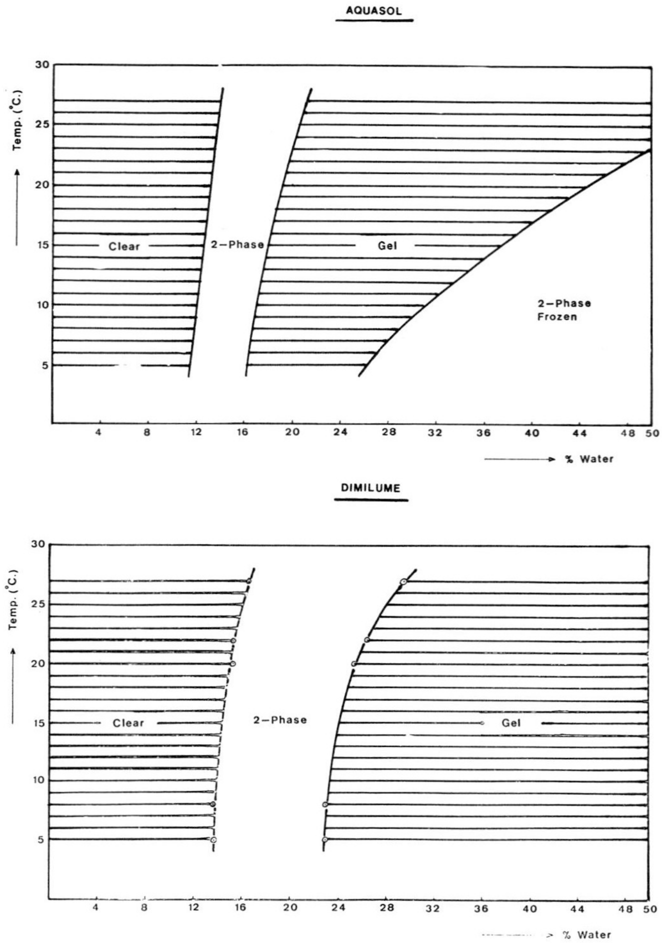

Fig. 14. Dependence of physical characteristics on temperature and sample sample load (water) for Aquasol and Dimilume.

sample in Insta-Gel, measured after 30 days storage. Figure 22 shows measurement of one regular and two other sample loads of water in Insta-Gel in terms of count rate versus time. Using temperature stabilised instruments it is of importance to allow the sample to reach the counting temperature of the compartment.

Chemiluminescence has been reported in the literature as one of the major problems when alkaline samples are to be counted with colloidal systems containing alkylphenol-polyethoxy-ethanols. Although the origin of the luminescent reaction has not been investigated, purification methods involving possible removal of these interfering compounds have been studied. None of these has been satisfactory. A method describing storage treatment of sodium carbonate for possible elimination of luminescent phenols, has been patented.[21] In contrast, I have not found this method to be of significant value. The phenomenon interfering luminescence, investigated by our laboratory has been found to be a direct function of the peroxide content of the tenside molecule. The ethylene oxide chain of the tenside molecule is known to form relatively stable hydroperoxides, which are attributable to chemiluminescence under alkaline conditions. The alkylphenol

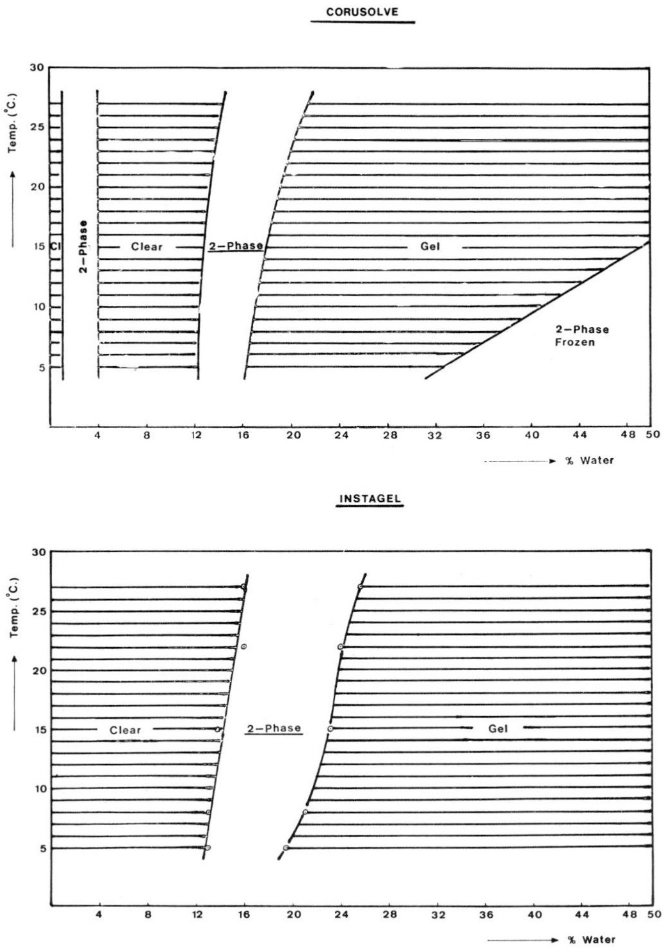

Fig. 15. Dependence of physical characteristics on temperature and sample load (water) for Corusolve and Insta-Gel.

derivatives as such have reasonable transfer energy and compete for spectral emissions, thus interfering with the scintillation measurement in the form of counts from spurious luminescence. Alkaline samples yield thousands of spurious counts with colloidal scintillators and decay of these counts can require several days. If alkaline samples are to be counted with colloidal scintillators the pH should be adjusted to neutral or slightly acid in order to obtain a normal background level within reasonable time.

The only colloidal scintillator having the general usability similar to the most universal colloidal systems with the advantage of rapidly reducing luminescence is Dimilume. The data in Fig. 23 show that using Dimilume as scintillator for a sample containing strong alkaline solubiliser with digested tissue, only a 45 min background stabilisation period is required compared with a much longer time for the usual type of colloidal scintillator.

The sensitivity of colloidal scintillators is best illustrated by studies of counting of low levels of environmental tritium in rainwater by Schell.[22] The minimum sensitivity level was as low as 193 picocuries per litre of water for

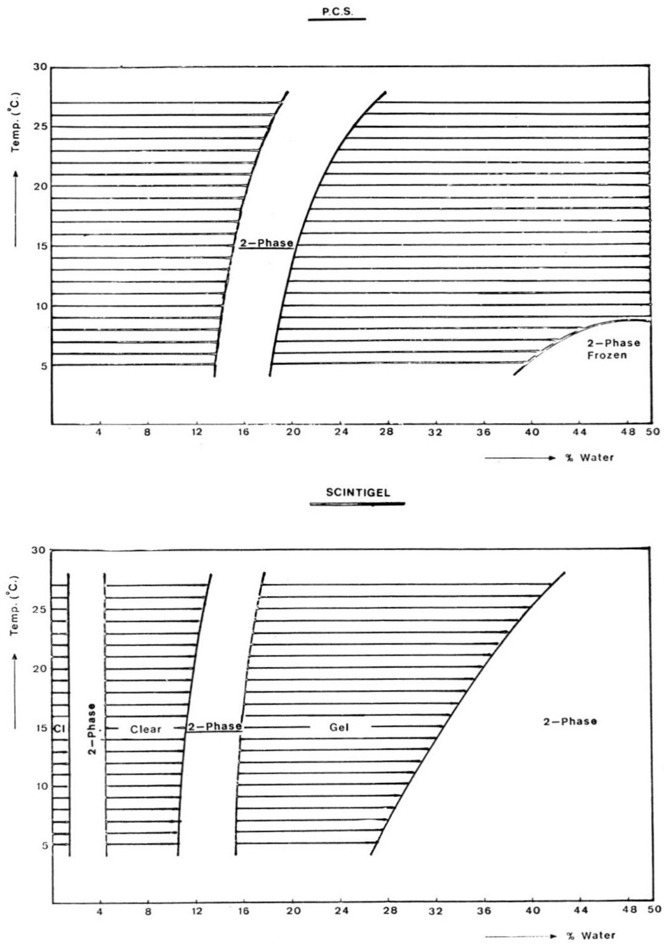

Fig. 16. Dependence of physical characteristics on temperature and sample load (water) for P.C.S. and Scintigel.

Insta-Gel, which represents a sensitivity improvement factor of 3.6 compared with dioxan.[23] There is no way to make similar comparisons for dioxan based scintillation fluids containing biological samples, because the organic solvent as such, as described earlier, is not suitable for scintillation counting of such samples.

The Figure of Merit, as a product of the counting efficiency (in percent) multiplied by the percentage of the sample load, has been used in comparative studies of counting systems of aqueous samples. Figure 24 shows such comparisons for tritium in water. It appears clearly from this study that the system Insta-Gel shows the highest Figure of Merit and the best universal usability as is demonstrated in the studies described earlier in this paper.

In conclusion, we can say that colloidal counting today represents the most universal system. It allows experimental measurements of low level radioactivity in environmental monitoring and routine determination of activity in body fluids for any particular purpose with high efficiency and accuracy. Together with sample preparation by automatic combustion techniques I see these counting systems

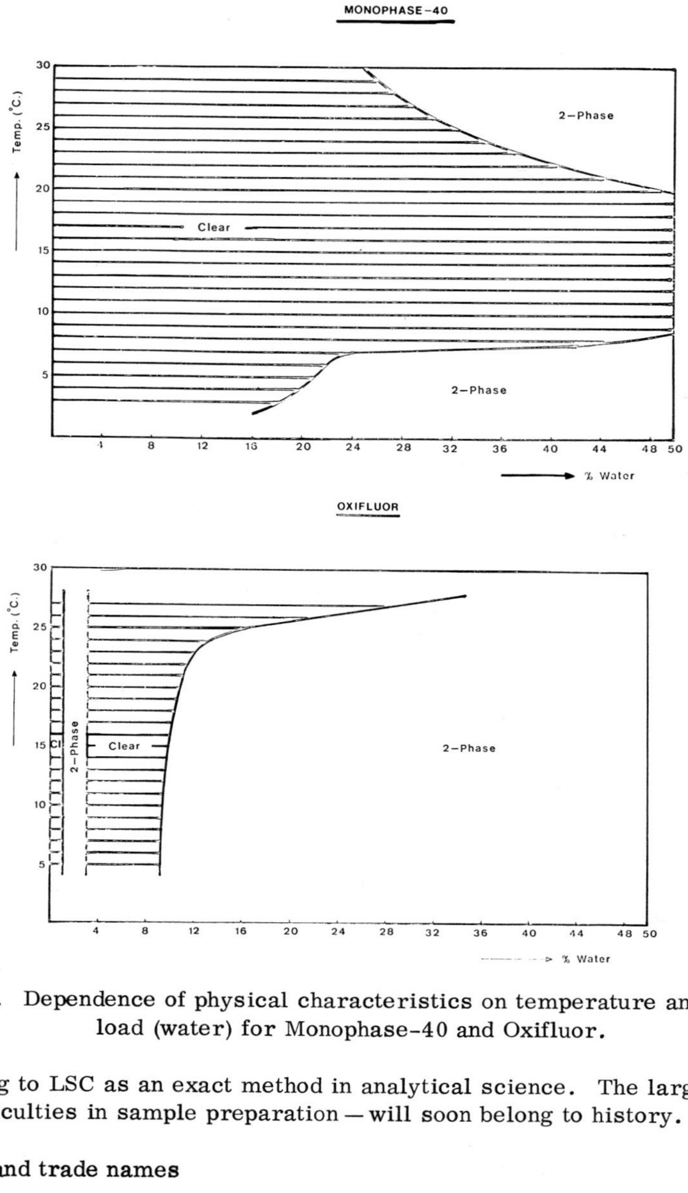

Fig. 17. Dependence of physical characteristics on temperature and sample load (water) for Monophase-40 and Oxifluor.

contributing to LSC as an exact method in analytical science. The largest drawback — difficulties in sample preparation — will soon belong to history.

Materials and trade names

The commercial products referred to were usually obtained through the owner of the trademark; if not, as otherwise specified as follows: Aquasol, Oxifluor: NEN Chemicals GmbH, Germany; Corusolve: ICN-Tracerlab, Belgium; Isolab: Isolab Inc., U.S.A.; Dimilume, Insta-Gel, Monophase: Packard Instrument Inc., U.S.A.; Multisol: Euratom, Centre Nuclear, Mol, Belgium; P.C.S.: Amersham-Searle Inc., U.S.A.; Scintigel: Karl Roth KG, Germany; Unisolve: Koch-Light Laboratories, U.K.; Triton N–101, X–100: Rohm & Haas, U.S.A. The vials used for these measurements were of the lowest background type glass (Packard Instrument Part No. 6000128).

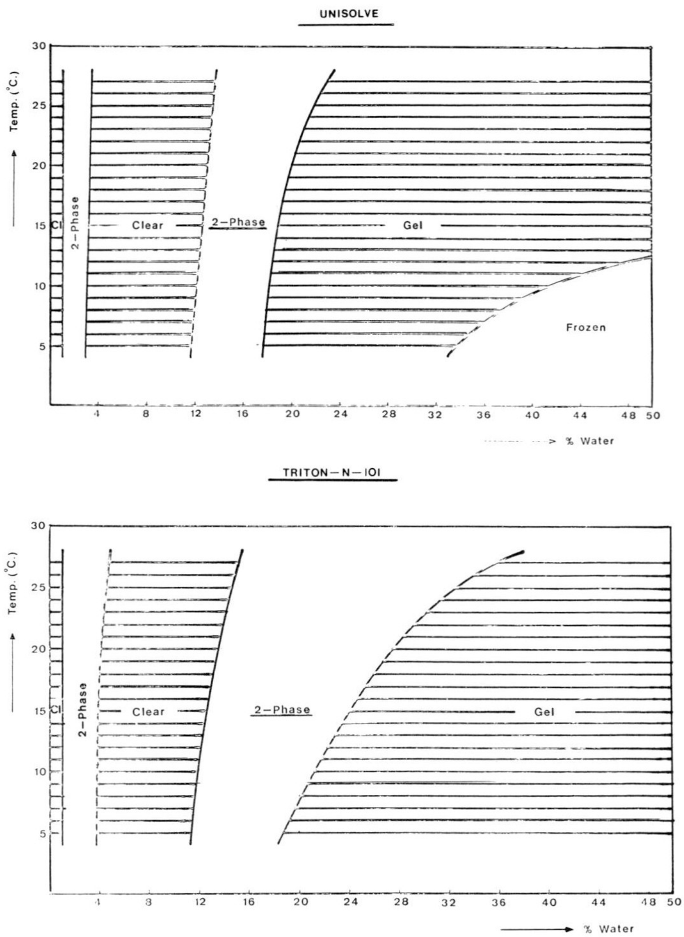

Fig. 18. Dependence of physical characteristics on temperature and sample load (water) for Unisolve and Triton N–101.

REFERENCES

1 Milton Furst and Hartmut Kallmann, Phys. Rev. **97**, 583 (1955).

2 G.T. Okita, J. Spratt and G.V. Leroy, Nucleonics **14** 76 (1956).

3 Frank Kinard, Science 28, 293 (1957).

4 G.A. Bray, Anal. Biochem. **1**, 279 (1960).

5 Frank E. Butler, Anal. Chem. **33**, 409 (1960).

6 G.A. Bruno and J.E. Christian, Anal. Chem. **33**, 1216 (1961).

7 G. Erdtmann and G. Herrmann, Radiochem. Acta **1**, 98 (1963).

8 R. Meade and R. Stiglitz, Intern. J. Appl. Radiation Isotopes **13**, 11 (1962).

9 M.S. Patterson and R.C. Greene, Anal. Chem. **37**, 854 (1965).

10 J. Shapira and W.H. Perkins, Science **131**, 414 (1960).

11 C.F. Gordon and A.L. Wolfe, Anal. Chem. **32**, 574 (1960).

12 Cooke et al. in Surface Chemistry, Munksgaard, Copenhagen 1965, p.231.

13 S. Srivastara and D. Haydon, Proceedings IVth International Congress of Surface Activity, Brussels, 1964.

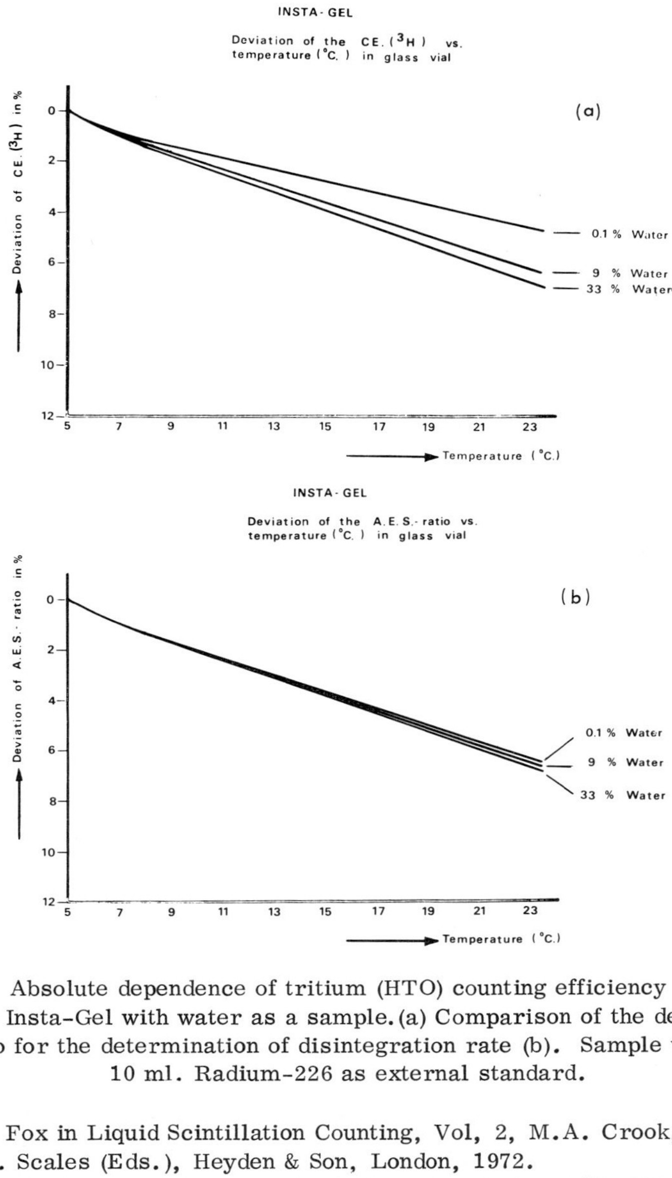

Fig. 19. Absolute dependence of tritium (HTO) counting efficiency on temperature for Insta-Gel with water as a sample. (a) Comparison of the deviation with AES ratio for the determination of disintegration rate (b). Sample volume 10 ml. Radium-226 as external standard.

14 B.W. Fox in Liquid Scintillation Counting, Vol, 2, M.A. Crook, P. Johnson and B. Scales (Eds.), Heyden & Son, London, 1972.

15 J. van der Laarse, Intern. J. Appl. Radiation Isotopes 18, 485 (1967).

16 P. Williams, Intern. J. Appl. Radiation Isotopes 19, 337 (1968).

17 R.H. Benson, Anal. Chem. 38, 1353 (1966).

18 Tetsuo Iwakura, Aiko Maebayashi and Yoshiko Kasida, Radioisotopes 18, 506 (1969).

19 P. Rauschenback and H. Simon, Anal. Chem. 256, 119 (1971).

20 L.H. Handler and J.A. Romberger, Intern. J. Appl. Radiation Isotopes 24, 129 (1973).

21 Royal H. Benson, U.S. Patent 3, 573, 219, 30 March 1971.

22 G. Sauzay and W.R. Schell, Intern. J. Appl. Radiation Isotopes 23, 25 (1972).

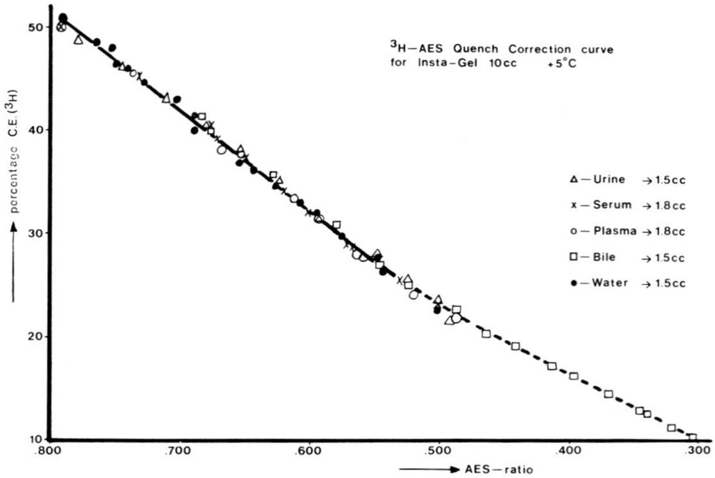

Fig. 20. AES quench correction curve for Insta-Gel with various biological samples labelled with Tritium (Battig).[24]

23 P.H. Williams and T. Florkowski in Proceedings of the Symposium on Radioactive Dating and Methods of Low-Level Counting, International Atomic Energy Agency, Vienna, 1967, p.703.

24 F. Battig, personal communication.

DISCUSSION

K. Painter: I hope everyone enjoyed the Packard commercial as much as I did. Certainly, Dr. Tarkkanen, somewhere there must exist a sample in the universe which works better in something other than Insta-Gel? Also, all your graphs show Insta-Gel taking aqueous samples from zero percent water on. Yet your literature states one should not use samples less than 1 ml (ref. Packard catalogue).

V. Tarkkanen: As to your first comment, this should be possible. The referred catalogue notice claims 0.1 ml, not 1 ml. This is easily understandable; the minimum amount of water is needed in order to obtain equal distribution of a sample within the detecting liquid.

B.W. Fox: As a person who has no connection with an instrument firm, I feel that this discussion demonstrates that with secret recipes there is no possibility of rational scientific discussion.

K. Painter: Could I ask Dr. Fox if he has worked out the MSQ value of Insta-Gel?

B.W. Fox: I have, but owing to the fact that I was limited to the basic composition of colloid scintillant, I could only work out the maximum MSQ value for that composition, which came to 2.70. I recognise, however, that the requirements of manufacturers are for a wide spectrum application, and they are prepared to sacrifice some efficiency for this parameter, which for these purposes is a correct approach. My aim is to find conditions where I can observe the maximum number of counts with the greatest counting stability. Dr. Tarkkanen, over what period do you regard counts as being stable?

V. Tarkkanen: If we cannot detect visible, physical instability like formation of two layers and freezing, the observed period is a minimum of 170 h.

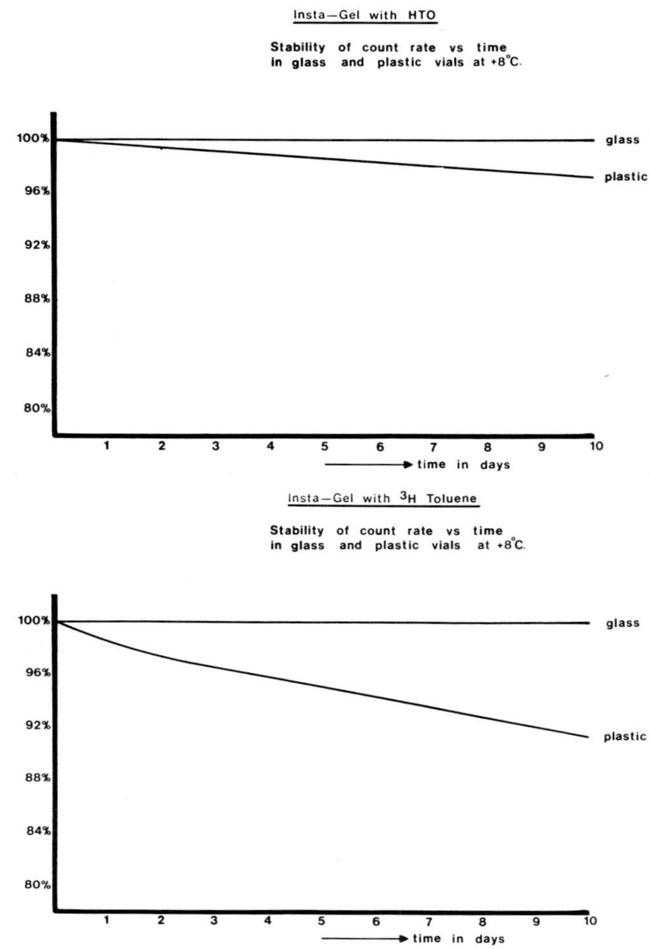

Fig. 21. Influence of the penetration of radioactivity through a plastic vial. Sample volume 15 ml.

E. Gunther: In dioxane-based scintillator some nuclides are adsorbed on the walls of the vials. Therefore one adds silica to have a controlled adsorption. What about the case of Insta-Gel? Is it also useful to introduce silica in such cases?

V. Tarkkanen: Addition of silica has only a negative influence in the form of self-absorption in Insta-Gel in the case of organic polymer samples. We have not enough experience concerning nuclides.

E. Gunther: Is the shape of the pulse spectrum in Insta-Gel the same as in the scintillator systems? For example, have you experience in determining the β-activity by an extrapolation method?

V. Tarkkanen: The pulse height spectrum is similar to a normal fluorescence emission spectrum. We do not have special experience using extrapolation for the determination of β-activity in Insta-Gel.

R. Dierckx: I have samples of highly concentrated lithium acetate $[(CH_3COOLi)\,nH_2O]$; would they be soluble in Insta-Gel?

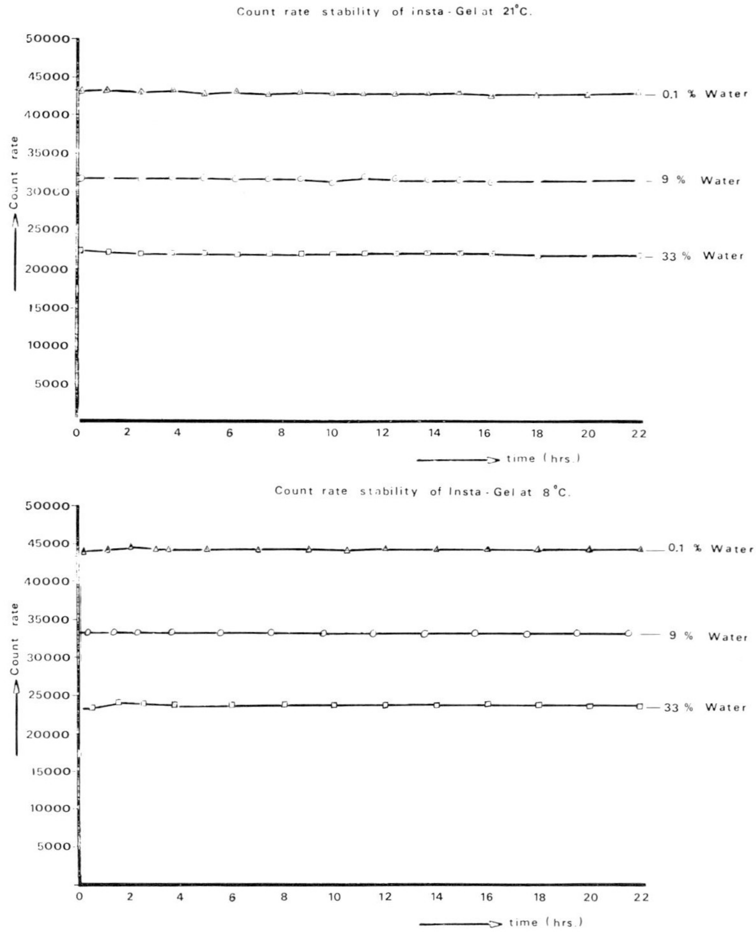

Fig. 22. Count rate stability of Insta-Gel with varying amounts of water at 8°C and 21°C. Final sample volume 15 ml.

V. Tarkkanen: I don't know, but if you like to count neutrons, the necessary formulation is available.

D.A. Ginger: You drew attention to an alleged product of my company, called Unisolve 3. Can you amplify your remarks regarding this product, as it has never been released for sale?

V.Tarkkanen: We tested Unisolve 1, 2 and 3 of which the Unisolve 2 was not usable due to abnormal low counting efficiency. Unisolve 3 gave better results than Unisolve 1. Therefore the presented results are most favourable for Unisolve 3.

D.A. Ginger: I must emphasise that only 1 litre of Unisolve 2 and 3 was ever manufactured at the express wish of a customer, not Packard-Becker, and comparisons obtained in this way are odious.

E. Mueller: Do you get the same efficiency for ^{3}H-water and ^{3}H-toluene in all the composition regions you describe as stable?

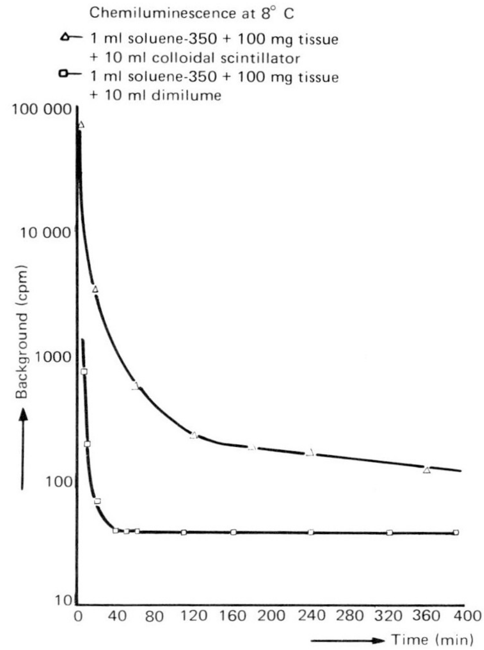

Fig. 23. Chemiluminescence decay in colloidal counting systems. Final sample volume 11 ml.

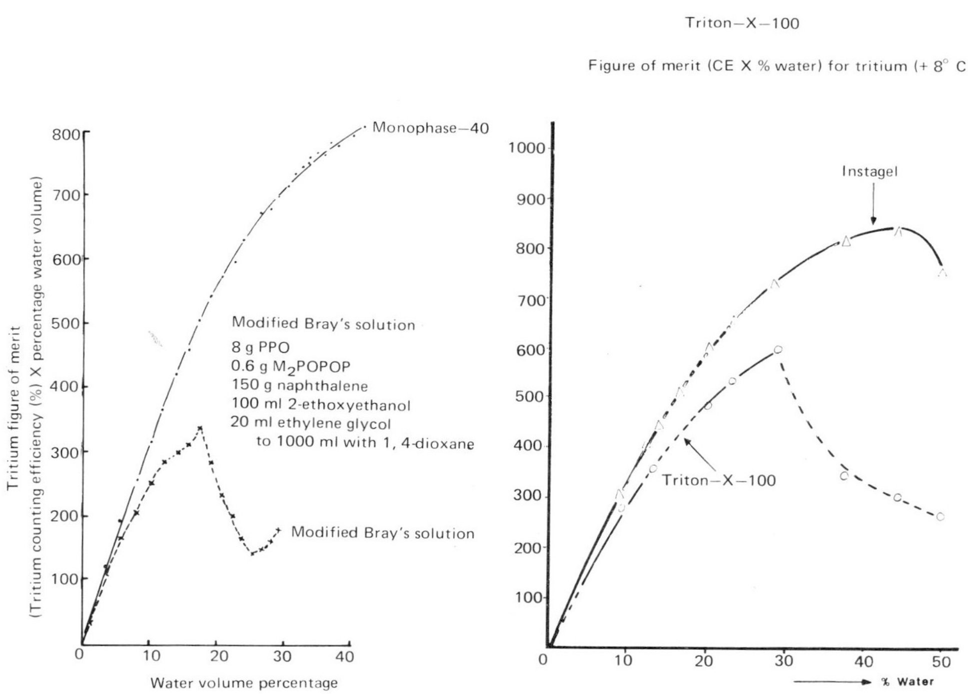

Figure 24.

V. Tarkkanen: We did this experiment only in the Insta-Gel and PCS solutions, and we did get the same efficiency in those parts of the stable regions which we tested.

M.W. Hollingsworth: You showed the difference between the chemical and colour quench curves for toluene systems. In the Insta-Gel system the points lay on the same curve. Is this generally true for these colloidal systems; for example, is PCS similar to Insta-Gel in this behaviour?

V. Tarkkanen: Insta-Gel shows similar correction curves with $CHCl_3$ and dinitro phenylhydrazine (yellow, the most usually occurring optical quench band) for tritium, but not for Carbon-14 when the instrument settings are optimal for counting efficiency. PCS shows different correction also for tritium when measured in the absence of an aqueous sample.

D.E. Bowyer: Insta-Gel showed little difference between chemical and colour quenching as determined by the external standard channels ratio method as given by Dr. Tarkkanen. This was presumably determined with a high energy external standard (Radium-226). Is this similarity also seen with a low energy external standard?

V. Tarkkanen: This experiment has not been performed by us using other than Radium-226.

B.W. Fox: Does Dr. Bowyer have any evidence, with respect to Insta-Gel, that there is a difference between standardisation by high energy external standard compared with low energy external standard. There is a difference in the Triton system.

D.E. Bowyer: No, I have no such evidence.

K. Painter: Users of commercial cocktails should be worried when counting tritiated plasma or serum samples that the weak β-particle may be self-absorbed by globular proteins and inaccurate d.p.m. values may result. I believe there is a publication in the literature critical of Insta-Gel on this very point. Although the data presented here is very complimentary to PCS, we would recommend to our users that digestion or combustion be used to test for self-absorption when counting tritiated serum or plasma samples.

V. Tarkkanen: As I mentioned, the term solubiliser in connection with colloidal scintillators is correct only regarding terminology within surface activity, not liquid scintillation counting: the sample must be in solution or in a colloidal form or in some cases so finely homogenised, i.e. in water or in alcohol–water, that the activity is eluted into the solution. There is one publication in which an attempt was made to count a concentrated, coarse homogenate in Insta-Gel. This is, of course, not correct and against our recommendations.

B. Scales: I am amazed that you obtain virtually identical calibration curves with chemical (thermal) and coloured quenching agents. In my experience even different chemical quenchers, let alone different colour quenchers, can give rise to different E.S.R./efficiency curves. I would like to ask if any members of the audience have any experience in the use of Insta-Gel and the production of quench correction curves? Also, how would Dr. ten Haaf explain the results?

V. Tarkkanen: Yes, different chemical quenchers can certainly give rise to different calibration curves. One should not make calibration curves with, for example, chloroform or acetone, but use water or other authentic materials as quencher. Also various instruments give different quench curves, depending on which external standard is utilised. The curves of colour and chemical quenching can be fairly close to each other if the settings of the instrument are adjusted accordingly.

F.E. ten Haaf: I should like to point out that numerous parameters have influences on the potential difference between colour and chemical quench curves and that it would not be correct to connect this only with the sample solvent. For instance, colour and chemical quench curves for Carbon-14 diverge much less if a rather narrow measuring channel is used, than in the case of a wide or integral channel. What channel width did you use?

V. Tarkkanen: We used optimal settings regarding counting efficiency, which were not optimal for separation of colour and chemical quench. For tritium, 55% gain, discriminator 50 to 1000; and for carbon, 6% and 8 to 1000 respectively.

Chapter 17

Optimisation in Colloid Scintillation Counting

B.W. Fox

Paterson Laboratories, Christie Hospital and Holt Radium Institute, Manchester, England

The colloid counting system has been shown to be a potentially useful means of assaying those solutions derived from biochemical experiments which contain many solutes other than the ones being measured and in which the labelled material is present in very low specific activity. Many such solutions have been examined by this general procedure (for review see Ref. 1), but the composition of the colloid system used has been the result of an empirical application of a composition, often used for a completely different solution than for the one intended.

In an earlier communication,[2] it was shown that the type of system used depended to a very considerable extent on the composition of the solution and that the final choice of mixture was highly dependent on the composition of this solution.

I wish to make it quite clear at the outset that my intentions are quite different from those of the many manufacturers of these colloid scintillant mixtures. I am attempting to measure very low specific activity solutions, given the fact that several millilitres of the aqueous solution is available and I want to be able to take full advantage of the concept of 'merit value'[2] to be able to measure the few counts present, where other methods of assaying, by aliquot sampling, precipitation, etc., fail. This aim differs from that of the industrial exploitation of the colloid system in that they require a water accepting mixture, which will also accept a varying type and quantity of other solutions and count with a reasonably high efficiency. However, in all these systems, there is inevitably considerable loss of the counting potential of the system, as the region of maximum value is nearly always confined only to a small area in the total contour diagram. It is clearly useful to be able to compare one ternary system with another from this point of view and to be able to judge the general properties of a scintillant system with regard to its general counting and stability characteristics at a glance.

In order to prepare a contour diagram, a triangular board, cut out and marked as in Fig. 1, is used. In the first instance, 0.5 ml of a concentrated scintillant solution in the primary solvent under study is added to each vial. Starting from the detergent:aqueous sample edge, 0.5 ml of pure primary solvent is added to the first row of vials (representing the 10% primary solvent level). 1.5 ml is added to the next row and each row with an incremental increase of 1 ml, until the last

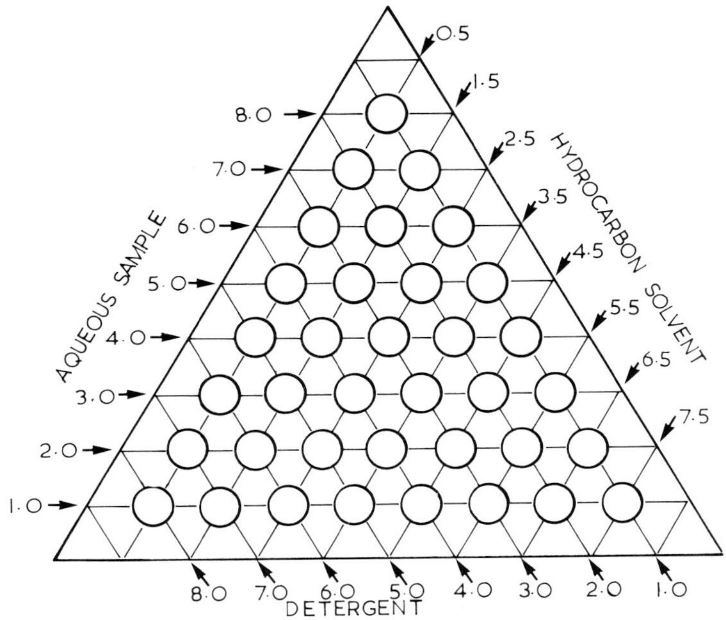

Fig. 1. The construction of a contour diagram. It is usually convenient to regard the lower left-hand corner as 100% detergent, the lower right-hand corner as 100% primary hydrocarbon solvent and the apex as 100% aqueous sample. The additions indicated are those of the pure components following addition of the concentrated scintillant solution (0.5 ml) to each vial.

bottle at the hydrocarbon solvent corner contains 8 ml of primary solvent (i.e. 7.5 ml pure solvent + 0.5 ml concentrated scintillant solution). Then, starting from the primary solvent:aqueous sample edge, 1 ml of detergent is added to the first two, 2.0 ml to the second and so on until the bottle at the detergent corner is reached when 8.0 ml is dispensed. A similar sequence of fillings with aqueous sample is conducted from the detergent:primary solvent border to the aqueous sample corner. The bottles are thoroughly shaken and the backgrounds taken to see if there is any contamination or unnecessary chemiluminescence interference.

Following spiking 10 µl of tritiated water to all the samples and thorough mixing, they are again counted after assessing for opacity by an empirical system (from 0 to 5 indicating clear to completely opaque). The degree of viscosity is also assessed on a 0 to 5 scale varying from completely fluid, like toluene, to a solid gel which remains rigid on inversion of the vial. Colloid stability (not counting stability) is assessed by a similar empirical set of numbers from completely stable for the duration of the experiment (1) to a rapid separation into two phases (4). If these values are independently obtained and are then plotted on triangular co-ordinate papers, it can be seen that contours may be constructed around the figures. Where apparently great changes in character occur in an unusual region, sub-division around that region may reveal a local change in this parameter.

The resulting diagrams for the system Triton X-100:xylene:water (PPO and POPOP as scintillants) are shown in Figs. 2, 3 and 4 for appearance, viscosity and colloid stability respectively. The resulting contours are by no means

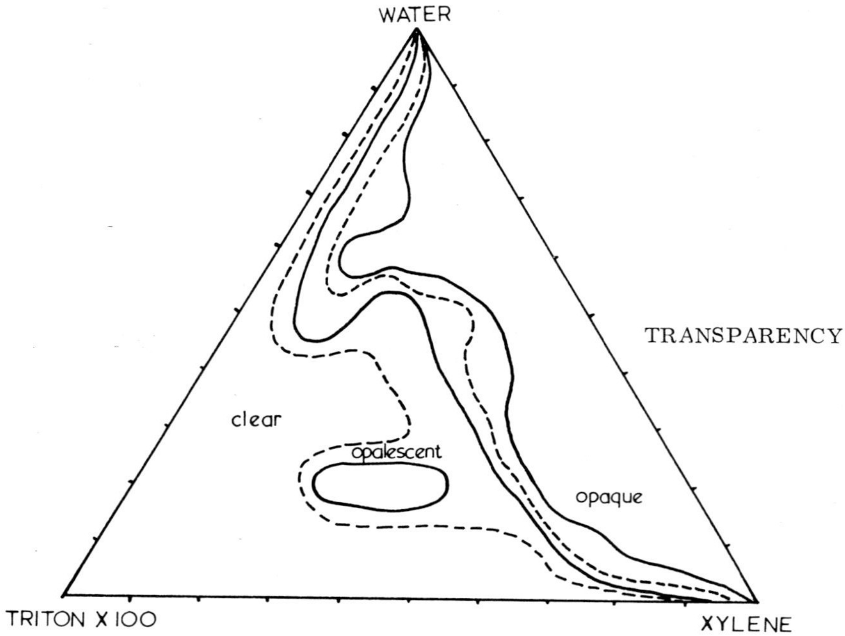

Fig. 2. The contour diagram of the appearance of the system Triton X-100: xylene:water at 21°C. The contours are constructed to enclose areas of similar appearance (see text).

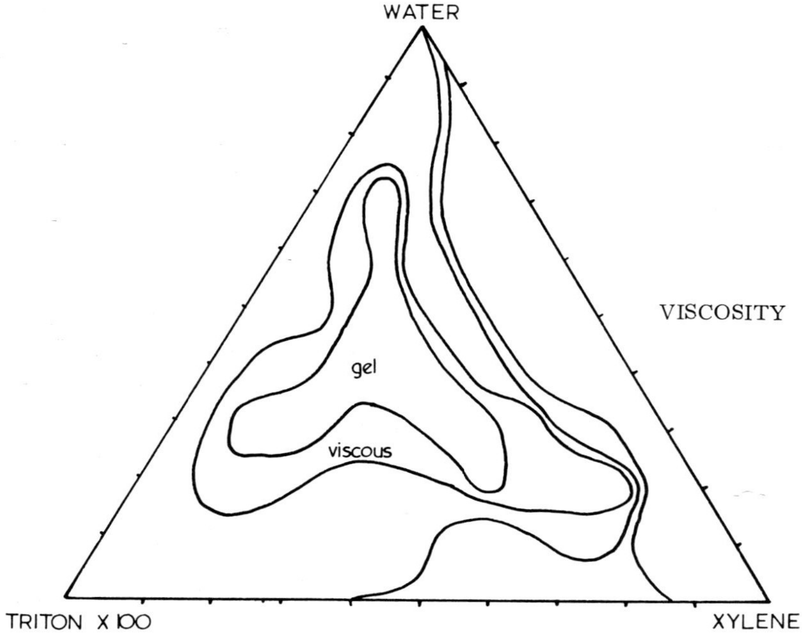

Fig. 3. The contour diagram describing the viscosity of the samples (see text)

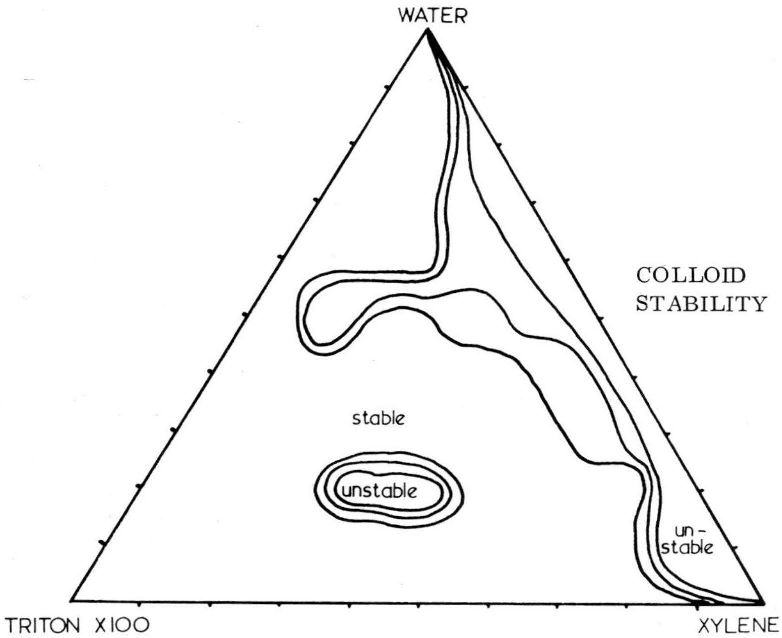

Fig.4. The contour diagram illustrating the regions of colloid stability (not counting stability) ranging from an unstable region, separating very soon after mixing, to a stable composition which remains the same after several days.

superimposable, and bear no obvious relationship to one another, but are clearly continuous within one parameter consideration. It is therefore not possible to construct such diagrams on vague descriptions employing two or more of these parameters, such as a cloudy gel or transparent fluid.

However, none of these parameters are really important from the point of view of liquid scintillation counting. Only two criteria are primarily important: (a) the merit value (i.e. the product of percentage efficiency and the percentage of sample) and (b) the counting stability. For the purpose of assaying this latter quantity, at least five recycled samples spread evenly over 48 h were taken. The standard deviation of the repeated counts were computed and in order that each sample could be compared with each other on an equal basis, the coefficient of variation was determined, i.e. the quotient (expressed as a percentage) of the standard deviation to the mean counts. Plotting both these parameters gives the contour diagram combination shown in Fig. 5. It can be seen that it is possible to select either a point which has a high merit value with some sacrifice in counting stability (A) or the reverse, high counting stability with some loss of merit value (B). The choice will depend on the operator's particular requirements and tolerances. When the appropriate point is chosen, the level of scintillant most suitable may be determined by varying the proportion of the concentrated scintillant solution to that of the primary solvent in that particular sample. The secondary solute appears to play a much more important role in this system than in an equivalent homogeneous system but the reason for this difference is not yet clear.

In an attempt to describe such a system with a unique value, a 'merit stability

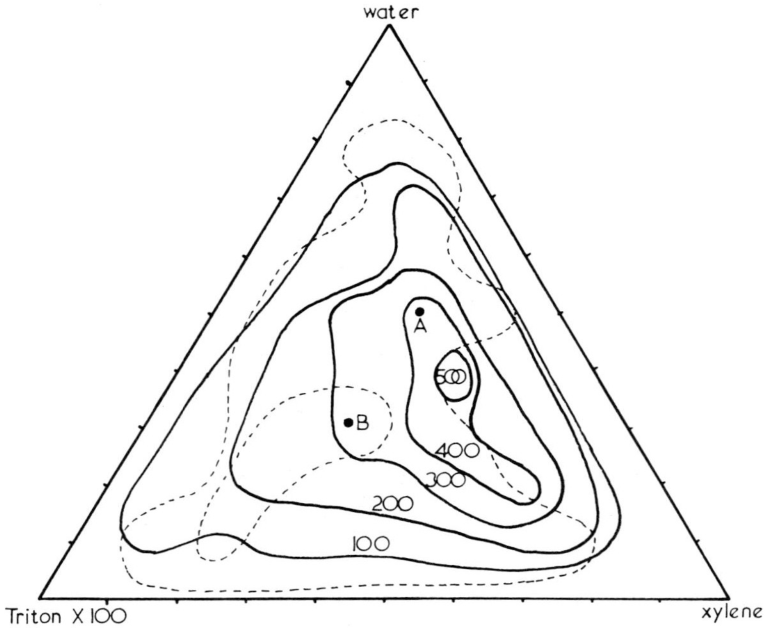

Fig. 5. A composite contour diagram for the system Triton X-100:xylene: water, showing merit values in contour steps of 100(——) and coefficient of variation (----). Inner contour less than 0.5%, outer contour encloses the area less than 1.0%. Position 'A' has a merit value of 447 and a coefficient of variation of 0.98%, whereas position 'B' has a merit value of 329 with a coefficient of variation of 0.33%. The highest merit value of 552 has a coefficient of variation of 7.5% and is clearly unsuitable for counting.

quotient' (MSQ) is suggested which is simply the merit value/coefficient of variation corrected for machine efficiency, i.e.

$$\text{MSQ} = \frac{\text{Merit value}}{\text{Coeff. of variation}} \times \frac{100}{\text{Machine eff. (Ref.)}}$$

The machine efficiency could be the reference standard for that particular isotope supplied with the machine or a 4 g/litre solution of PPO in toluene, with or without 0.02 g POPOP depending on photomultiplier type.

In view of the considerable sensitivity of this figure to small changes in coefficient of variation, it is tentatively suggested that the logarithm of this figure could be used. The figure is still not satisfactory, however, since it is only meaningful if there is always some instability. When the log values are plotted on the triangular co-ordinate system, there is a clear distinction between two systems, one showing instability over a wide area (Fig. 6) and the other (Fig. 7) showing a relatively larger area of stability, and would be presumably more favourable to the requirements of the manufacturer. However, the pattern can only give a broad impression about the system, and to obtain the more exact conditions the superimposition of the two parameters as indicated in Fig. 5 must remain the most useful method of selection.

However, Table 1 shows some maximum $\log_{10}$ MSQ values for some systems tried. It must be emphasised, however, that these conditions and values are

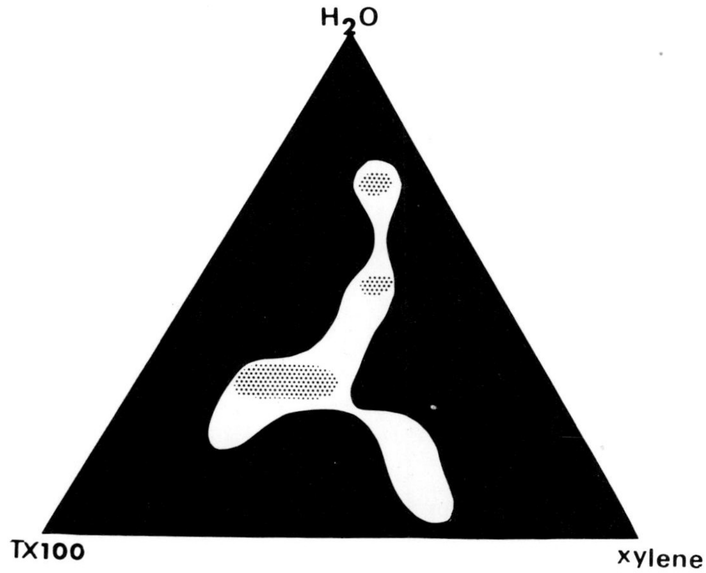

Fig. 6. Plot of $\log_{10}$ MSQ for the system Triton X-100:xylene:water, with PPO/POPOP as scintillant. The three small areas enclosed within the clear area are regions where the value is greater than 3. 0, whereas the black outer area is less than 2. 5 and represents either low stability or poor merit values.

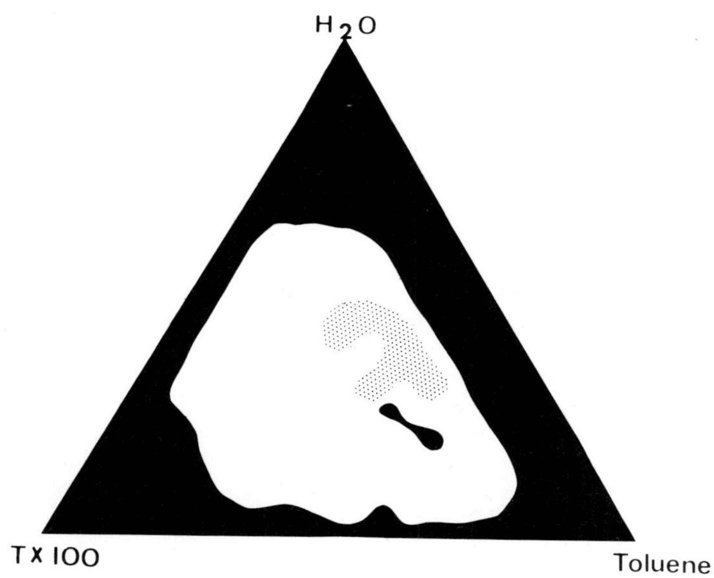

Fig. 7. Plot of $\log_{10}$ MSQ numbers for the system Triton X-100:toluene:water (PPO and POPOP as scintillants). It can be seen that the unshaded area representing $\log_{10}$ MSQ between 2. 5 and 3. 0 indicate that a greater region of the area is considered to be potentially useful for counting.

Table 1. The relative merit stability quotients of a number of ternary systems.

Aqueous solution	Detergent	Primary solvent	Secondary solvent	Primary solution	Secondary solution	MSQ	$\log_{10}$ MSQ
Water	Tx 100	Toluene	—	PPO	POPOP	2395	3.38
Water	Tx 100	Xylene	—	PPO	POPOP	2073	3.32
Water	Tx 114	Xylene	—	PPO	POPOP	1436	3.14
Water	TN 101	Xylene	—	PPO	bisMSB	1333	3.13
Water	Tx 57	Toluene	—	PPO	POPOP	960	2.98
Water	Tx 45	Toluene	—	PPO	POPOP	593	2.77
Water	Tx 100	Xylene	Naphth	PPO	POPOP	542	2.73

valid in these cases only for water containing tritiated water standard. Recently, the factor MSQ has been modified to take into account its considerable sensitivity to low coefficients of variation. A more realistic figure, abbreviated to MISQ, is defined as follows:

$$\text{MISQ} = \frac{\text{Merit value}}{1 + \%\ \text{coeff. variation}} \times \frac{100}{\text{Ref. efficiency}}$$

When the coefficient of variation is zero, the MISQ becomes an instrument efficiency corrected merit value and will thus be capable of being used as a comparative figure to describe the potential efficiency of a colloid scintillation mixture. The maximum MISQ for toluene-water detergent systems would thus be

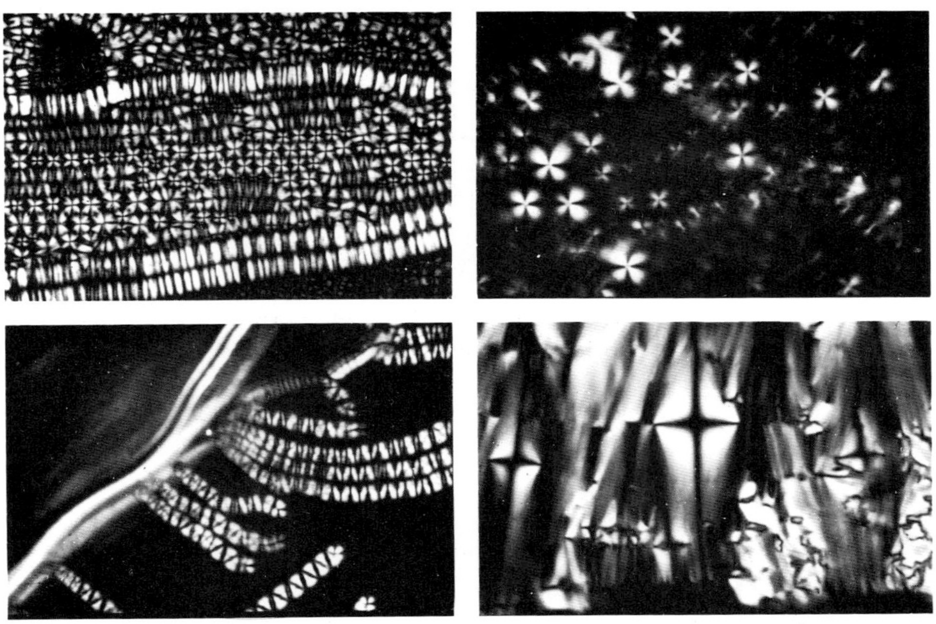

Fig. 8. The plates show four of the many different appearances under crossed polaroid optics of regions of the birefringent zone of the system Triton X-114:xylene:water (approx. 5:1:4). The fluctuating temperature of the microscope stage can change the complex structure dramatically in short periods of time (magnification approx. 600).

related to the detergent as follows, Triton X-100 (1014), Triton N101 (507), Triton X57 (368), Triton X45 (297) and clearly indicates that amongst these detergents, with PPO/POPOP as solutes, Triton X100 shows the best potentialities for the most sensitive detection of ^3H-H_2O.

As an added interest, a number of selected regions from a ternary combination of Triton X-114:xylene:water were examined through crossed polaroid filters at approximately 600 magnification and a number of these are indicated in Fig. 8. It can be seen that a remarkable ordered arrangement of these colloid systems exists which exhibits a great variety of liquid crystalline structures. The variation of such structures clearly go a long way in accounting for the extreme versatility of the system and the potential value of the system in studying membrane and similar structures can be envisaged.

REFERENCES

1 E. Rapkin, 'Gel and Emulsion Counting of Aqueous Solutions', Digitechniques published by Intertechnique.

2 B.W. Fox, in Liquid Scintillation Counting, Vol. 2 (Eds. M. Crook, P. Johnson and B. Scales), Heyden & Son, London, 1972, p.189.

DISCUSSION

J.F. Stoutjesdijk: You add 0.5 ml of a concentrated PPO-solution and so the PPO-concentration in the organic phase is not constant. Cannot this give differences in counting efficiency?

B.W. Fox: When you are undertaking this kind of analysis using phase diagrams, you must decide whether to have the same level of PPO in each vial or the same concentration of PPO per ml toluene in each vial. From a practical point of view, therefore, we decided on the first approach to attempt to modify the level of PPO in the chosen vial to determine final optimal working conditions. If the second method is chosen, it would be very difficult to interpret the results due to self-quenching effects in the toluene-rich regions.

J.F. Stoutjesdijk: Did you investigate Insta-Gel with different amounts of H_2O?

B.W. Fox: Yes.

F.E.L. ten Haaf: You mentioned a particular sample which even showed phase separation but nevertheless had good counting stability. Wouldn't this depend rather strongly on the phase in which the marker is dissolved?

B.W. Fox: Yes, but I included it as an example to show that good stability may sometimes be achieved even under such conditions. It also demonstrates that the visual appearance is no index of the true structure, since it is likely that both so-called phases are probably colloidal.

K. Painter: We have noted variations in efficiency and stability of Triton systems based upon the order of addition of components. Have you noticed any similar variations?

B.W. Fox: No. We always add the aqueous component last and ensure that the whole is thoroughly mixed, either on a rotary mixer or by warming and cooling, before counting. We have not examined possible effects due to order of mixing.

P.E. Stanley: Batches of Triton X-100 vary somewhat. Do your MSQ values change with different batches of the detergent?

B.W. Fox: We buy our batches of Triton X-100 in bulk and consequently we have not been able to compare batches since the same batch has been used throughout. This is an interesting point and it should be done.

Chapter 18

The Reliability of Estimating Carbon-14 in Biological Samples

B. Scales

I.C.I. Pharmaceuticals Division, Alderley Park,
Macclesfield, Cheshire, England

INTRODUCTION

Although very many methods now exist for the preparation of samples to be assayed by liquid scintillation counting, most laboratories use relatively few. The reasons for using any one sample preparation technique are often very basic ones, such as cheapness, familiarity, universality and possibly ignorance of other more appropriate methods. On occasions, the reliability of accepted methods is questioned[1] and large discrepancies can be found between results obtained using different methods of sample preparation.

In most laboratories it is common practice to 'normalise' the raw counting data, i.e. to attempt to correct for the unavoidable variables which occur in liquid scintillation counting, by computing d.p.m. values relative to (a) a standard which may be purchased for that purpose or (b) a sample of radioactivity, a portion of which is used in the actual experiment. However, some of the variables cannot be readily assessed, and it is this difficulty which gives rise to the discrepancies which may be found when different methods are used to analyse any one sample.

Further problems can arise when vastly different types of samples have to be analysed, and the results totalled, or compared to give ratios. In these circumstances, partly because of the necessary use of different methods of sample preparation, reproducible, but highly inaccurate final results, can be obtained. The problem can even be magnified and errors increased if different individuals combine their efforts.

The purpose of this paper is to report the results of a study which was done in order to check the agreement between different methods of sample preparation and between different people using the same method. To this end, 22 individuals with different degrees of familiarity with liquid scintillation counting were each asked to assay two different urine and two different plasma samples for their Carbon-14 content, each person using his preferred methods of sample preparation. They were asked to report the results as d.p.m. Carbon-14/ml sample. The only way that the measurements could hopefully differ from normal procedures was that individuals would, where possible, do replicate analyses rather than just one or two. This would enable a more acceptable statistical analysis of the results to be carried out.

METHODS

Preparation of radioactive biological fluids

A high activity urine sample was obtained from a human dosed with ^{14}C-propranolol, and a low activity sample was obtained from a marmoset dosed with ^{14}C-ICI 55, 897. The radioactivity in both these samples was present as stable, highly water-soluble metabolites. The high and low activity human plasma samples contained ^{14}C- ICI 54, 594, a highly protein-bound carboxylic acid.

The approximate specific activities of these samples and the volumes given to each investigator are shown in Table 1. The high activity samples were prepared at about 50, 000 d. p. m. /ml since this is, in general, a typical high level found in urine and plasma samples after a dose of Carbon-14 labelled drug. The low activity samples were of such a specific activity that the background contribution to the count-rate would be significant.

Table 1. Samples prepared for analysis.

Sample	Code	Specific acitivity (approx. d.p.m./ml)	Volume supplied to investigators (ml)
Human urine	H.U.	50, 000	2.0
Marmoset urine	L.U.	500	17.0
Human plasma	H.P.	40, 000	2.5
Human plasma	L.P.	1, 000	12.0

All the specific activities were such that a 10-min count would produce reasonable accuracy, and our standard ^{14}C-hexadecane spiking fluid (nominally 10, 000 d. p. m. /50 μl) would be adequate for efficiency determination after a further 1 to 10-min count.

Sample preparation for liquid scintillation counting

In the event, the Carbon-14 activity in the samples was measured using the three different techniques listed in Table 2.

Table 2. The composition of scintillation media used to assay the samples.

System	Scintillator composition
Emulsion counting:	
Toluene–Triton X-100	Butyl-PBD (0.8% w/v) in toluene 2 parts + Triton X-100 1 part
Solution counting:	
Dioxane–naphthalene	Butyl-PBD (0.8% w/v) + naphthalene (10.4% w/v) in dioxane
Combustion Techniques:	
Packard 305	CO_2 absorbed in a mixture with final composition, ethanolamine:methanol:butyl-PBD (2.0% w/v) in toluene (4:9:6).
Kalberer–Rutschmann	Ethanolamine:methanol:butyl-PBD (0.8% w/v) in toluene (3:22:25)

The emulsion counting system was essentially that described by Turner,[2] and the solution counting system was that described by Werbin et al.[3] The Packard 305 is a semi-automatic flask combustion system which can be operated in either the Hydrogen-3 or the Hydrogen-3 plus Carbon-14 mode, thereby allowing the separation of the two isotopes prior to counting. The second combustion system was essentially that described by Kalberer and Rutschmann,[4] with minor modifications. The scintillation medium for each of these systems had the composition given in Table 2.

Counting procedures

Samples were counted in either a refrigerated Philips Liquid Scintillation Analyser, refrigerated Packards (models 3340 or 3003) or in room temperature Packards (models 2240). All samples were allowed to reach temperature equilibrium in the counters before the Carbon-14 activity was measured. Samples spiked with ^{14}C-hexadecane (50 μl) were recounted using the same instrument settings.

Categorising results for analysis

The biological samples were counted for from 10 to 100 min and the spiked samples for 1 or 10 min. Although the accuracy of the d.p.m. calculations depends on the net counts accumulated, the variations in calculated d.p.m. caused by the different times used in this study were insignificant. For uniformity, the d.p.m. were calculated from the maximum counts accumulated in any one counting operation. Assuming that the prepared samples were stable, neither the type of instrument used nor the temperature of the counting chamber should affect the d.p.m. calculation. In almost all instances, after examining the prepared samples, these variables could be ignored in categorising the results for analysis.

The most important aspects of the study which affected the d.p.m. calculations were the actual methods of sample preparation. These were categorised for the analysis after due consideration of the physical principles involved and a knowledge of their possible effects.

Thus, all investigators who used the toluene–Triton X-100 system used 10.0 ml of scintillation fluid, but the total aqueous volume ranged from 0.025 to 4.0 ml. The physical properties of this emulsion system were so clearly defined that the prepared samples were readily sub-divided into the categories listed in Table 3.

Table 3. The analysis of urine and plasma samples using the toluene– Triton X-100 system.

Category	Total aqueous volume added to 10 ml scintillation fluid
T-T, 1	Less than 0.25 ml
T-T, 2	From 0.5 to 1.0 ml
T-T, 3	4.0 ml
T-T, 4	4.0 ml of L.P.
T-T, 5	as T-T, 2, but no b.g. subtract for L.P. and L.U.

A number of variations in the use of the dioxane– naphthalene (Table 4) system were considered for categorisation. However, the results obtained using 12.5% aqueous volume in a total of from 14 to 20 ml at refrigerated or ambient temperatures were statistically identical to those obtained using 5% aqueous volume at ambient temperature. The addition of fumed silica (Cabosil, Degussa Aerosil) was also without effect. All these methods are listed as D - N, 1. The only misuse of the method was in the failure by one person to add sufficient water to prevent crystallisation in the 0 - 4° C machines. This method was therefore considered separately (D - N, 2). The range of sample volumes used in the two combustion systems is also given in Table 4. Both systems were relatively free from obvious investigator error. Only one investigator used the Kalberer - Rutschmann method, and because of the obvious differences in the absorption systems, his results were considered separately (C, 2).

Table 4. The analysis of urine and plasma samples using the dioxane-naphthalene system and the combustion techniques.

Category	% Aqueous volume and conditions		
D–N, 1	12.5% at 0°C – room temp.	)	± fumed silica
	5% at room temp.	)	
D–N, 2	5% at 0–4°C		
C, 1	Packard 305	)	0.02 to 0.50 ml
C, 2	Kalberer–Rutschmann	)	on low ash paper

RESULTS AND DISCUSSION

Statistical analysis of the results

All the individual figures cannot be shown. A summary of results for each of the nine categories of sample preparation is given in Table 5.

The first line of figures in each compartment (a) gives the number of investigators, the number of batches and the total number of samples.

The second line (b) gives the calculated mean specific activity (d.p.m. Carbon-14/ml) and an asterisk indicates that it is significantly different ($P < 0.05$) from the overall mean at the foot of the table.

The third line (c) gives the 'within person' coefficient of variation (C.V. %). This is a measure of the repeatability of the method on that sample when used by one person. It is the standard deviation of the observations expressed as a percentage of the mean, where the standard deviation is based only on the variation encountered when one person makes repeat determinations.

The fourth line (d) gives the 'between person' C.V. %. This is a measure of the reproducibility with which different people using the same method on one sample can obtain similar results. It is calculated in the same way as the previous figure, except that the standard deviation is based on the variations encountered when different individuals make repeat determinations. An asterisk after this figure indicates that the value is significantly greater ($P < 0.05$) than the previous one; this shows that there are significant differences between people using the same method.

A question mark in each compartment indicates that the values were obtained from relatively little data and should be interpreted with care.

A visual examination of the prepared samples in the counting vials, and of the raw counting data obtained from them, showed that certain results listed in Table 5 could be sensibly ignored in the calculation of the overall results. Thus the results from method T - T, 4 on L.P. and D - N, 2 on H.U. could be discarded because of gross inhomogeneity of the samples, resulting in very unstable and irreproducible counting rates. Also, the results from T - T, 5 could be eliminated because of failure to subtract background counts.

The statistical effect of discarding these obviously unreliable results was a minor one, the result being to slightly raise the overall mean specific activity and to slightly lower both of the C.V. %'s. The high overall 'between person' C.V. % was therefore caused by less obvious errors than these.

Individual and method differences

The mean results obtained by each individual were tested to see if they were statistically different from the overall mean for that sample. Of the 22 investigators in the study, 10 produced results on all 4 samples which were indistinguishable from the overall means. These 10 investigators used mainly methods T - T, 1,

T - T, 2 and D - N, 1.

Table 5. Mean specific activities and coefficients of variation for each sample and each method.

Method		Sample			
		L.P.	H.P.	L.U.	H.U.
T–T, 1	(a)		4–5–24		6–6–24
	(b)	–	39, 085*	–	50, 652
	(c)		2.0		1.9
	(d)		3.5*		2.3
T–T, 2	(a)	9–15–71	6–8–41	9–11–50	7–8–44
	(b)	1405*	39, 218*	546	52, 536*
	(c)	5.2	1.7	3.6	2.1
	(d)	13.3*	2.6*	5.9*	4.1*
T–T, 3	(a)	3–3–6	4–6–15	4–4–9	1–1–2
	(b)	1462*	30, 924*	440*	39, 800*
	(c)	? 3.7	3.6	10.5	? 4.2
	(d)	3.9	10.4	16.7	–
T–T, 4	(a)	1–1–3			
	(b)	832*	–	–	–
	(c)	3.0			
	(d)	–			
T–T, 5	(a)	1–1–2		1–2–8	
	(b)	1481*		629*	
	(c)	0.9	–	? 3.1	–
	(d)	–		28.3*	
D–N, 1	(a)	5–8–20	6–7–19	7–8–22	7–8–28
	(b)	1532	37, 697	550	51, 131
	(c)	1.4	3.0	3.4	4.5
	(d)	4.8*	13.9*	18.3*	8.5*
D–N, 2	(a)				1–1–4
	(b)				? 42, 875*
	(c)	–	–	–	12.2
	(d)				–
C, 1	(a)	5–5–17	5–6–14	3–3–7	3–3–6
	(b)	1230*	34, 386*	413*	46, 034*
	(c)	6.0	12.1	? 6.9	? 1.7
	(d)	16.5*	14.3	14.6	11.2*
C, 2	(a)	1–1–3			
	(b)	1422*			
	(c)	? 0.4	–	–	–
	(d)	–			
Overall	(b)	1367	37, 235	535	50, 868
	(c)	4.9	4.3	4.6	3.6
	(d)	12.2*	10.2*	14.4*	6.6*

(a) = No. of investigators — batches — samples.
(b) = Mean specific activity (d.p.m. Carbon-14/ml).
(c) = Within person C.V.%.
(d) = Between person C.V.%.
? = Few data available, interpret with care.
* on line (b) means that the result is significantly different ($P<0.05$) from overall values.
* on line (d) means that this value is significantly greater than result in line (c).

However, 8 other investigators each produced 3 or 4 results which were significantly different from the overall means ($P < 0.05$), and they used mainly methods T - T, 3, D - N, 1, C, 1 and to a lesser extent T - T, 2. As can be seen from Table 5, these methods produced the highest 'between person' C.V. %, although it is important to note that method D - N, 1 was the only method to give mean specific activities for all 4 samples which were indistinguishable from the overall means. After careful examination of the raw data and discussion with the participants it was decided that in most, but not all, cases it was the analytical systems employed, rather than the manipulative techniques of the individuals, which were at fault.

Examination of raw data from emulsion counting

Method T - T, 1. This method of sample preparation was only used with the high activity samples, Table 5. It was differentiated from T - T, 2 because it had been observed in preliminary studies that the use of small volumes of water could result in a separation of droplets containing water-soluble salts which assayed with a different efficiency to solvent-soluble materials. It was hoped that the use of highly water-soluble ^{14}C-Inderal metabolites and ^{14}C-54, 594 would amplify this point. In the event, the mean result for H. P. was significantly above the overall mean, and that for H. U. was indistinguishable from the mean. Only with the high activity urine sample did different people obtain essentially the same result.

Method T - T, 2. This was the most commonly used method. Only with the low activity urine sample did the mean specific activity agree with the overall mean (Table 5); the results from the other three samples were all significantly higher than the overall mean.

On a number of occasions, the samples in a batch showed marked differences in optical properties, either before or after spiking. This was very noticeable with diluted plasma samples, and counting efficiencies were found to range from 33 to 42%. These changes in physical properties of the samples were ignored, and there resulted a wide range of calculated specific activities.

In contrast, the use of 1.0 ml undiluted low activity plasma occasionally gave rise to an unexpectedly high and constant counting efficiency of from 61 to 62%. This resulted in a relatively low specific activity. These observations were disturbing because they showed that presumably inaccurate results could be obtained with good repeatability.

Method T - T, 3. This method was used on all samples. In the majority of cases the low activity samples were diluted at least four-fold and the high activity samples forty-fold, as part of the procedure of adding 4.0 ml aqueous phase to 10.00 ml scintillator.

With the low activity urine samples, three participants used 4.0 ml undiluted urine. The efficiency of one of these batches of samples was only 9% due to inadequate channel settings, but the result was acceptable.

The mean results by this method for H.P., L.U., and H.U. (Table 5) are 18 – 24% below the overall average values. An examination of the raw data showed that the counting efficiencies were quite reproducible and that it was variability in the initial 10-min count rate which was responsible for the large 'within person' coefficient of variation, i.e. the lack of repeatability.

In general, in all categories of emulsion counting, the high 'between person' C.V.% was due primarily to instability of samples. This resulted in unstable sample counts and variable counting efficiencies, which occurred in at least 30% of the prepared batches. In not one instance did an investigator assay different dilutions of sample to see if he would obtain the same result.

Examination of raw data from solution counting

Method D - N, 1. This was the second most common method used. The mean results for each of the four samples were indistinguishable from the overall means (Table 5), but as with most other assay methods the 'between person' C.V.% was relatively high. The 'within person' C.V.% was similar to that of method T - T, 2, and further examination of individuals' raw data was undertaken to try to explain the differences between people.

The method was used by some of the least experienced personnel who, in retrospect, used incorrect pipetting techniques. Also they spiked only one sample in each batch, even though the count rate of the samples varied by up to 8%. Moreover, one investigator failed to completely dissolve the hexadecane spike in the dioxane scintillator, thereby obtaining a false value for the counting efficiency.

These problems apart, the method gave very good reproducibility. Only one person assayed a range of volumes for each sample. The average specific activities were all very close to the overall means and the standard deviations of the results were as low as any. This illustrates that in this system, even with varying sample volumes, the repeatability is high.

Therefore, the large 'between person' C.V.% observed in solution counting was caused primarily by incorrect sample preparation due to poor training.

Examination of raw data from combustion counting

Method C, 1. This was the third most popular method, and as shown in Table 5 it was used for all samples. Not only are the mean values for each of the four samples significantly lower than the overall means (24% for L.U.), but the 'within person' and the 'between person' coefficients of variation are excessively high for almost all samples.

Although combustion to CO_2 and H_2O is hailed by many as the only reliable method for biological samples, many combustion systems leave much to be desired. The repeated blocking of the gas lines leading from the combustion vessel, with carbonaceous material, is ample proof that many combustions are incomplete.

On some occasions, very high blank values were noted when non-labelled samples were combusted immediately after a labelled sample; in these cases, carry over was equivalent to 5% of the Carbon-14 count of the previously combusted sample.

Although the machine did occasionally meter a variable volume of liquid into different vials, this should not seriously affect the d.p.m. calculations when, as was the case in this study, all samples were spiked.

Therefore, the gross variability and the overall low values were due to :

(a) the inadequate combustion of many samples, and
(b) the incomplete transfer and collection of combustion products.

Method C, 2. Only one individual used this method, and obtained the highest specific activity and the lowest 'within person' C.V.% for all the combusted samples of L.P. Such a result, in contrast to method C, 1, is probably due to complete dissolution of combustion products in the scintillator, whether they were fully oxidised to CO_2 or not.

SUMMARY

For each sample, it has been shown that there are highly significant differences between methods ($P<0.005$) and between people using the same method ($P<0.005$). The particular differences vary from sample to sample.

Methods which rely on the solubility of the aqueous sample in dioxane gave results for each of the four samples which were indistinguishable from the overall means. The majority of the results obtained with toluene– Triton X-100 after the addition of no more than 10% aqueous solution were significantly higher than the overall means, whereas when the water content was greater than 10% the majority of the results were significantly lower.

Variations in solution counting were chiefly due to inaccurate pipetting, leading to large 'between person' coefficients of variation, whereas variations in most of the emulsion systems were due largely to sample instability under counting conditions.

The combustion of samples by the Packard 305 automatic system gave very unreliable results in that, besides giving results well below the overall means, the machine produced the largest 'within person' and 'between person' coefficients of variation of any of the methods tried. The Kalberer – Rutschmann variation appeared to be the most reliable of the two combustion procedures at present available.

On the whole, these results are very disturbing, and are all the more difficult to assess because no absolute standard exists. It is probable that the most reliable and acceptable method is that which gives the smallest 'within person' and 'between person' coefficients of variation.

Table 6. Reliability of methods.

Method	No. of batches assayed	% of batches giving mean d.p.m. statistically indistinguishable from overall means	Reliability of method
T–T, 1	11	100	good
T–T, 2	42	95	good
D–N, 1	31	77	good
T–T, 5	3	66	bad
T–T, 3	14	50	bad
C, 1	17	30	bad
C, 2	1	0	good
T–T, 4	1	0	bad
D–N, 2	1	0	bad

As shown in Table 6, and when the above considerations are taken into account, methods T - T, 1, T - T, 2 and D - N, 1 are to be preferred. Although we always set ourselves the highest standards for accuracy and reliability in our experiments, it seems that, as carried out above, all leave something to be desired.

ACKNOWLEDGEMENTS

I would like to thank all those who contributed to this study. I am also grateful to Mr. J.A. Lewis for the statistical analysis of the results.

REFERENCES

1 W.A. Cope, and J.A. Double, in Liquid Scintillation Counting, Vol. 2, (Ed. M.A. Crook, P. Johnson and B. Scales), Heyden and Son, London, 1972, p.223.

2 J.C. Turner, Intern. J. Applied Radiation and Isotopes **19**, 557 (1968).

3 W. Werbin, I.L. Chaikoff and M.R. Imada, Proc. Soc. Exp. Biol. and Med. **102**, 8 (1959).

4 F. Kalberer and J. Rutschmann, Helv. Chim. Acta **44**, 1956 (1961).

DISCUSSION

F.R. Jacobsberg: I presume scintillation counting is still carried out at ICI Pharmaceuticals. What are your own conclusions regarding techniques?

B. Scales: Yes, we continue to do a great deal of liquid scintillation counting. In my talk I have painted the very worst picture, but this extreme case can exist. Our problem is possibly due to the fact that we do not offer a radioisotope analytical service, but let everyone loose on our many machines, each person taking full responsibility for his own results. One possible outcome of this is shown in the above results and we are attempting to alter this situation. My own beliefs are that when assaying any sample, replicates should be assayed, either by using different volumes of sample or by using different scintillation systems. In these circumstances, the agreement of replicate specific activities is indicative of a reasonably reliable analysis.

Chapter 19

Micellar Scintillators: a Rational Approach to the Design of Stable Assay Solvents for Liquid Scintillation Counting

M.J. Ewer[a] and N.G.L. Harding

Postgraduate Medical School, Hills Road, Cambridge, England

Liquid scintillation counting involves the efficient transfer, in solution, of emission energy from a decaying isotope to a transducing molecule and from thence, as photons, to a quantitative photomultiplier system. With the increasing use of liquid scintillation techniques there has been a parallel demand for solvents able to accommodate appreciable amounts of widely differing samples. Contemporary samples often contain low isotopic activity above background. Ideal solvents should therefore have the following properties:

1. Large volumetric capacity for aqueous samples.
2. Low background count-rates at the emission energies of the commonly used beta-emitting isotopes.
3. Temperature-dependent noise should be minimal.
4. Stable background count-rate with increasing sample volume.
5. Minimal change in counting efficiency as sample volume increases.
6. Resistance to quenching effects of a wide range of samples.
7. Dose-response curves suitable for determining counting efficiency by both internal and external standardisation techniques.
8. The light output should be a function of the energy of the sample isotope.
9. No chemical reaction between sample and scintillator or solvent molecules.

Despite the abundance of information about the scintillation properties of organic solvents and solutes in aqueous and non-aqueous environments, there is no coherent theory which will permit both the rational design of 'ideal' scintillation solvents and explain their function. Accordingly, some years ago we initiated a programme broadly designed to examine theories upon which we could base design and predict function of systems possessing features 1 to 9 above. One practical benefit envisaged was that one type of solution would cover most laboratory requirements. In the present communication we briefly review the evolution of this programme; specific details of the several aspects of the work will be reported elsewhere.

Before discussing the development of micellar scintillators, some clarification

---

[a] On leave of absence from Nuclear Enterprises Ltd.

of terms is necessary. 'Scintillator' refers to solute molecules able to transduce radioactive decay energy to photons. This term often loosely refers to a complete system of solvents and solutes to which radioactive samples are added for assay. Such cocktails are referred to as 'scintillants' or 'liquid scintillators'. 'Fluor' is a loose term usefully describing molecules which emit photons in response to radioactive and non-radioactive stimuli such as ultraviolet or visible light, chemical or thermal stimuli. A micelle is an aggregate of molecules bound together in an ordered structure by non-covalent forces. Micellogenic molecules referred to in this review generally consist of two regions containing three parts: an hydrophilic region comprised of charged and hydrophilic moieties, and hydrophobic areas built from hydrocarbons.

EVOLUTION OF SCINTILLANTS

Several transducing systems have evolved for converting decay energy to photons with usable efficiency. The simplest preparations consist of aromatic solvents (toluene, xylene; about 9 molar) containing primary scintillators (about 25 mM) and sometimes a secondary wavelength emitter (about 0.5 mM). Energy from the beta-particle elevates some solvent molecules to their lowest excited pi-electron singlet state. These molecules are among the 350 solvent molecules per molecule of primary solute, and can translocate their energy to excite the latter into a similar state. Excited primary solute molecules emit fluorescence photons for detection by the photomultipliers of the electronic counter.

Reduced to its simplest terms the fundamental problem, then, is to incorporate water containing various solutes into toluene which contains the fluors. There are several milestones in the history of attempts to solve this problem. First, there has been extensive use of short-chain hydrophilic solvents such as 1,4-dioxane,[1,2] ethanol[4-7] or mixtures[8] admixed with toluene. In some, the freezing point is depressed by polyethers.[1,3,9] For many purposes these solvents are still satisfactory but they suffer the disadvantages of restricted sample capacity, chemiluminescence and sometimes low efficiency. The former difficulty is inherent in the type of solvent used whilst the latter effects can be partially offset by careful purification and storage of solvents. Also in this group, long-chain hydrophilic solutes such as Triton are effective but are chemiluminescent and unsuitable for assay by external standard techniques.[10]

In a second approach, particles of sample are gelled or suspended in the solvent by aluminium stearate,[11-13] Thixcin[11,14-16] and silica.[17,18] Apart from difficulties in sample preparation, self-quench is appreciable when particle diameters approximate to the range of the beta-particle. Thirdly, samples have been absorbed by plain or derivatised cellulose or glass-fibre papers, desiccated and counted. Such membrane techniques often have low efficiency for tritium and the assay is dependent upon orientation of the strip in the counting vial.

A central factor is that the efficiency with which a fluor emits photons depends upon its environment. This efficiency is a complex function, containing statements about the solute concentration, efficiency of transfer of energy from solvent to solute, fraction of the excited solute molecules emitting photons, efficiency of transmission of photons by the solvent, and match between the energy of the emitted fluorescence photons and spectral sensitivity of the photomultiplier system. For some fluors environmental effects are very important. For example, emission from fluorescein is pH-dependent; the quantum yield from aniline derivatives increases markedly as the molecules move to non-polar environments. Therefore prime requisites for scintillators are that the environment of the molecule should possess excellent energy transfer characteristics yet should also be correctly modulated by changes in sample composition so that existing methods of internal or external standardisation can be used. In this way the fluor must be bifunctional, signalling not only the radioactive content of a sample but also reporting any quenching effects arising from that addition. At the outset, therefore, it was important to examine micellar systems both for their ability to monitor sample isotope efficiently and for a faithful response to systems of external standardisation.

APPLICATION OF MICELLAR THEORY TO THE SCINTILLATION PROCESS

It would be desirable to remove contaminants such as salts, protein or solvents prior to assay of the sample isotope. This would minimise quenching effects of inert molecules upon the scintillation process. Could such a separation be achieved in the liquid scintillator itself? A solution with an ordered structure ideally is required for presenting the isotope, at intermolecular distances, to the scintillator. At the same time it would be beneficial to place any irrelevant, but quenching, molecules in a region of the solution structure where quench effects would be minimal.

Davson—Danielli membrane models[20] offer structures with some of the desired

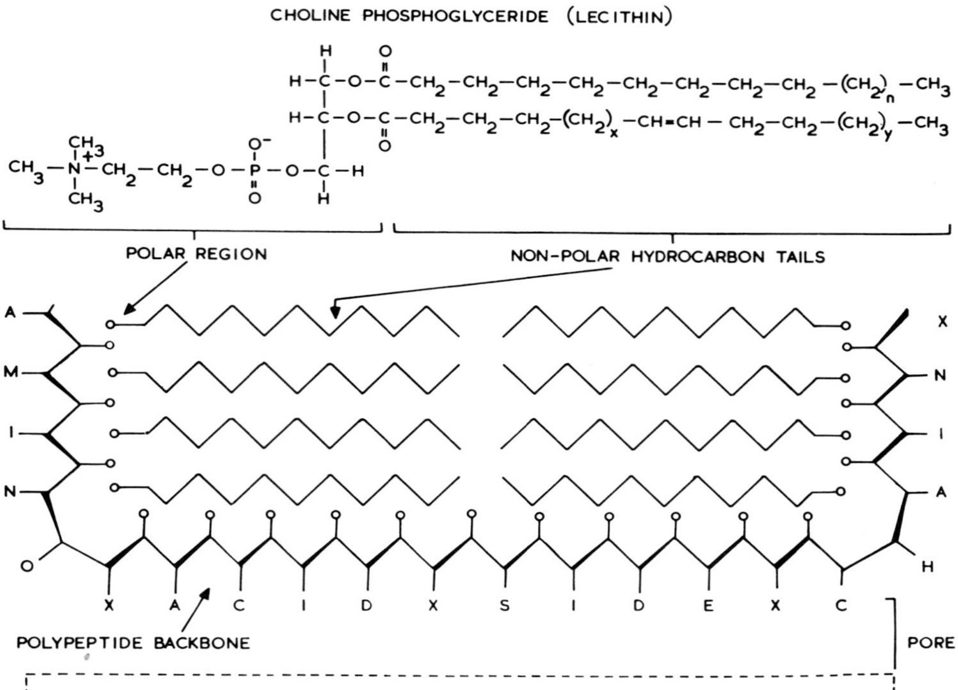

Fig. 1. A Davson-Danielli model of membrane structure. Though various models of membrane structure have been proposed, this diagram illustrates the features pertinent to the development of micellar scintillators. The hydrophobic chains of lecithin molecules are aligned to form a liquid hydrocarbon core of the bilaminar structure. The polar regions have formed a coherent structure with the backbone and side chains of a structural protein. The pore represents a boundary in the membrane, analogous to the periphery of the micelle in aqueous micellar systems. The number of carbon atoms in each chain can vary as indicated.

properties. Figure 1 illustrates the model in terms of structural principles underlying this programme. Choline phosphoglyceride possesses two hydrocarbon 'tails' of varying length depending upon the source of the molecule. The ethanolamine and phosphate moieties contribute both an hydrophilic character and zwitterionic charge to that region of the molecule. The hydrocarbon tails are held together by hydrophobic interactions which may total several kcal binding energy per mole. Such forces adequately maintain the hydrocarbon chains in a relatively constant and hydrophobic environment, which seems ideal for the scintillator molecule.

At the periphery of the structure the hydrophilic and zwitterionic regions are free to interact with, and hence form relatively stable (but probably dynamic) structures with charged molecules present in the aqueous phase of the sample. In Fig. 1 a protein molecule is shown performing this function but the protein could be replaced in the model by organic or inorganic structures. Into this ordered structure it is necessary to insert the extra molecules required to render the system suitable for use as a liquid scintillator.

SIMPLE MICELLAR SCINTILLATORS

We have indicated that the properties of the micelle depend upon the structure of the hydrocarbon region and the hydrophilic charged regions of the molecules forming the micellar skeleton. At the outset of this work we had no basis for predicting which structures would yield micellogenic molecules suitable for our application. A simple system was therefore examined in order to ascertain whether the principle was valid and to gain some insight into the structural aspects of the problem.

Particular attention has been given to the properties of micelles formed in aqueous systems. For several reasons, however, we wished to examine 'inverted' micelles which form in non-aqueous solvents. The lecithin–benzene[21] system was the obvious starting point.

A first complication was that unless the lecithin was rigorously purified there was a progressive rise in the background count-rate (tritium window). Even with purified material, oxidative yellowing was associated with an increased count-rate and seemed to account for much of the quenching observed. The progressive rise in background counts could be eliminated by treatment with borohydride under photolysing conditions. Despite the increasing background count-rate of the untreated system (Fig. 2, left-hand panel) there was little change in the titration curve relating lecithin concentration to the external standard ratio (Fig. 2, right-hand panel). In dissociating the rise in background count from the origin of the quench, the latter profile confirmed the high degree of purity required for efficient function of the system. Also, this type of study, together with aqueous titration data, indicated that the zwitterionic configuration may not be an optimal charge configuration for the hydrophilic region of the molecule. An interesting feature was that addition of water and scintillator components to the lecithin—toluene solution (Fig. 3) decreased the background count rate and marginally improved the quenching characteristics (Fig. 3, right-hand panel).

Micellar structure in the lecithin—benzene system depends upon the lecithin concentration[22] so it was important to determine the effects of micelle size upon the counting characteristics of the system. In the ternary system (water—lecithin—toluene), Fig. 4 implies that the counting of tritiated water best occurs when aggregates of large micelles are present. A feature of this system is that the micelles required ultrasonic energy for reasonably rapid production (left-hand panel), and that if produced in this way the ternary micelle required appreciable

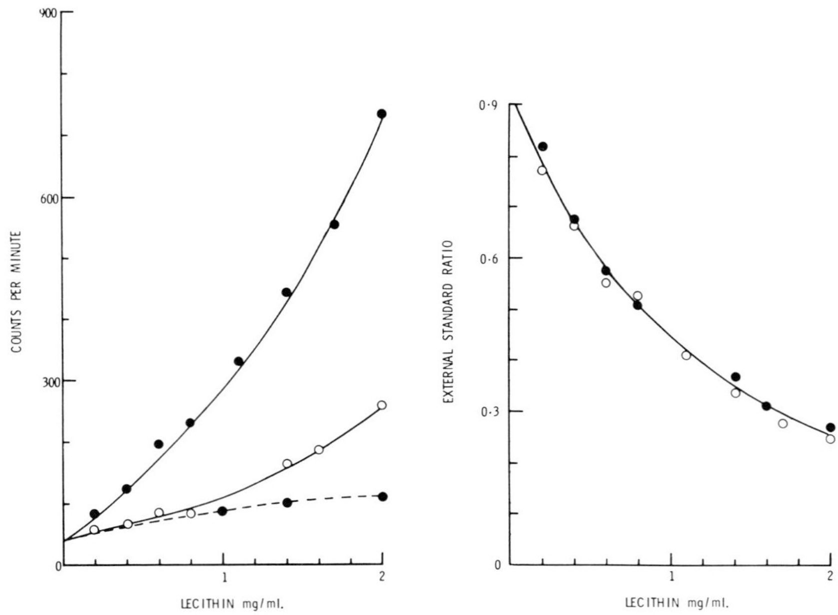

Fig.2. The instability of lecithin associated with background count-rate in hydrocarbon micelles. Vials contained 10 ml toluene, 85 mg butyl-PBD and stated amounts of lecithin. A'wide' tritium window was used to obtain the background count-rate in a Nuclear Enterprises 8312 Scintillation Counter (—●—●—). Vials were left for 24 h under aerobic conditions at room temperature in the dark and recounted (—○—○—). To vials, 25 mg potassium borohydride was added and the lecithin degradation products reduced for 24 h prior to adding butyl-PBD and counting (---●--●---). The latter vials are not shown on the external standard profile, but gave a similar response curve.

time to stabilise. The slight second peak at 10 h indicates that there may be an intermediate heterogeneous micelle population able to equilibrate more quickly than other classes of micelle. Stabilisation was marked by a gradual increase in count-rate and transition in the external standard ratio profiles from a rather flat response to a useful linear profile (right-hand panel).

In the binary (lecithin—toluene) system, the amount of water imbibed depends upon both the nature and concentration of the micellogenic molecule and upon the organic solvent used. As water is progressively added an end-point is reached marked by formation of macroscopic aggregates. Solution structure at this point is not precisely identified but for convenience may be viewed as two phases. The upper phase, rich in organic solvent, is often of good optical clarity. The lower phase, however, contains opaque hydrated aggregates of lecithin. We would therefore expect both phases to show different counting characteristics.

A common finding in later experiments has been that the end-point in many binary systems is preceded by formation of a clear gel. Some of these gels are unstable and subject to separation of as many as four phases so that in practise the end-point is less clearly identified in such cases. Consequently the term 'gelation—precipitation point' is used to describe a region in which phase separation occurs.

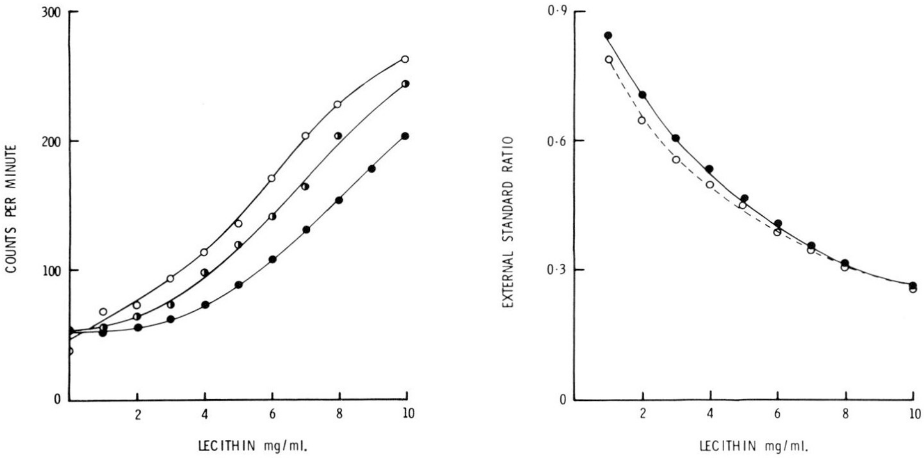

Fig.3. Effects of scintillator and water upon the background count-rate of the lecithin–toluene micellar system. The composition of vials is given in the legend to Fig.2, except that 200 µl water was added (—○—○—) to vials containing butyl-PBD and lecithin (—●—●—). The latter were prepared by adding 85 mg butyl-PBD to each vial after the first assay (—◑—◑—). Counting conditions are as for Fig.2.

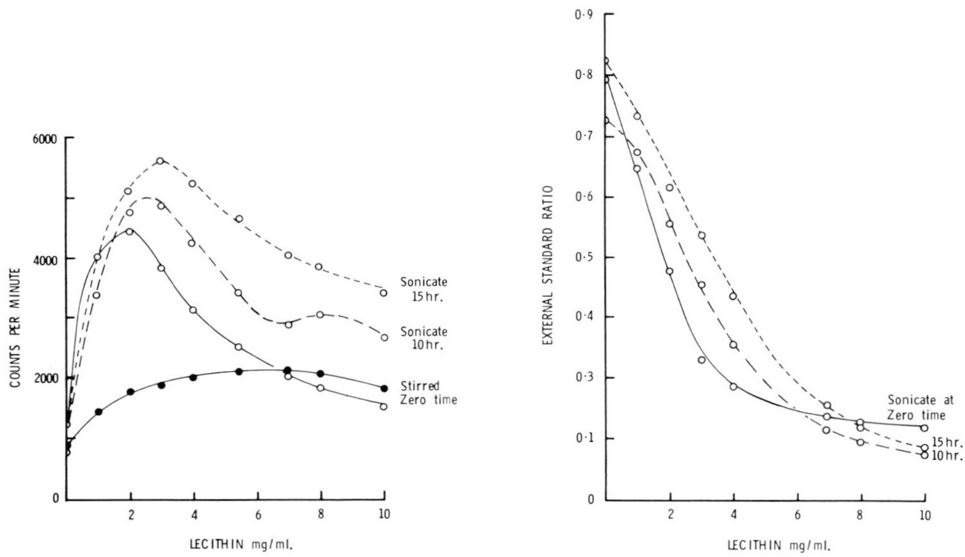

Fig.4. Efficiency of counting tritiated water in ternary lecithin micelles produced by ultrasonic energy or mechanical stirring. The composition of vials is given in the legend to Fig.2, except that each vial contained 3.6 x 10^4 d.p.m. tritium as tritiated water (100µl). After adding the isotope, vials were stirred mechanically for 30 min and counted (—●—●—). Duplicate vials were sonicated at 28 kc/s, probe amplitude 5µ for 1 min, and counted immediately (—○—○—), at 10 h (–––○–––○–––) and finally at 15 h (---○---○---). A wide tritium window was used in the 8312 Counter.

Figure 5 confirms that both phases have different counting characteristics and that optimal counting in each phase occurs at different lecithin concentrations of the initial-binary solution as might be expected from partition effects. A prime feature of these experiments is that at the gelation—precipitation point the sum of

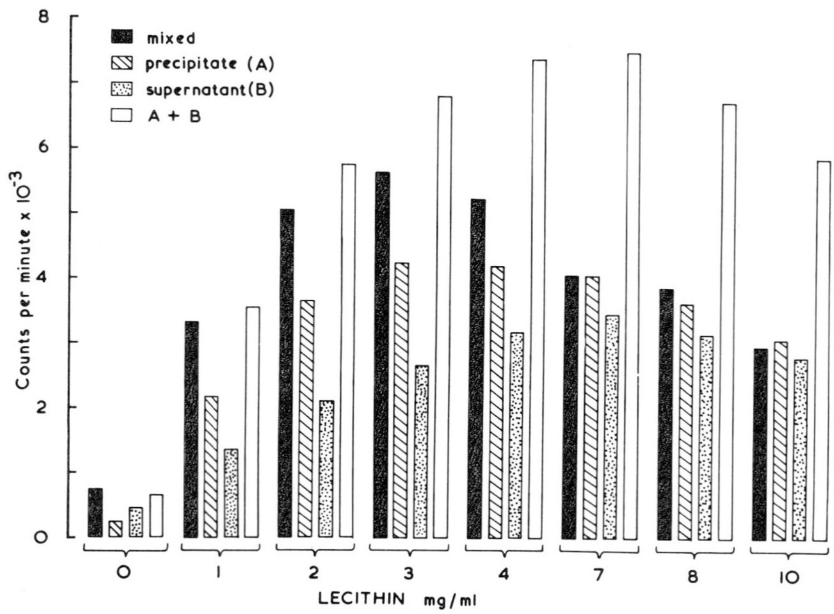

Fig.5. Partition effects and aberrant counting at the gelation—precipitation point. Vials similar to those described in the legend to Fig.4 were titrated with unlabelled water until the end-point was reached (see text). Each vial was thoroughly mixed and assayed (= 'mixed'). Phases were then separated under gravity for 20 h. The supernatant and precipitate of each vial was assayed separately as shown, and the separate count-rates summed (A+B).

the counts in the two individually assayed phases yields improved counting efficiency when compared with assays of the physically admixed phases. Figure 5 shows that there can be as much as 75% difference when the system is evaluated in this way.

In addition to demonstrating the validity of the concept, experiments such as these showed several encouraging features:

1. Aqueous samples could be assayed efficiently in a micellar scintillant.
2. The background count-rate could be eliminated by purification of the lecithin.
3. Acceptable dose-response profiles were obtained by external standardisation techniques.
4. Profiles obtained by external standard techniques were in good accord with those obtained by internal techniques using either tritiated water or toluene.
5. Information was obtained about the molecules required for forming the necessary micellar structures in the solvent. Stability of lecithin was a particular limiting factor in the further development of this binary system. A second difficulty was that the lecithin structure possessed characteristics which required appreciable expenditure of energy in order to establish the ternary (water—lecithin—toluene) micelles.

6. Binding energies in the micelle were sufficient to sustain stable ternary structures. Destruction of the micellar structure occurred when the aqueous capacity of the system was exceeded. This was associated with phase separation and aberrant counting.

It seemed likely that two solvent structures occurred during conversion of the system from binary (lecithin—toluene) to ternary (water—lecithin—toluene) micellar modes by addition of an aqueous sample. Figure 6 illustrates the transition from a phased solution of toluene and water (panel A) to micellar mode. When lecithin is added to the upper (toluene) phase, the bifunctional lecithin molecules tend to arrange randomly (panel B) prior to formation of micelles (panel C). Admixture of water from the lower phase doubtless produces some hydrated lecithin molecules (panel B1), but is more likely to yield an ordered ternary micelle (panel C1). At present we know very little about the static or dynamic features of these micelles. Answers to such questions are important for the rational design of micellogenic molecules for use in future systems.

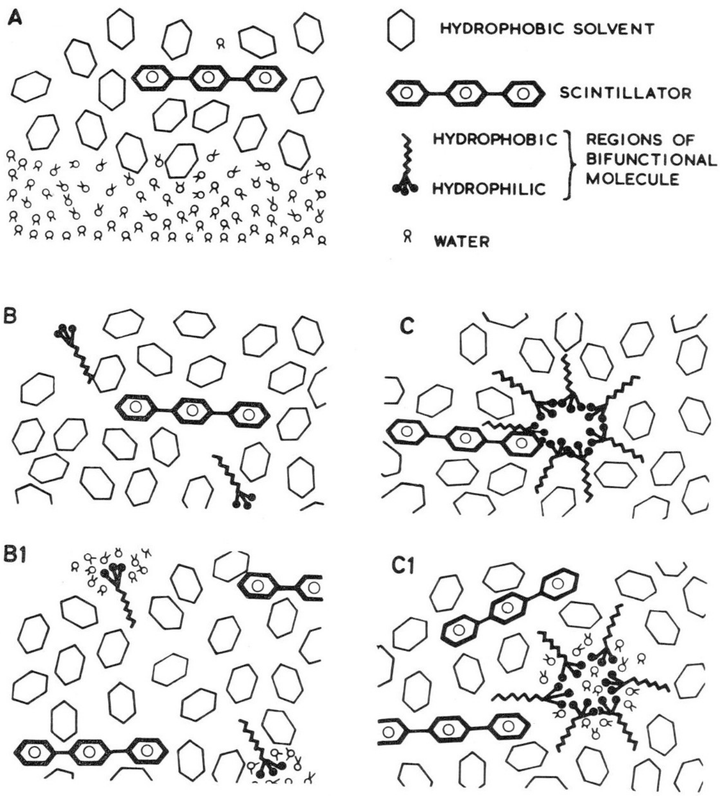

Fig. 6. Models illustrating the tendency of bifunctional micellogenic molecules to form binary and ternary micelles. The effects of micellongenic reagents upon a phased solution of toluene and water (A) are described in the text.

It was apparent from the lecithin—toluene system that in order to realise the potential of this approach it would be necessary to examine synthetic micellogenic reagents as replacements for the naturally occurring choline phosphoglyceride used in the first-generation experiments.

SECOND-GENERATION MICELLAR SCINTILLATORS

The initial problem was to dissect the lecithin molecule to determine the structural features required for optimal function in the ternary micellar system. Several questions arose: How many hydrocarbon chains are required per molecule and how long should the chains be? Are aromatic structures advantageous? Which hydrophilic groups are most appropriate? What type of charge, if any, is required at the hydrophilic terminus? These five questions gave rise to a wealth of possible experiments and organic syntheses, the number approaching Avogadro's Number in magnitude!

At the outset the experimental approach was therefore confined to studying simpler molecules and restricted to questions concerning the length of the hydrocarbon chain, the type of charge and requirements for the hydrophilic region of the molecule. As the ground was cleared we were able to examine other questions such as: What are the effects of aromaticity in the hydrocarbon moiety?

Findings from these experiments may be briefly summarised: In many situations a terminal cationic charge was particularly useful. Other solvent and sample conditions demanded charge of opposite polarity. A well-defined hydrophilic terminus is obligatory; the type of charge depends upon the samples to be added and the length of the hydrocarbon chain is to some extent dictated by the nature of the sample and the solvent of the organic phase.

Some of these features are illustrated by the data in Table 1. Here, alkylphosphate esters have been examined as micellar reagents in conjunction with different organic solvents. Similar data are obtained for certain cationic derivatives but the alkylphosphate esters are shown for they demonstrate interesting features when examined in fluorinated solvents.

In hydrocarbon solvents the major determinant of water capacity of the binary system is the structure of the organic solvent rather than the structure of the incoming water. In fluorocarbon solvents, however, there was greatly increased capacity for water, and the structure of the incoming water now contributes noticeably to the capacity of the system. Although data is not shown, structure of the alkylphosphate is also important for determining the aqueous solvent properties.

When the binary solutions are examined as scintillation reagents, hydrocarbon solvents produce high efficiency when assessed with either tritiated toluene or tritiated water. Although efficiencies are similar, there is perhaps some reduction in efficiency using tritiated water as the standard. The previous models (Fig. 6, panel B versus panel C1) indicate that there may be some less efficient energy transfer from tritium in the hydrated core of large micelles to scintillant molecules in the hydrophobic domain around such micelles.

Methyl substitution of the aromatic solvent decreases water capacity and increases counting efficiency. This is true for both the hydrocarbon and fluorocarbon series, though anomalous behaviour is observed with the latter as will be seen below. The basis for such effects may lie in non-covalent interactions between fluor, alkylphosphate esters and solvent, though this remains to be explored in detail. The ternary solutions have similar optical clarity in the visible range so an as yet unidentified modification of micellar structure, in turn dependent upon solvent structure, may be responsible for improved energy transfer in the system.

Fluorocarbon solvents show markedly improved water capacity when compared with hydrocarbon analogues (Table 1) but show appreciable quench for tritium. Similar reduction of efficiency for both standard tritiated toluene and tritiated water implies an appreciable impedance to energy transmission through the

Table 1. Relationship between hydrophilic properties and tritium counting efficiency in alkylphosphate micellar systems.

Solvent	End point by titration with ml		Background	Tritium added as				Ratio
				Toluene		Water		
	H_2O	D_2O	(c.p.m.)	A (c.p.m.)	Efficiency (%)	B (c.p.m.)	Efficiency (%)	B/A
ØH	0.25	0.22	50	9,557	22	47,625	23	4.98
ØMe	0.155	0.152	45	11,890	28	55,386	27	4.66
m$Ø Me_2$	0.071	0.067	50	12,645	30	54,553	27	4.31
ØF	1.2*	0.9*	38	2,246	5	12,501	6	5.57
$ØCF_3$	0.55	0.43	45	3,535	8	10,833	5	3.06
m$Ø(CF_3)_2$	T	T	48	4,789	11	2,615	1	0.55

Alkoxydecylphosphate esters were examined at a concentration of 75 g/litre in the solvents shown. Ø refers to the C_6H_5 structure except where prefixed by m, in which case it refers to the C_6H_4 nucleus in the meta-configuration. Me refers to methyl. Assays were performed with a wide tritium window in order to demonstrate the quenching effects of the fluorocarbon series upon the background count-rate. All solutions contained butyl-PBD 8.5g/litre. The end-point in titrations with water and deuterium oxide is the gelation—precipitation point of the system. T refers to an indefinite, turbid end-point occurring soon after starting the titration. * indicates that the titration end-point had not been reached. Efficiencies were determined in 7 ml solution containing water (first column) plus the stated isotope.

hydrophobic domain of these systems. An important exception to this seems to be trifluoromethylxylene which shows some eleven-fold greater counting efficiency for toluene compared with water at the end-point.

Several proposals can be advanced to explain such findings (Fig. 7). Stabilised layers, composed of scintillator- and toluene-rich hydrophobic regions, could be sited upon a scintillator-poor lamina. As shown in the diagram, the micelle may invert and produce increased light output despite the increased amount of water. Exploration is necessary to understand and perhaps take advantage of this curious phenomenon.

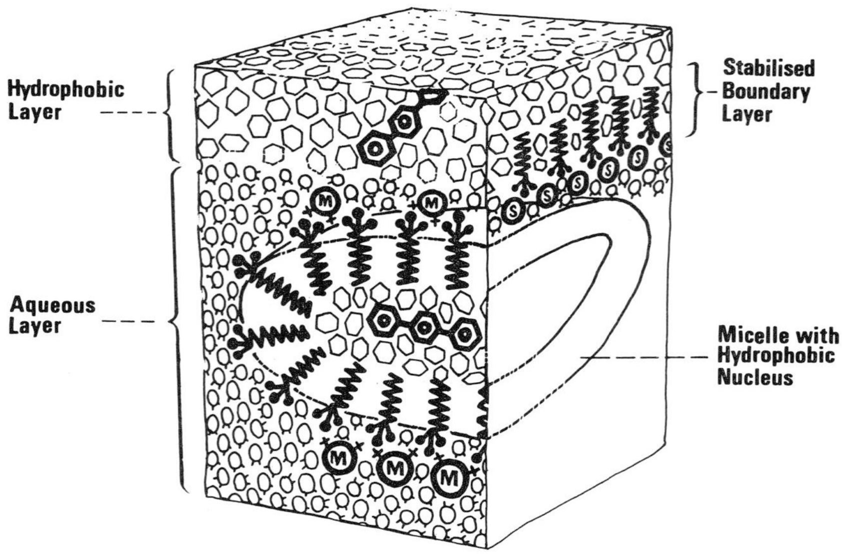

Fig. 7. Some stable structures of the ternary (water–lecithin–toluene) system at the gelation–precipitation point. An inverted micelle is shown with the hydrophilic charged regions stablised by a complementary cation (M) from the sample. At the boundary between the upper toluene and lower aqueous layers the micellogenic reagent is shown to be stabilised by molecules of high affinity (S) derived also from the sample. In both cases the scintillator is located in the toluene phase. Symbols are identical with those shown in Fig. 6.

CONTEMPORARY AND FUTURE MICELLAR SCINTILLATORS

During evolution of this programme it was evident that the micellar concept not only provided highly efficient scintillants with good capacity for a wide range of samples, but also offered a useful theoretical approach to the design of functional systems. Experiments of the type outlined above delineated the following important design principles.

The structure of the micelle is an important determinant of both counting efficiency and the fidelity of external standardisation. Binary and ternary micellar systems are relatively free from unwanted thermal noise components of the background count-rate. Macroscopic phase transitions are undesirable because of their association with anomalous counting and unreliable response to standardisation by both internal and external methods. To determine isotope dose with absolute veracity in such a situation would require an internal standard of isotope molecules identical with those in the sample. Kinetic characteristics of molecules forming the micelle matrix should be such that stable ternary micelles rapidly form on addition of sample. The scintillant must be stable for use and storage at room temperature.

A series of contemporary micellar scintillators has evolved[23] from these studies. The solutions show extensive tolerance to a wide range of samples produced in current biochemical and clinical practice. For example, protein solutions such as human plasma, or chaotropic reagents such as urea are absorbed and counted with ease. An important feature is that good efficiency is sustained as sample volume is increased. Figure 8 (left-hand panel) shows the titration profiles for plasma in a micellar scintillant compared with a commercial product currently available. Identical values of quench are obtained when tritiated water and tritiated toluene are used as internal standards. Hence the external standard can be used to advantage to monitor efficiency (Fig. 8, right-hand panel). Other samples seem to follow this example.

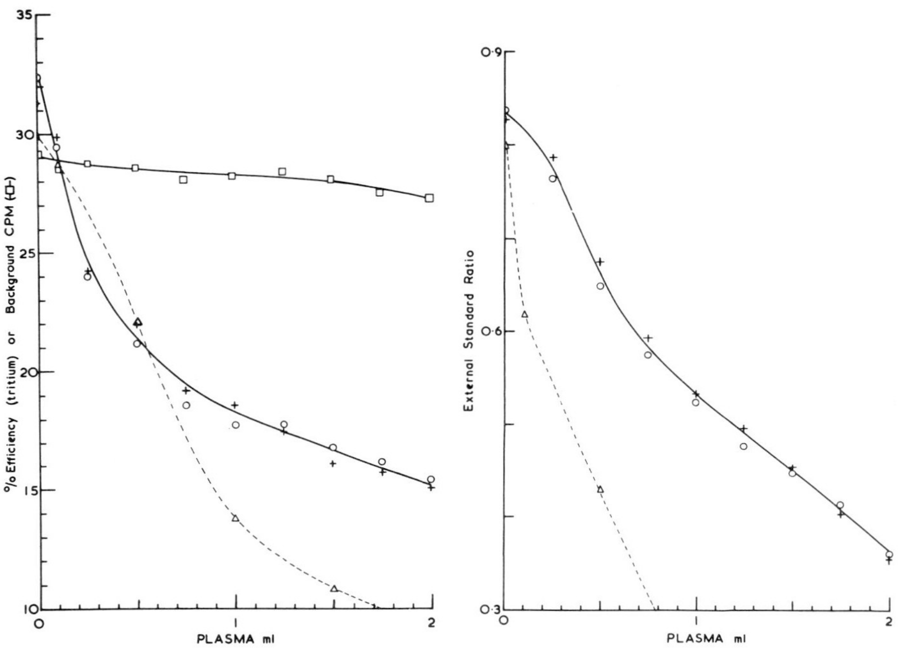

Fig. 8. The assay of human plasma in a micellar scintillant. The indicated volumes of fresh plasma were added to a series of vials containing 10 ml of a micellar scintillant[23] and efficiency determined by addition of standard tritiated water (0) or tritiated toluene (+), and the External Standard Ratio determined for each set of vials. Tritium window, ambient temperature, N.E. 8312 Counter. A currently available gel scintillator is shown for comparison (-----), the efficiency in this case being determined by tritiated toluene.

There seem to be several directions which future work may take. Some of the unexplained findings with experimental micellar scintillants need to be analysed in detail to improve our understanding of the structure and function of these systems. These avenues lead towards the design of micellogenic molecules possessing an intrinsic fluor region. At the same time a generation of Reticular Scintillants is envisaged. In these, the solution structure appears rather like a liquid crystal (Fig. 9) containing scintillators and isotopes in their traditional domains. Large sample capacity and high light output are attractive theoretical features of these structures.

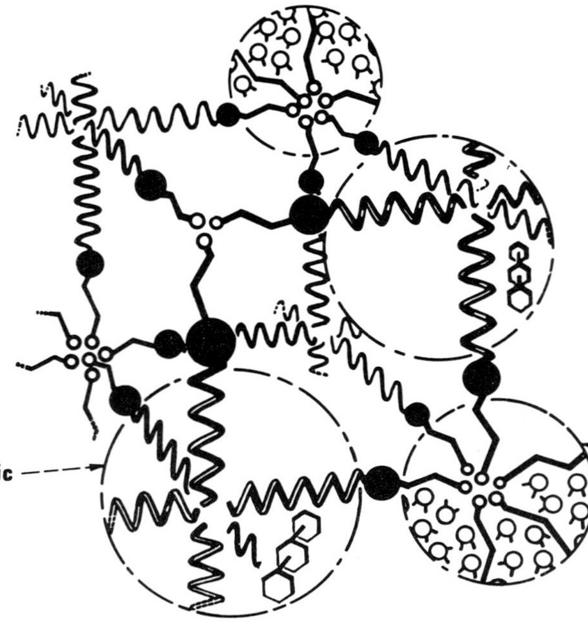

Fig. 9. A model of the solution structure of a Reticular Scintillant. Symbols are similar to those used previously (Fig. 6), except that the molecules of the lattice are shown with three defined regions rather than the two regions shown in the diagrams of the micellogenic reagents.

ACKNOWLEDGEMENTS

N.G.L.H. gratefully acknowledges financial support from the Medical Research Council. Drs. Jocelyn Dow and Derek Robinson are thanked for their critical discussions during the course of the work, and Dr. Philip Wraight is thanked for evaluating some of the micellar systems. N.G.L.H. and M.J.E. thank Nuclear Enterprises Ltd. for support of the programme.

REFERENCES

1 J.D. Davidson and P. Feigelson, Intern. J. Applied Radiation and Isotopes 2, 1 (1957).
2 M. Furst and H. Kallman, Physical Reviews **97**, 583 (1955).
3 J.D. Davidson in Liquid Scintillation Counting (Ed. C.G. Bell and F.N. Hayes), Pergamon Press, New York and London, 1958.
4 C.A. Ziegler, D.J. Chleck and J. Brinkerhoff, Analyt. Chem. **29**, 1774 (1957)
5 G.T. Okita, J. Spratt and G.V. LeRoy, Nucleonics **14**, 76 (1956).
6 J.C. Bateman, H.N. Carlton, R.C. Calvert, and G.E. Lindenblad, Intern. J. Applied Radiation and Isotopes **7**, 287 (1960).

7 F.N. Hayes and R.G. Gould, Science **117**, 480 (1953).
8 F.E. Kinard, Rev. Sci. Instr. 28, 293 (1957).
9 E.A. Peets, J.R. Florini and D.A. Buyske, Analyt. Chem. 32, 1465 (1960).
10 B.W. Fox, Intern. J. Applied Radiation and Isotopes **19**, 717 (1968).
11 S. Helf in Liquid Scintillation Counting (Ed. C.G. Bell and F.N. Hayes), Pergamon Press, New York and London, 1958.
12 B.L. Funt, Nucleonics **14**, 83 (1956).
13 B.L. Funt and A. Hetherington, Science **125**, 986 (1957).
14 C.G. White and S. Helf, Nucleonics **14**, 46 (1956).
15 S. Helf, C.G. White and R.N. Shelley, Analyt. Chem. 32, 238 (1960).
16 J. Shapira and W.H. Perkins, Science **131**, 414 (1960).
17 C.F. Gordon and A.L. Wolfe, Analyt. Chem. 32, 574 (1960).
18 D.G. Ott, C.R. Richmond, T.T. Trujillo and H. Foreman, Nucleonics **17**, 106 (1959).
19 T.B. Ryves, J. Sci. Instr. 37, 201 (1960).
20 H. Davson and J.F. Danielli, (Ed.), in The Permeability of Natural Membranes, Cambridge University Press, 1943.
21 P.H. Elworthy, J. Chem. Soc. 1951 (1959).
22 P.H. Elworthy, J. Chem. Soc. 813 (1959).
23 Data Sheet, Nuclear Enterprises NE 260.

DISCUSSION

C.J. Palais: External standard ratio was correlated to lecithin concentration. Did the authors also try to correlate this ratio to counting efficiency for the various concentrations? If so, what were the results?

N.G.L. Harding: As I have indicated, lecithin provided us with a means for exploring the scintillant properties of inverted ternary micelles. Tritium efficiency was between 15–25% depending upon the purity of the lecithin and mechanism by which the micelles were produced. If the lecithin was rigorously purified under anaerobic conditions, counting efficiency was independant of lecithin concentration in the working range. However, I have shown our profiles with the external standard values to emphasise first that quench can arise from instability of the micellogenic reagent, that this quench is concentration-dependant and, finally, that in the micellar system such quench can be competently quantified by the external standardisation technique. I should perhaps add that during our work we have been concerned to examine the response of micellar scintillants to the external standard for a couple of reasons. First, as an isotope user I am very conscious of the value in not having to dispense aliquots of standard calibrated isotopes and then recount one's vials. Secondly, as a simple-minded computer user I find that the less curvilinear are response profiles, the more easy it is to devise programs to analyse the data.

SHORT COMMUNICATION

Activity Determination of Phosphorus-32 Sources with High Solid Content

Eberhard Günther

Physikalisch-Technische Bundesanstalt, Braunschweig

Activation probes are frequently used for fast neutron flux density determinations. With the aid of Californium-252 as neutron source cross-sections for a number of activation probes were to be determined at the Physikalisch-Technische Bundesanstalt. To do this a quick, reliable and accurate method of activity determination was desired; its error should be less than 2% as other uncertainties are in the same order of magnitude. At last liquid scintillation counting was chosen, because it fulfilled best the condition of being more or less independent of the solid content of the sample.

One of the possible activation probes is NaCl, which results in samples with a relatively low specific activity of Phosphorus-32. As the use of $4\pi\beta$-counting for the activity determination had to be excluded because of the large self-absorption, the following experiments were performed.

At first a mixture of 65% scintillator Insta-Gel and 35% water was shaken with a well calibrated carrier-free Phosphorus-32 solution. The count rate of weighed portions of this gel was measured in the balance point mode. For that purpose the gain was adjusted so that the count rate for a chosen window of 20–1000 divisions gave a maximum. The counting efficiency for these samples was about 97% with the standard deviation 0.4%. The contribution of the statistical error amounts here to 0.2%. In further determinations with increasing NaCl quantities up to 2 g per sample (or up to 1 g $NaH_2PO_4.H_2O$) nearly the same efficiency value was obtained, whilst the standard deviation rose to about 1%. It can be assumed that the main source of error beyond the counting statistics is the instability and inhomogeneity of the gel.

The activity determination of the samples described above with the extrapolation method led to results 5% in excess.

It can be concluded that liquid scintillation counting using the internal standard method is a suitable procedure of activity determination in cases of high solid content.

DISCUSSION

J.F. Stoutjesdijk: Why didn't you use Cerenkov measurement?

E. Günther: The properties of our measuring equipment (e.g. no bialkali multipliers) have not been optimal for this type of measurement. We plan, however, such measurements in the near future. Thank you for this suggestion.

SECTION V

SPECIAL TOPICS

Chapter 20

Cerenkov Counting and Other Special Topics

R.P. Parker

Physics Department, Institute of Cancer Research
and Royal Marsden Hospital, Sutton, Surrey, England

INTRODUCTION

Cerenkov light is electromagnetic radiation, normally in the near ultraviolet and visible parts of the spectrum, which is emitted when charged particles of sufficient energy pass through a dielectric medium. It was first observed as a bluish-white luminescence by Mme Curie in 1910 and studied experimentally during 1926–29 by the French radiochemist Mallett.[1] Further detailed experimental work was performed by the Russian Cerenkov,[2] whose results were explained theoretically by Frank and Tamm.[3]

Practical use of the effect awaited the invention of the photomultiplier in the late 1940's. The first uses were in nuclear physics, and the application to radioassay came surprisingly late. In 1950 Belcher[4] observed a luminescence of low order from aqueous solutions of radioactive isotopes which he later[5] showed to be primarily Cerenkov emission. Little use appears to have been made of his work until the mid-1960's when, following a suggestion by Rapkin[74] and stimulated by the development of sensitive automatic liquid scintillation counting equipment and the subsequent demand for improvements in sample preparation, there were some sporadic reports of applications.[6–12] The method has since come into routine use for the assay of high-energy β-emitters and has been the subject of a number of fairly recent reviews.[13–16] In the present paper the more important features of the technique are summarised with emphasis on possible sources of error. More extended attention is given to those aspects which have received most attention during the past 3 years, namely, quench corrections and the use of wavelength-shifters, and finally the increasing number of applications are reviewed.

NATURE OF THE CERENKOV EFFECT

When a charged particle, such as an electron, passes through a dielectric medium, local polarisation is produced along its path. Immediately after its passage the polarised molecules return to their quiescent state with the emission of electromagnetic radiation. Normally the light pulses will interfere destructively but if the velocity of the particle is greater than that of light in the medium it is possible for the wavelets to interfere constructively, in a similar manner to the production of a shock wave from a supersonic aircraft. The resulting pulse of light is called 'Cerenkov emission'.

The light is highly directional, being emitted as a cone of radiation whose half angle ϕ is given by :

$$\cos \phi = \frac{c}{vn} \tag{1}$$

where c is the velocity of light in vacuo, v the velocity of the particle in the medium and n the refractive index of the medium. Thus there is a threshold for emission given by :

$$v_{min} = \frac{c}{vn} \tag{2}$$

An inverse relationship between v_{min} and n exists, and for water v_{min} corresponds to 0.263 MeV for electrons. The number of photons N emitted per unit path length x of an electron over a selected spectral region[3] is :

$$\frac{dN}{dx} = \frac{2\pi}{137} \left(\frac{1}{\lambda_2} - \frac{1}{\lambda_1}\right) \left(1 - \frac{c^2}{v^2 n^2}\right) \text{ photons/cm} \tag{3}$$

where λ_1 = upper limit of selected waveband (cm) and λ_2 = lower limit. Values have been tabulated for water by Ross.[17] The intensity of the emitted light increases towards shorter wavelengths and extends into the near ultraviolet, where many photocathodes are very sensitive.

The pulse of radiation is very short (<1 ns) and is quite distinct from fluorescence and bremmstrahlung. A comprehensive treatment is given by both Jelley[18] and Marshall.[19]

The number of photons emitted in water by Phosphorus-32 is similar to that encountered in the measurement of Hydrogen-3 by liquid scintillation counting and for that reason standard liquid scintillation counting apparatus is suitable. Figure 1 shows experimentally determined Cerenkov spectra from Iodine-131 (β_{max} 0.61 MeV) and Phosphorus-32 (β_{max}1.71 MeV) compared with the Hydrogen-3 spectrum using a typical scintillator cocktail. They are clearly comparable, but it is also obvious that the measurement of the lower-energy β-emitters by the Cerenkov method makes severe demands on the equipment.

FACTORS AFFECTING COUNTING EFFICIENCY

Factors will now be summarised under convenient headings. Variations in them between samples, if not allowed for, will affect the accuracy of the method.

1 Basic factors

a. Isotope energy. The intensity of Cerenkov emission, and hence the detectability, is governed by the number of electrons emitted above the threshold value. For particular nuclides a theoretical maximum efficiency may be calculated[16] and these are given in Table 1 later in this paper. For reasons outlined below it is most unlikely that such high values will be obtained in practice.

2 Instrumental factors

a. Optical. The directional nature of Cerenkov emission results in inefficient detection of the light by the normal twin-photomultiplier arrangement. This is particularly true for low-energy β-emitters, where the cone of emission is narrow and the number of available photons small. Experimental studies have been carried out by Ross.[13] In practice the collection of light may be improved by using plastic vials

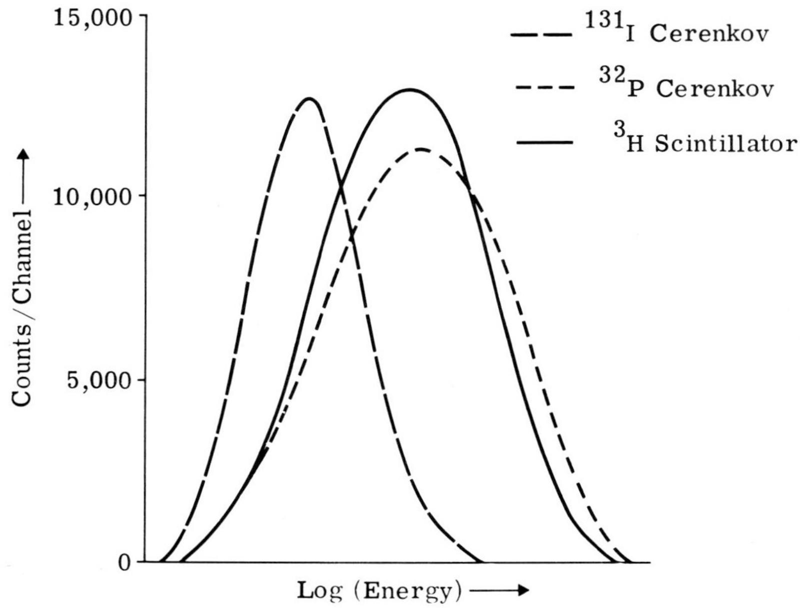

Fig. 1. Cerenkov spectra of unquenched aqueous solutions of Phosphorus -32 and Iodine-131, compared with the liquid scintillation spectrum of Hydrogen-3.

which cause greater light scattering than glass[14,20,21] and by the use of wavelength shifters (see Section 3 (d) below).

Greater efficiencies can be obtained if the coincidence condition is removed[22,23] but the background is considerably increased, as is the detection of chemiluminescence. Since a significant proportion of the Cerenkov light is in the near ultraviolet, the number of photons detected depends on the optical absorption characteristics of the vial, and quartz photomultiplier windows and plastic vials are better than glass. Variations between batches of vials can affect results.

b. Type of photocathode. As noted above, the number of photons detected is very dependent on the quantum and spectral sensitivity of the photocathode. Recent improvements have shown corresponding gains in counting efficiency.[14] Further advances are to be expected, but it must be noted that photocathodes matched to the peak emission of liquid scintillators are not necessarily best for the Cerenkov spectrum.

c. Instrument settings. In all types of counting systems the amplifier, E.H.T. and channel settings will considerably affect the performance. For the Cerenkov counting of most radionuclides the Hydrogen-3 settings will be satisfactory (see Fig. 1), but if the β-energy is particularly low or high, adjustments should be made.

3 Factors affected by sample preparation

a. Refractive index of solvent. As shown by Eqns. (2) and (3) the refractive index affects both the threshold energy and the number of photons emitted. An increase in refractive index will improve the efficiency, particularly for low-energy β-emitters. Using S-11 photocathodes a change in refractive index of 0.01 increases the efficiency for Tellurium-204 (β_{max} 0.77 MeV) by as much as 10%, whilst for Phosphorus-32 (β_{max} 1.71 MeV) the increase is less than 1%.[24] The detection of a β-emitter as low in energy as Carbon-14 (β_{max} 0.156 MeV) by Cerenkov emission in β-bromonaphthalene has been reported by Ross,[17] presumably due to anomalous refractive dispersion.

b. Density of solvent. A decrease in density of the medium will result in a longer path length for the electrons with a corresponding gain in the number of photons emitted. A decrease of 7% in density has similar affects to an increase of 0.01 in refractive index, described above.[24] In the case of both density and refractive index the material of the vial will also affect results since some of the Cerenkov emission is due to electrons travelling through the walls. Variations between samples may cause significant errors.

c. Other characteristics of solvent. Certain organic solvents exhibit weak scintillator effects which may not be reproducible due to variable degrees of impurity quenching. The effect of a number of different solvents has been studied by Haviland and Bieber.[25]

Absorption bands in the solvent will considerably reduce the number of photons reaching the phototubes, and may occur in both the visible and near ultraviolet parts of the spectrum. Colour quenching of this type can be very important and is considered in detail below.

d. Wavelength shifters. Certain substances have the property of absorbing light of short wavelength (e.g. in the near ultraviolet) and re-emitting it isotropically in the visible region. Thus the effects of both ultraviolet absorption and the direction of Cerenkov light are reduced.

Heiberg and Marshall,[26] who examined a number of possible substances, suggested the use of 2-amino-6, 8-naphthalene-disulphonic acid, disodium salt. Other workers recommend 2-naphthylamine-6, 8-disulphonic acid, sodium potassium salt[21,22,24] and 7-amino-1, 3 naphthalene-disulphonic acid.[27] Ross[23] studied a number of compounds and favours 4-methylumbelliferone at a concentration of about 100–200 mg/litre since it is stable and relatively independent of pH over a wide range. Some idea of the improvement obtained can be seen from Table 1.

Table 1. Detection efficiency in water.

Nuclide	βmax (MeV)	Counting efficiency % of disintegrations			
		S-11 cathode	S-11 with wavelength shifter	RCA 4501 -Ve photomultiplier	Max. theoretical value
^{36}Cl	0.71(98%)	2.3	4.7	5.3	58
^{40}K	1.32(89%)	13	31	35	82
^{32}P	1.71(100%)	25	50	43	86
^{144}Pr	2.98(98%)	54	75	59	95
^{42}K	2.0 (19%) 3.6 (82%)	61	85	76	95

e. Volume of sample. The sample volume will affect light collection and hence counting efficiency. Again the effect is more pronounced at low β-energies.[15] Counters using high sensitivity photomultipliers do not exhibit such marked dependence.[21] Note that the whole volume can contain sample, not just 20–30% as in liquid scintillation counting.

EXPERIMENTAL VALUES OF COUNTING EFFICIENCY

A literature search[7,8,14,21,22,28-30] reveals considerable agreement in experimental values for different radionuclides, indicating that the make of counter has little effect. Parker and Elrick[14] report substantially higher values when high sensitivity photomultipliers are used. Typical values are given in Table 1, together with results using S-11 photocathodes with the wavelength shifter 2-napthylamine-6, 8-disulphonic acid[24] and the maximum theoretical efficiency[16] which is based on the number of β-particles exceeding the Cerenkov threshold.

It is interesting to note that at low energies it is more important to detect the small numbers of photons (i.e. use a high quantum efficiency phototube) than re-emit them isotropically. At high energies there is no problem in actually detecting the majority of the photons, so redirection is more important. A combination of the two would presumably give an even greater effect.

An additional improvement is given by using polyethylene vials.[9,14,20,21]

BACKGROUND COUNT RATE

One of the advantages of the Cerenkov technique is that the background counting rate is lower than when a scintillator is used. Furthermore, a number of workers[8,9,14,27] have reported that plastic vials (notably polyethylene) have lower backgrounds than glass. Table 2[15] lists the background count-rates obtained for various types of vial in the pre-set counting channels using Intertechnique SL-40 equipment. The total count-rate is not too dependent on the volume of water, the addition of which causes a widening rather than an increase in the background pulse-height spectrum. It is clear that the radioactive content of the vial material is an important component of the background and some correlation with potassium content is apparent. Care must be taken with some materials (such as quartz and polystyrene) which can exhibit long-term phosphorescence.[24,31]

Table 2. Background count-rates obtained in the various pre-set counting channels using Intertechnique SL–40 equipment.[15]

Type of vial	Background (c.p.m.)					
	^{3}H channel		^{14}C channel above ^{3}H channel		^{14}C channel	
	empty	20 ml water	empty	20 ml water	empty	20 ml water
Polyethylene	11.2	12.4	5.6	12.2	24.1	30.2
Wheaton	19.2	19.4	5.2	12.0	31.9	37.8
Low K glass	36.2	36.4	4.6	11.3	48.4	55.2
Soda glass	100.2	93.0	6.2	11.9	123.1	118.3

Other causes of background include thermal and electronic noise, and Laney[32] has described an electronic method by which cross-talk may be reduced. Cerenkov emission in the photomultiplier, counting vial and sample due to environmental γ-radiation, cosmic rays and nearby γ-emitting samples contribute to the remaining background, as mentioned elsewhere in these Proceedings.

QUENCHING

In normal liquid scintillation counting, chemical quenching, in which molecules are de-excited without the emission of visible radiation, is often serious. It does not exist with Cerenkov counting due to the nature of the effect. Many workers have substantiated Cerenkov's original observation[2] in this respect. Thus there is considerable freedom with regard to sample preparation, which is normally limited only by chemiluminescence[33] and colouration.

Colour quenching can be serious, particularly if the absorption bands lie in the ultraviolet or blue parts of the spectrum, thus yellow or red colouration is worse than blue. Figure 2 shows the pulse-height spectrum of quenched and unquenched Iodine-131 and Phosphorus-32. In the case of the higher-energy nuclide the pulse-height spectrum shifts to lower energies, together with a reduction in intensity; for the lower-energy nuclide there is little energy shift but considerable intensity reduction. S-11 photocathodes were used.

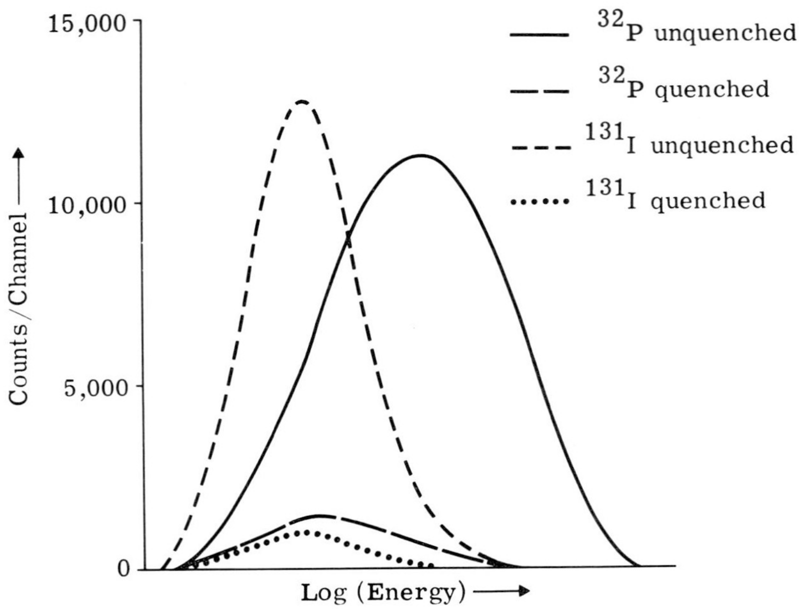

Fig. 2. Cerenkov spectra of unquenched and heavily quenched Phosphorus-32 and Iodine-131.

During the past 3 years there have been a number of references in the literature to quench correction methods for Cerenkov counting, together with some disagreement as to the best one to choose. The following are the methods considered:

1 Decolouration of sample

This is the most direct method, and has been used extensively for urine samples by Gomm.[34] If the counting rate is adequate it is often useful to dilute the samples in order to decrease the colouration.

2 Internal standardisation[35]

The disadvantages are similar to those encountered in liquid scintillation counting, namely the recounting necessary, the difficulty in sample recovery and the extra labour involved. It does not appear to have been widely used in Cerenkov counting.

3 Internal channels ratio[36,37]

Since this is based on the degree of shift in the pulse-height spectrum, Fig. 2 indicates that it will be more suitable for Phosphorus-32 and higher-energy β-emitters than for those of about 1 MeV β_{max} and below. High quantum efficiency photocathodes should improve the situation. For this reason it was not favoured in the earlier report by Elrick and Parker,[24] although Stubbs and Jackson[38] reported favourably in the case of Phosphorus-32 and Moir[39] has shown it to be successful for Potassium-42. Wiebe et al.[40] carried out a comprehensive study in the case of Phosphorus-32 and showed that the nature of the shift of the pulse-height spectrum is the same for all colours, but the degree of shift at a given absorbancy is dependent upon the colour involved. The errors involved were estimated to be up to 5%.

Figure 3 shows the calibration curves we have obtained for S-11 photocathode equipment using blue and yellow dyes for three different energy radioisotopes. The greater accuracy at high energies is clearly demonstrated and with modern apparatus the method should be satisfactory for Phosphorus-32 and above, but must be used with caution for lower-energy β-emitters. The results are also, of course, dependent on the counting statistics obtained.

4 External standardisation

An external γ-ray source will produce a Compton electron distribution in the sample and vial; if the electrons are of sufficient energy, Cerenkov light will be emitted which will be subject to quenching in a similar way to the sample, provided the spectra are not too different.

Figure 4 illustrates the pulse-height spectra obtained for water using both Radium-226 (10 μCi) and Caesium-137 (20 μCi) external standards. It is clear that only the Radium-226 is suitable. Both Elrick and Parker[24] and Brownell and Lauchli[41] suggest that the change in total count-rate from the Radium-226 source is a reliable indicator of the amount of quenching. A typical calibration curve for the low-energy β-emitter Iodine-131 is shown in Fig. 5. One curve is sufficient for all colours, but the total count-rate can vary from one vial to the next due to differences in wall thickness.

5 External standard channels ratio

This employs changes in the channels ratio due to the external standard, and a calibration curve obtained with Iodine-131 and Radium-226 external standards is shown in Fig. 6. One curve is again sufficient for all colours except at very high pigmentations. It has the advantage over the straightforward external standard techniques in that the results are relatively volume-independent, but it should be noted that the optimum channel settings are very different from those used with a scintillator.

6 Spectrophotometric methods

Here the counting efficiency is correlated with the optical transmission at a fixed wavelength. They have been used for Cerenkov counting[24,40] but are more time-consuming than the above methods, which can be made automatic.

7 Cerenkov inserts

A novel method of avoiding colour quenching effects is to contain the sample in an opaque, thin-walled cylinder which is immersed in a colourless liquid contained in a standard counting vial.[42] Whilst the β-rays penetrating the central tube can be detected by Cerenkov emission it is more effective to surround the tube with a

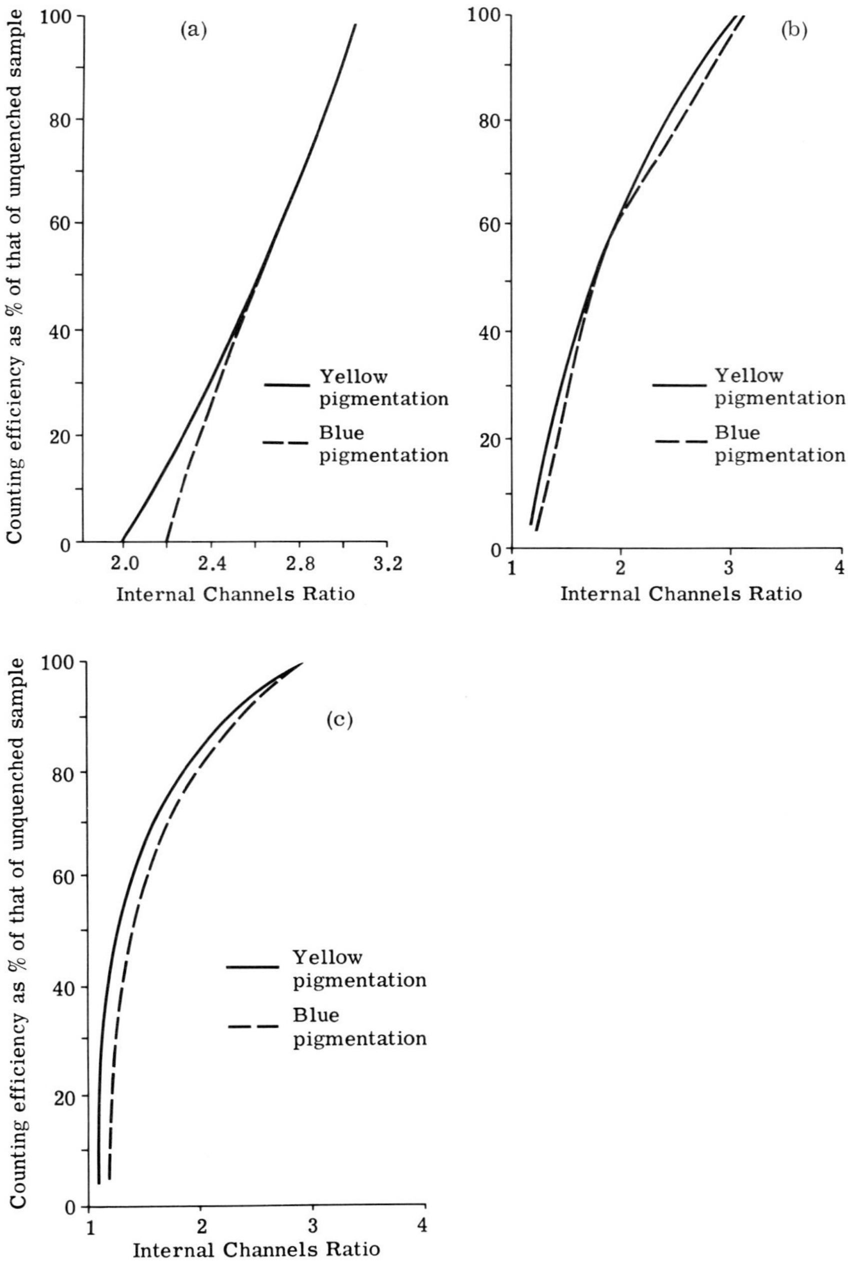

Fig. 3. Internal channels ratio calibration curves for (a) Chlorine-36, (b) Phosphorus-32 and (c) Ruthenium-106–Rhodium-106.

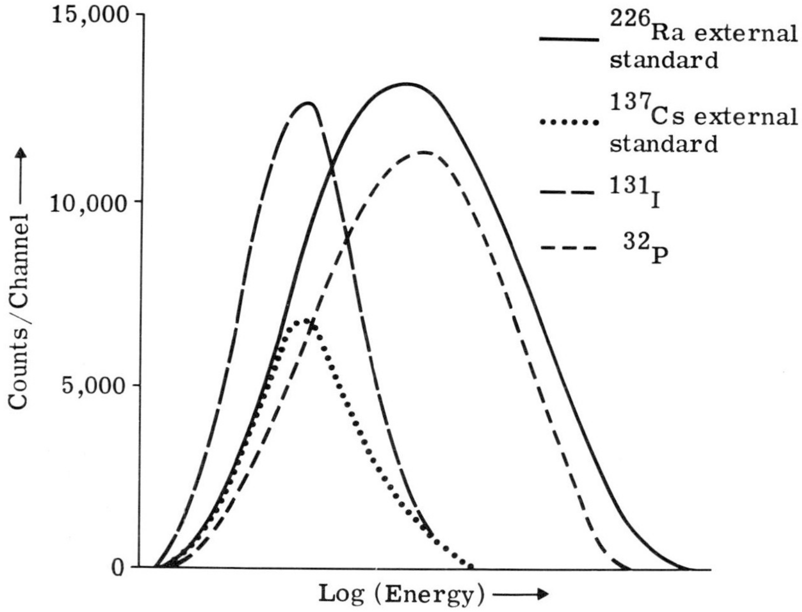

Fig.4. Cerenkov spectra of unquenched Iodine-131 and Phosphorus-32 compared with the Cerenkov spectra due to external sources of Radium-226 and Caesium-137.

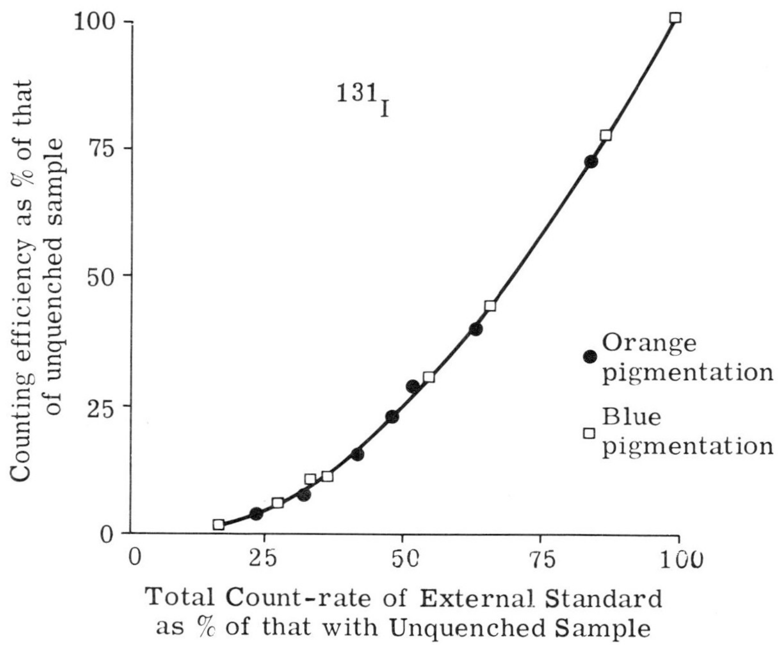

Fig.5. Variation in total count-rate due to an external Radium-226 source with counting efficiency for quenched samples of Iodine-131.

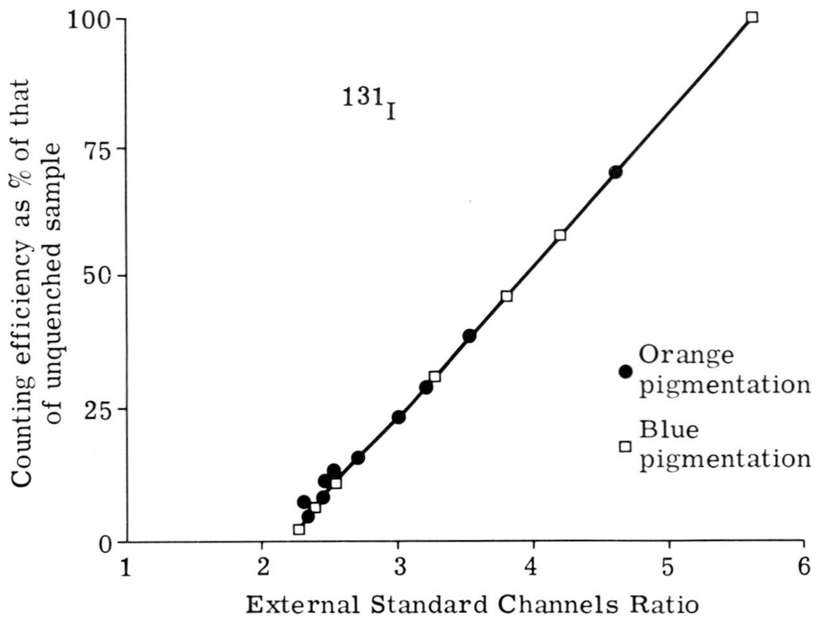

Fig. 6. Quench calibration curve for Iodine-131 using the external channels ratio technique with Radium-226.

liquid scintillator, a method also reported by Braunsberg and Guyver.[6] Colour quenching problems are thus avoided, but at the expense of efficiency.

8 Conclusion

For most applications the internal channels ratio method would appear satisfactory, provided the pigmentation is not too intense or the β-energy too low. In these cases a Radium-226 external standard method is preferable. A useful comparison of quench correction methods for liquid scintillation counting has been given by Noujaim et al.[43]

APPLICATIONS

Cerenkov counting has found widespread application primarily due to the simplicity of sample preparation, the fact that the sample can be recovered for further treatment and the saving in cost compared with scintillator methods. Since the whole of the sample is the test solution the figure of merit — (efficiency) x (sample volume) — may well be greater than if the liquid scintillation method were used.

There are now well over 40 publications specifically describing applications of Cerenkov counting in radioassay and a number of these will be summarised in order to show the diversity of uses. The divisions used below are somewhat arbitrary, and are chosen primarily for convenience.

1 Reaction effluent and protection problems

Following Belcher's early work, Yamada[12] developed an 8-litre detector for the estimation of $^{90}Sr-^{90}Y$, the sensitivity being adequate for the determination of an 8×10^{-14} Ci/cm^3 sample in a counting time of 5 min. A large volume can be used because the path length of the Cerenkov radiation is large. De Volpi and Porges[7] also measured reactor effluents using a specially designed counter, this time with only 12-ml samples, and determined Manganese-56 and Sodium-24 as well as $^{90}Sr-^{90}Y$. Iwakura[21] used a standard liquid scintillation counter for measuring low-level aqueous waste with a detection limit of 5×10^{-13} Ci/ml for 1 MeV β-emitters. The general problem of Cerenkov light-monitoring for evaluating fission-product contamination in reactor effluent streams is considered by Rippon.[44] Recently a continuous flow measuring system has been developed[73] in which the higher-energy β-emitters are measured by Cerenkov counting after the continuous separation of the individual radionuclides. An interesting variation is that of Vaninbroukx[45] who used the Cerenkov effect for the accurate determination of the activity of extended reactor probes, such as the fast neutron monitors based on the Phosphorus-31 (n, p) Silicon-31 and Sulphur-32 (n, p) Phosphorus-32 reactions. An accuracy better than 1% is claimed. This is primarily due to the simple sample preparation and the absence of errors in correcting for chemical quenching.

2 Animal and human studies

The majority of these applications concern the determination of Phosphorus-32. Media include ashed biological materials,[11] supernatants from soft tissues,[47] phospholipids from rats,[48,49] labelled RNA[50] and DF-Phosphorus-32 platelets.[51] Palmer[52] and Schneider[75] describe the measurement of Phosphorus-32 following colorimetric phosphorus analysis, the blue reduced phosphomolybdate complex being decolourised before counting. In this way the specific activity may be determined, and the method is especially recommended for the analysis of samples separated by thin-layer chromatography. The Phosphorus-32 effluents from a column chromatogram have also been counted using a modified fraction collector.[53] In plasma proteins Potassium-42 may be determined.[28] The Cerenkov method is attractive for those applications because of the ease of sample preparation and availability for subsequent analyses; even inhomogeneous samples such as Sodium-24 labelled tissue slices can be counted directly in buffer solution.[6]

3 Plant studies

These are interesting since many workers have counted dry material and reported satisfactory results. In a series of papers Lauchli[27,54] describes the determination of Rubidium-86 and Chlorine-36 in plant roots and similar samples, using a sophisticated technique employing a wavelength shifter and correcting for colour quenching using a Radium-226 external standard. Efficiencies of 60% for Rubidium-86 and 13% for Chlorine-36 are reported. A similar quench correction method was used by White and Ellis[55] for Phosphorus-32 in dry ashed plant tissues. Algae may be counted directly on glass fibre filters with an efficiency for Phosphorus-32 of 12% in glass vials and 24% in polyethylene vials.[56] An alternative method[57] is to place the plant tissues in a vial containing water or hexane, the latter providing a greater counting efficiency due to scintillator action, particularly if leakage of the Phosphorus-32 is possible. In agrochemical studies, low specific activities are often encountered, but wet ashing with decolouration using activated charcoal can be satisfactory.[58,59] The use of Cerenkov inserts containing plant material on glass fibre filters is another technique used,[42] but is really a variation on liquid scintillation counting.

4 Multi-nuclide studies

A combination of Cerenkov counting (to measure high-energy β-emitters such as Phosphorus-32, Iodine-131 and Potassium-42) and liquid scintillation counting (to measure low-energy β-emitters as well, including Hydrogen-3, Carbon-14 and

Phosphorus-33) permits rapid determinations of suitable isotope mixtures.[60-62] It has been extended[63] to a true two-channel Cerenkov technique for Strontium-89–Strontium-90 using a ratios technique modified from Hydrogen-3/Carbon-14 mixture analysis. The build-up of the Yttrium-90 daughter permits estimation of Strontium-90 even in the presence of long-lived high-energy β-activity. Giubileo and Camera[64] report a quick method for the measurement of $\beta-\gamma$-activity in biological samples, using liquid scintillation counting for β-emitters below 1 MeV, Cerenkov counting for those above and gamma spectrometry where applicable. A combination of Cerenkov counting and decay studies has been used to determine radon and thoron in environmental samples.[8,21] Bromine-80g and Bromine-80m (via its Bromine-80g daughter) were measured with high efficiency in organic solutions by the Cerenkov radiation from the high-energy β-rays emitted in the Bromine-80g decay.[76] The efficiency for the contaminating Bromine-82 activity was less than 1%.

5 Activation analysis

Although Cerenkov counting has not found great use in this field, three papers are of interest. In the first[65] the phosphorus content of micro samples of nucleic acids is determined quantitatively by neutron activation, separation of the phosphorus on a chromatographic column and subsequent Cerenkov counting. Hoch et al.[66] have utilised the Cerenkov technique to count Iodine-128, formed by neutron activation of Iodine-127 used as a biological tracer. Nanogram quantities of iodine may be measured. Lithium has been determined in the 3–400 μg range by detecting the 13 MeV (max) β-particles from the 0.84 s Lithium-8 produced by the reaction Lithium-7 (n, γ) Lithium-8.

6 Special applications

There have been some novel applications in unrelated fields which are worthy of mention. Wilson et al.[67] have used Cerenkov light to detect radioactive sources beneath the sea: the main problem was apparently bioluminescence. Cerenkov light emitted by Phosphorus-32 incorporated in the eye has been detected by a photomultiplier as a test for ocular lesions.[68] Sealed Cerenkov sources may be used as standard light sources for testing equipment[69] and as a sensitive spectrophotometer when incorporated with a narrow-band optical filter in a liquid scintillation counter.[70] The substance under investigation is placed around the rod-like light source, and the absorption measured. Liquid scintillation counters have also been used for the determination of particle concentration (for example 0.5 μm polystyrene spheres or suspensions of E. coli) by the extinction of the Cerenkov emission from a solution of Chlorine-36. It should be of value when the particle concentration is too high for normal turbidimetry or manipulation is hazardous, as in the case of pathogenic bacteria

Finally, mention should be made of a flow counter developed by Eyre[72] in which a scintillation fluid is automatically mixed with the column effluent of an automatic amino acid analyser. Counting efficiencies of 60-65% for Carbon-14 and 7–10% for Hydrogen-3 were obtained. By omitting the scintillator it should be possible to use it for high-energy β-emitters by the Cerenkov technique, as in the work on fission products described by Colomer et al.[73] This has also been successfully carried out for Krypton-85 in a water stream. The Krypton-85 can be separated from the water for Cerenkov counting by diffusion into polyethylene.[20]

CONCLUSIONS

The large and varied number of applications is clear evidence that Cerenkov counting is now receiving widespread application in the assay of high-energy β-emitting radioisotopes, primarily due to the simplicity and cheapness of sample preparation and the absence of chemical quenching. Workers have experimented with solid and inhomogeneous samples with considerable success and have thus exploited the simplicity of sample preparation. Quench corrections are now being applied with confidence, avoiding the need for decolouration. There have been only a few attempts at routinely measuring β-emitters of less than 1 MeV, but the use of wavelength shifters and high quantum efficiency phototubes should improve the situation.

It is hoped that this trend will be encouraged by the manufacturers, who could usefully provide pre-set Cerenkov counting channels similar to those already supplied for Hydrogen-3 and Carbon-14. Correct settings should be available for use with Radium-226 external standardisation techniques, and there is little doubt that counting efficiencies would be considerably increased using more senstive photomultipliers, or if more than two photomultipliers were used in order to minimise the direction effect of the Cerenkov emissions. Whilst increasing the complexity, it should add relatively little to the cost of the sophisticated equipment now being marketed. Certainly the design of the reflecting chamber of a liquid scintillation counter should be examined in order to optimise it for Cerenkov techniques.

REFERENCES

1 L. Mallett, C.R. Acad. Sci. (Paris) 188, 445 (1929).
2 P.A. Cerenkov, C.R. Acad. Sci. (U.S.S.R.) 2, 451 (1934).
3 I. Frank and Ig Tamm, C.R. Acad. Sci. (U.S.S.R.) **14**, 109 (1937).
4 E.H. Belcher, Brit. J. Radiol Suppl. No. 2, p.282 (1950).
5 E.H. Belcher, Proc. Royal Soc. **A216**, 90 (1953).
6 H. Braunsberg and A. Guyver, Anal. Biochem. **10**, 86 (1965).
7 A. De Volpi and K.G.A. Porges, Intern. J. Applied Radiation and Isotopes **16**, 496 (1965).
8 K. von Haberer, Atomwirtschaft **10**, 36 (1965).
9 V.W. Hoffman, Radiochim. Acta. **4**, 117 (1965).
10 J. Narrog, Atompraxis **11**, 373 (1965).
11 V.H. Vemmer and J.O. Gutte, Atompraxis, **10**, 475 (1964).
12 S. Yamada, J. Phys. Soc. (Japan) **17**, 865 (1962).
13 H.H.Ross, Analyt. Chem. **41**, 1260 (1969).
14 R.P. Parker and R.H. Elrick, in The Current Status of Liquid Scintillation Counting (Ed. E.D. Bransome) Grune and Stratton, New York, 1970, p.110.
15 R.P. Parker, Cerenkov Counting, Digitechniques Tech. Rev. No.4, Intertechnique, Paris, 1970.
16 D. Santoso, Ph.D. Thesis, Univ. of Maryland (Univ. Microfilms 72–21, 131), 1972.
17 H.H. Ross in The Current Status of Liquid Scintillation Counting (Ed. E.D. Bransome) Grune and Stratton, New York, 1970, p.123.
18 J.V. Jelley, Cerenkov Radiation and its Applications Pergamon, London, 1958.
19 J. Marshall, Phys. Rev. **86**, 685 (1952).
20 J.D. Buchanan, Radiochim. Acta **9**, 218 (1968).
21 T. Iwakura, Radioisotopes (Tokyo) **18**, 500 (1969).
22 R.P. Parker and R.H. Elrick, Intern. J. Applied Radiation and Isotopes **17**, 361 (1966).
23 H.H. Ross, in Organic Scintillators and Liquid Scintillation Counting (Ed. D.L. Horrocks and Chin-Tzu Peng), Academic Press, New York, 1971, p.757.

24 R.H. Elrick and R.P. Parker, Intern. J. Applied Radiation and Isotopes **19**, 263 (1968).
25 R.T. Haviland and L.L. Bieber, Anal. Biochem. **33**, 323 (1970).
26 E. Heiberg and J. Marshall, Rev. Sci. Instr. **27**, 618 (1956).
27 A. Lauchli, Intern. J. Applied Radiation and Isotopes **20**, 265 (1969).
28 J.E. Johnson and J.M. Hartsuck, Hlth Phys. **16**, 755 (1969).
29 B. Francois, Intern, J. Applied Radiation and Isotopes **18**, 525 (1967).
30 P.J. Garrahan and I.M. Glynn, J. Physiol. **186**, 55P (1966).
31 E.H. Belcher, Nature, **170**, 571 (1952).
32 B.H. Laney in Organic Scintillators and Liquid Scintillation Counting (Ed. D.L. Horrocks and Chin-Tze Peng), Academic Press, New York, 1971.
33 D.S. Kearns, Intern. J. Applied Radiation and Isotopes **23**, 73 (1972).
34 P.J. Gomm, AERE—PR/HPM **11**, 47 (1967).
35 F. M. Hayes, Intern. J. Applied Radiation and Isotopes **1**, 46 (1956).
36 L.A. Baillie, Intern. J. Applied Radiation and Isotopes 8, 1 (1960).
37 E. T. Bush, Analyt. Chem. **35**, 1024 (1963).
38 R.D. Stubbs and A. Jackson, Intern. J. Applied Radiation and Isotopes **18**, 857 (1967).
39 A.T.B. Moir, Intern. J. Applied Radiation and Isotopes **22**, 213 (1971).
40 L.I. Wiebe, A.A. Noujaim and C. Ediss, Intern. J. Applied Radiation and Isotopes **22**, 463 (1971).
41 J.R. Brownell and A. Lauchli, Intern. J. Applied Radiation and Isotopes **20**, 797 (1969).
42 P.E. Ballance and S. Johnson, Intern. J. Applied Radiation and Isotopes **20**, 447 (1971).
43 A. Noujaim, C.Ediss and L. Wiebe, in Organic Scintillators and Liquid Scintillation Counting (Ed. D.L. Horrocks and Chin-Tze Peng), Academic Press, New York, 1971, p.705.
44 S.E.H. Rippon and Nuc. Instr. Meth. **21**, 185 (1963).
45 R.L. Vaninbroukx, Proc. Symp. Standardization of Radionuclides, Vienna: IAEA. 1967, p.443.
46 V. Camera and H. Tanguy, EUR—4084, 1 (1968).
47 M.K. Johnson, Anal. Biochem. **29**, 348 (1969).
48 J. Plesums and W.H. Bunch, Anal. Biochem. **42**, 360 (1971).
49 G. Threlfall, D.M. Taylor and A.T. Buck. Am. J. Pathology, **50**, 1 (1967).
50 S. Mizuno, H. Eguehi, K. Yano and H. Yamaguchi, Radioisotopes (Tokyo) 18, 19 (1969).
51 R.D. Smith, J.J.B. Anderson and Ristic Miodrag, Intern. J. Applied Radiation and Isotopes **23**, 513 (1972).
52 F.B. St. C. Palmer, Anal. Biochem. **31**, 493 (1969).
53 H.R. Matthews, J. Chromatogr. **36**, 302 (1968).
54 A. Lauchli, in Organic Scintillators and Liquid Scintillation Counting (Ed. D.L. Horrocks and Chin-Tzu Peng), Academic Press, New York, 1971, p.771.
55 R.P. White and B.G. Ellis, Soil Science Am. Proc. **32**, 740 (1968).
56 W. Hülsen and U. Prenzel, Anal. Biochem. **26**, 483 (1968).
57 T. Awerbuch and Y. Avnimelech, Plant Soil **33**, 260 (1970).
58 F.J. Haasbroek and E. Lotz, S. Afr. J. Agr. Sci. **10**, 285 (1967).
59 D. Sipos and L. Koch, Atomtech. Tajek **12**, 313 (1969).
60 W.E. Kisieleski, A.N.L. — 7635, 152 (1969).
61 J.R. Robinson, Intern. J. Applied Radiation and Isotopes **20**, 531 (1969).
62 L.C. Brown, Analyt. Chem. **43**, 1326 (1971).

63 M. Goldman and E. Beckman, U.C.D. — 472 — 116, 100 (1969).
64 M. Giubileo and V. Camera, EUR — 4066, 1 (1968).
65 F. Giradi, V. Camera and E. Sabbioni, Radiochem. radioanalyt., Lett. 2, 195 (1969).
66 F.L. Hoch, R.A. Kuras and J.D. Jones, Anal. Biochem. 40, 86 (1971).
67 A.E. Wilson, J.F. Tinney and J.J. Koch, UCRL — 50007 — 68 — 2, 8 (1966).
68 W.M. Burch, Nature 234, 358 (1971).
69 W. Anderson and E.H. Belcher, Brit. J. Appl. Phys. 5, 53 (1954).
70 P. Jordan and P. Koberle, Experientia 25, 335 (1969).
71 J. Ashcroft, Intern. J. Applied Radiation and Isotopes 20, 555 (1969).
72 D.R. Eyre, Anal. Biochem. (in the press).
73 J. Colomer, M. Cousigne and G. Metzger, in Liquid Scintillation Counting Vol.2. (Ed. M.A. Crook, P. Johnson and B. Scales), Heyden, London, 1972, p.181.
74 E. Rapkin, Intern. J. Applied Radiation and Isotopes 15, 69 (1964).
75 P.B. Schneider, J. Nuc. Med. 12, 14 (1971).
76 G.P. Gennaro and K.E. Collins, J. Radioanalytic Chem. 5, 387 (1970).

DISCUSSION

E. Gunther: What can you say about the uncertainties of Cerenkov counting in comparison with LSC (toluene scintillator)? Are measurements with the same reproducibility of, for example 0.1% in the case of Phosphorus-32, possible?

R.P. Parker: Yes.

J.F. Stoutjesdijk: The count-rate of a ^{32}P-solution in 30–70% H_2SO_4 is about 5% higher than in H_2O; in 90% H_2SO_4 it is about 5% lower. Can you explain this? (The refractive index of H_2SO_4 is about the same as that of H_2O: the density of pure H_2SO_4 is 1.8.)

R.P. Parker: In fact the refractive index of H_2SO_4 varies from 1.339 at a density of 1.028 to 1.437 at a density of 1.811. There are two competitive processes: as the density increases the count-rate relative to H_2O will decrease, whilst as the refractive index increases the count-rate relative to H_2O will increase. The photon yield relative to water is not directly proportional to the refractive index at a given density, and careful calculation shows that experimental results are borne out by the theory. It is not a chemical quenching effect.

D.L. Horrocks: Chemical quench may not be as important in Cerenkov counting using wavelength shifters as in ordinary liquid scintillation counting since the transfer to the wave-shifter is by a radiative process and the process of chemical quenching occurs by a non-radiative competition with the excited molecule. There is a possibility of molecular competition for the excitation energy once it is associated with the wavelength shifter.
Also the use of quartz-faced, high efficiency MPT's will provide a higher detection efficiency but these advantages may be negated if a glass bottle is used for the sample vial. It is necessary to use a plastic vial or a quartz vial to see the maximum effect.

R.P. Parker: Chemical quenching occurs in Cerenkov counting, but I agree it is possible (although somewhat unlikely) when wavelength shifters are used. I fully agree with your point concerning choice of sample vial, which is very important and which can alter the amount of colour quenching observed, if, for example, there is an absorption band in the near ultra-violet.

J.F. Stoutjesdijk: When wavelength shifters are added, chemical quenching may influence the counting efficiency and so the composition of the sample may become important. This possible source of error is not stressed sufficiently when the use of wavelength shifters is advocated.

R.P. Parker: This is true, but it is not normally an important effect. The efficiency of a wavelength shifter will depend on the concentration and on the pH of the solution. Also the addition of a wavelength shifter is likely to affect the degree of colour quenching. All these possible effects must be taken into account, and the user may feel that the added complexity and need for additional sample preparation is not justified. My own view is that it is only really worthwhile when low count-rates are encountered and there are many samples of the same type to be measured.

Chapter 21

Analytical Bioluminescence Assays Using the Liquid Scintillation Spectrometer: A Review

P.E. Stanley [a]

Department of Agricultural Biochemistry, Waite Agricultural Research Institute, The University of Adelaide, Glen Osmond, South Australia

INTRODUCTION

Bioluminescence is the enzyme- or protein-mediated production of light and is a form of chemiluminescence since it arises from the excited states of the products of exothermic reactions. The quantum yield, i.e. the total quanta emitted divided by the number of reacted molecules, is high for many bioluminescent species. Thus a value of close to one has been recorded for firefly luminescence,[1] whereas those of many chemiluminescent systems are frequently much less efficient[2,3] with values as low as 10^{-15}. However, bioluminescent reactions differ thermodynamically from other biological reactions since the emission of an einstein (a mole of photons; 6.025×10^{23}) requires 63.5 kcal (2.75 eV) at 450 nm and 47.6 kcal (2.07 eV) at 600 nm. This is considerably more energy than is available in any single-step biological reaction. Thus the biochemistry and physical chemistry of bioluminescence has been of considerable interest. However, these aspects will not be discussed here and the reader is referred to recent books, reviews and symposia[1-17] for further information.

Since bioluminescence is generally a highly efficient process, it is extremely useful as an analytical tool, especially now that modern low-background photomultipliers are available to count photons with high quantum efficiency. An example will show the sensitivity that can ideally be attained for measuring adenosine triphosphate (ATP) using the luciferin-luciferase system of the firefly *Photinus pyralis*. When used to monitor such a reaction, a good photomultiplier with a background of 50 c.p.s. and an average quantum efficiency of 5% is capable of measuring approximately 10^{-19} moles ATP/s. Since light collection is not perfect and because the rate of the bioluminescent reaction diminishes with time, the estimate of sensitivity would have to be lowered by one or two orders of magnitude. In practice, the best sensitivities obtained are about 10^{-15} moles for ATP[18-19] and 10^{-16} moles for reduced nicotinamide adenine dinucleotide (NADH).[20]

---

[a] Present address: Department of Clinical Pharmacology, The Queen Elizabeth Hospital, Woodville, Adelaide, South Australia.

GENERAL ASPECTS

In the early 1950's analytical bioluminescence assays were in use in few laboratories and most measurements were made using the photomultiplier as a photometer (current producing) rather than as a quantum or photon counter (pulse producing). Morton,[21] Duquesne and co-workers,[22-24] James[25] and Tothill[26] have considered various aspects of photon counting with special reference to low light levels. Seliger[8] has compared the current and pulse procedures for detecting and measuring very low light levels and concludes that 'there is no a priori reason to use the more complex pulse counting techniques over the much simpler D.C. amplifier current measurements for optimum detection of very weak chemiluminescent reactions'. This is based on a photomultiplier with a figure of merit of one. Robben,[27] in a detailed review paper, has come to a similar conclusion when comparing the two techniques under conditions where the figure of merit approaches unity, but under the conditions employed by Duquesne et al.[22] pulse counting gives the best sensitivity. However, the dearth of suitable photometers and the ready availability in many laboratories of liquid scintillation spectrometers, which can be readily adapted for quantum counting, has made this instrument the one of choice for work in the biological and medical fields. The modern unit, fitted with photomultipliers having low-noise bialkali photocathodes capable of excellent signal-to-noise operation at ambient temperatures, enables measurements of ATP and NADH to be made routinely in picomole (10^{-12}) amounts.[28-41]

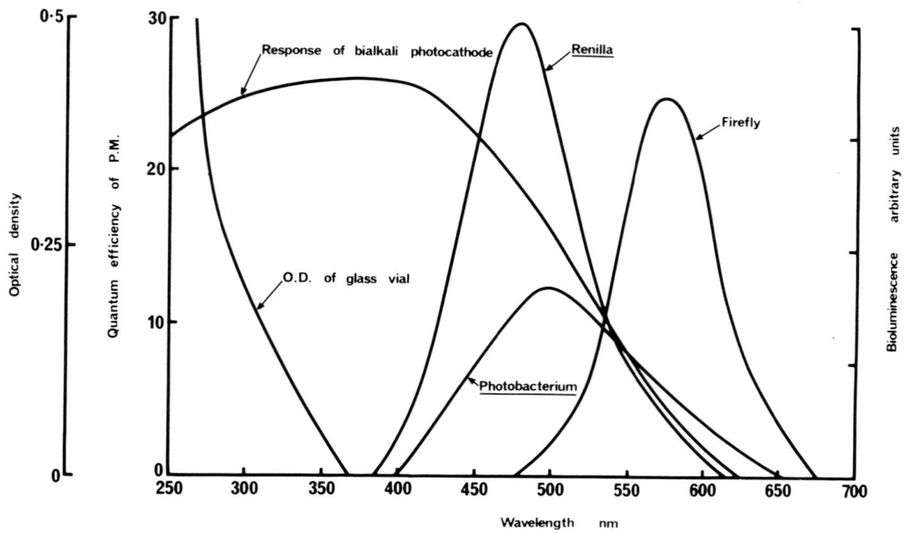

Fig.1. The quantum efficiency of a bialkali photomultiplier (EMI 9635QB 42a), the optical density of a glass scintillation vial together with emission spectra of the firefly, Photobacterium and Renilla. The emission spectra are not normalised.

Figure 1 shows the quantum efficiency curve for a bialkali photocathode (SbKCs) fitted to a photomultiplier of a type commonly used in spectrometers, together with the emission spectra of the firefly (ATP), Photobacterium (NADH) and Renilla reniformis (adenosine 3'-phosphate 5'-phosphate, PAP). Thus good quantum efficiencies are attainable. Frequently, however, the instruments have features which are superfluous for analytical bioluminescence, e.g. refrigeration, coincidence circuitry, external standardisation and several pulse height analysers, but until other less complex but equally reliable automatic pulse counting equipment becomes available commercially the conventional liquid scintillation spectrometers will continue to be used for bioluminescence and chemiluminescence assays.

Bioluminescence is composed of single discrete photons, whereas in liquid scintillation spectrometry the discrete scintillations consist of a burst of photons, some of which go to one photomultiplier and some to the other. Coincidence circuitry establishes the legitimacy of the pulses and electronically reduces the background derived from the thermal noise of the two photocathodes to negligible values. However, in order to count single photons efficiently in the spectrometer, the coincidence circuit must be switched off,[32, 33, 37-40] this, of course, results in a concomitant increase in background which with good modern bialkali photomultipliers is quite acceptable. Photons counted per unit time will thus be directly proportional to the ATP or NADH in the sample. In the coincidence mode a squared proportionality is observed [32, 37] and the useful range is smaller.[37] In the out-of-coincidence mode the pulse height spectrum for these single photons is in the lowest sixth of the pulse height range for unquenched tritium[37] and the discriminators and amplifiers on the pulse height analysers should be set accordingly.

Schram[36] and others[42a] have recently emphasised the importance of selecting optimal high-voltage and discriminator settings when maximum counting sensitivity is required. These settings are important in attaining the best signal-to-noise ratio and those used for the liquid scintillation counting of tritium are not adequate. Higher voltages do not necessarily give better separation of photomultiplier noise and signal, and these may give rise to memory effects at high counting frequencies.[36] On the other hand, Lin and Cohen[41] advocate the use of higher voltages. The individual characteristics (gain and thermal noise) of the photomultipliers will no doubt dictate their best operating conditions.[42a, b] The use of high counting rates causes another problem, which is associated with saturation of the pulse height analyser. The result is that individual pulses (from single photons) are not resolved and the apparent pulse height spectrum is distorted.[37] As a consequence, calibration curves deviate significantly from linearity at high counting frequencies. With modern equipment, about 2×10^6 random pulses/min is the upper limit for normal use. To employ closely set discriminators in order to diminish the observed counting rate does not overcome the problem since the analyser has to appraise <u>all</u> pulses entering it. Some equipment is fitted with live timing so that the real time clock is switched off during the processing of each pulse. The user must be aware that at high counting rates a considerable extra elapsed time will occur over and above the preset time and this will generally give false results because the bioluminescence rate usually changes with elapsed time.

All measurements of analytical bioluminescence are concerned with reaction rates. Thus on mixing ATP with firefly luciferin-luciferase (in the presence of O_2 and Mg^{2+}) there is a rapid increase in light output, reaching a maximum in about one second (the peak or flash height), and this is followed by a gradual exponential decrease in light production.[6] The rate of decay depends on the amount of ATP, enzyme, pH and ionic strength,[28] as well as buffer type,[28] inhibitory ions[30, 43] and temperature. Under proper conditions a decay period of 30 min or more can be observed and this rate remains directly proportional to the ATP concentration,[33-35] so a log-lin plot of counts per unit time versus time gives a series of parallel straight lines over a wide concentration range of ATP. A somewhat similar, although not identical, situation is found in the case of <u>in vitro</u> bioluminescence of <u>Photobacterium</u>, which can be effectively employed to measure either FMN (flavin mononucleotide) or NADH[34, 38, 39, 50] since the enzymes present catalyse the following reaction sequence :

$$NADH + FMN \xrightarrow{\text{Dehydrogenase}} NAD + FMNH^{+}$$

$$FMNH^{+} \xrightarrow[\text{Long-chain aldehyde}]{\text{Bacterial luciferase}} \text{light}$$

When using the liquid scintillation spectrometer, conditions are usually so arranged as to minimise decay, e.g. use of arsenate buffer for ATP assays,[28] since quanta are counted over a fixed time interval starting at a preset time after mixing the reactants outside the instrument. This preset time is generally 5-10 s and the fixed time 5-30 s.[28-41] When a reaction is to be followed over many minutes, the repeat count mode is employed and the use of spectrometers fitted with buffered output (such as the Searle Analytic Unilux III and Mark II instruments) is particularly advantageous for this purpose since the next counting sequence is underway again as print-out commences. Since print-out may take from 5-15 s on some instruments, it follows that a considerable loss of data occurs in these when monitoring a reaction sequence taking, say, 2 min. Should a continuous analogue representation be required, this can be obtained by coupling a ratemeter (time constant 0.3 s or less) and a chart recorder (full-scale deflection in less than 1 s) to the output stage of one of the pulse height analysers.[36, 37, 42] An alternative is to employ a multichannel analyser in multiscale. In this mode, channels are opened sequentially for a preset time, say 0.1 or 1 s, and the counts occurring in that time interval are stored in adjacent channels. This is especially useful for fast reactions.[39, 40, 44] Analogue representation is usually available on an oscilloscope display screen and digital information can be retrieved by interfacing to a computer or teletype.

As mentioned earlier, photometers are also used for assaying bioluminescence but here the measurement is usually of the peak or flash height since some workers consider this value to be less susceptible to interference from other substances.[49] St. John,[46] however, recommends that light emission be integrated over a 30 s time interval commencing at the time of mixing the reactants. This gives high sensitivity, high reproducibility, convenience and speed. He has evidence that measurements of the peak intensity do not necessarily give the most reliable results for ATP in complex systems. A number of photometers designed for bioluminescence assays have been described [18, 19, 47-54] and there are commercial units made by Du Pont (Biometer) and JRB Inc., San Diego, California (ATP photometer); the former is designed to measure the flash height and the latter integrates over a fixed time interval following mixing.[36] Prydz and Froholm[55] have described the use of Beckman DB spectrophotometer for the purpose and St. John[46] has used a bioluminescence sensing chamber (American Instrument Co.). Ausmus[56] has employed the highly sophisticated GeMSAEC Centrifugal Photometric Analyzer, designed to perform the analysis of ten samples, four standards and a blank simultaneously. The instrument is interfaced to a PDP8 computer for control of the unit and calculation of the results. Samples can be processed rapidly and accurately.

With the scintillation spectrometer, the use of a fixed cell within the counting vial gives more reproducible results when using small volumes than the vial produces alone.[33] A similar cell can also be employed to keep the reaction at a useful ambient temperature, using a cooled spectrometer,[36, 38, 39] when the scintillation vial is filled with water at the required temperature. A cell fixed in the detector improves the reproducibility of results and obviates the problem of phosphorescence of glassware.[35] Transfer of reagents in this unit occurs by means of a motor-driven micro-syringe. Stanley[40] has described a cell which fits into the detector chamber and is filled manually with syringes; it may also be used to measure flash heights.

When large numbers of samples have to be processed, automation is an advantage. Hammerstedt[57] has interfaced to a commercial spectrometer a Hamilton precision liquid dispenser, which adds a fixed amount of firefly luciferin-luciferase into the next sample at the time of printing data of the previous one; 1400 samples can be processed in a 7 hr session. Another approach has been to employ an auto-analyser on-line to a flow detector in a spectrometer.[58] The development of further automatic equipment is very important, especially for dynamic measurements when several reagents are to be added.[38, 44, 59, 60]

A problem which is common to many bioluminescence assays is that of phosphorescence of vials. Thus, scintillation vials which have been exposed to sunlight or unfiltered fluorescent lighting will phosphoresce (single photons) for up to several hours, giving rise to both variable and high backgrounds ($10^3 - 10^4$ c.p.s.), and require 'dark' adaption for several hours prior to use. A room lit only with conventional tungsten lights is essential for this type of work. Some workers obviate the problem by employing special cells fixed in the detector chamber.[33, 35, 40] Vial caps can also be a source of phosphorescence[61, 62] but need not be used in analytical bioluminescence assays.

Not generally recognised is the problem of contamination of glassware and solutions with dust and micro-organisms. Thus the luciferase system of Photobacterium is readily absorbed on dust particles and gives erroneous data, while the presence of micro-organisms results in degradation of compounds such as NADH and ATP. Solutions are best prepared and then filtered through a bacterial filter. Even then, very dilute solutions of ATP ($<10^{-9}$ molar) kept at $0^{o}C$ at neutral pH appear to break down after a few hours.[37]

SPECIFIC ASSAYS AND PROCEDURES

Beijerinck[63] in 1902 was the first worker to use bioluminescence as an analytical technique and assayed O_2 using luminous bacteria (Photobacterium). He detected oxygen produced during photosynthesis by crushed clover leaves covered with a suspension of luminous bacteria. McElroy[64] first used the firefly system to measure ATP in microgramme amounts. The development of sensitive photomultipliers in the early 1950's resulted in a rapid increase in the uses of bioluminescence assays in a few laboratories, and Strehler and Totter[28] published the first paper on the technique using quantum counting in 1952.

Assay of ATP (adenosine triphosphate)

The basis of the assay is the formation of an activated enzyme-substrate complex, specific for ATP :

$$LH_2 + E + ATP \xrightarrow{Mg^{2+}} E \cdot LH_2AMP + PP_i$$

(LH_2 = luciferin)

This complex then reacts with molecular oxygen to give an excited state enzyme-decarboxy keto luciferin product which, with a quantum yield of nearly unity, emits a yellow-green light (λ_{max} = 562 nm).[1] Thus with ATP the only reactant not in excess, the light emitted is proportional to the amount of this compound. The concentration of magnesium ions is important[43] under certain conditions but 5 mM is generally satisfactory. A cofactor requirement, D(-)luciferin, is apparent but this is present in crude firefly lantern extracts. About 1 mg of luciferin can be prepared from 2000 fireflies.[17] Inactive extracts prepared from commercially available lanterns (perhaps old stock) can usually be traced to a deficiency of this compound. Fortunately synthetic luciferin is now available commercially (Sigma Chemical Co.) and may be synthesised by two different procedures.[65, 66]

It is now standard practice to slow down the decay rate of luminescence by conducting the reaction in phosphate or arsenate buffers, thus making measurements on spectrometers easier to perform.[28, 37] A pH in the range 7.5 to 9.0 is usual for the reactions since the quantum yield exceeds 0.9 at these values. Below about pH 6.5 the yield is only 0.45 and the emission spectrum shifts further into the red (614 nm)[1] where the photomultipliers are considerably less sensitive (see Fig. 1). The use of internal standards is thus mandatory when measuring samples in acid solutions.[37]

The specificity of the reaction for ATP is very high but recently two other nucleotides have been shown to be active in light production. Thus, 3-iso-AMP (ribose attached to N-3 of adenine)[67-70] produces red light as does ε-AMP (but not ε-ATP).[69, 70] However, these two compounds are unlikely to be present in biological materials and thus the reaction can be considered to be a specific assay for ATP. However, the use of crude luciferase preparations does lead to other problems of specificity associated with the presence of adenylate kinase and nucleoside diphosphate kinases. The enzymes mediate the in situ formation of ATP in the presence of ADP and other nucleoside triphosphates. This is readily detected since under these conditions light output increases steadily for at least half a minute rather than staying steady or decreasing slightly. Rasmussen and Nielsen[45] have purified the luciferase by filtration on high porosity gels and have considerably reduced the level of these two enzymes. While the purification and crystallisation of luciferase has been described,[71] its use in pure form in these assays is not economical at the present time. Cormier and Totter[14] have pointed out that ATP assays could be readily carried out without depleting the firefly population by using the readily available bacterial enzyme to assay NADH as follows:

$$\text{ATP} + \text{NMN} \longrightarrow \text{NAD} + \text{PP}_i$$

$$\text{NAD} + \text{Malate} \longrightarrow \text{Oxalacetate} + \text{NADH}$$

$$\text{NADH} + \text{FMN} \longrightarrow \text{FMNH}^+ + \text{NAD}$$

$$\text{FMNH}^+ + O_2 + \text{R·CHO} \longrightarrow \text{Products} + \text{Light}$$

(NMN = nicotinamide mononucleotide)

The sensitivity attainable with a particular firefly preparation is of course dependent on the level of endogenous ATP. The latter can be reduced considerably by incubating the crude extract of 20°C for several hours. A better procedure to remove ATP is the use of Sephadex G-25 gel filtration.[45] Unfortunately this procedure also removes the luciferin cofactor, which must then be added back to the reaction mixture. About 10^{-15} moles ATP can then be detected by this method. Lyman and DeVincenzo[19] recommend the addition of small amounts of apyrase to the luciferase to hydrolyse endogenous ATP. The present author finds this procedure difficult to control.

A good deal of attention has been paid to the proper extraction of ATP from biological material. Thus, it is important to assess whether its level has changed during sampling, extraction or storage.[56, 72-75] Those extraction procedures tested have included: boiling water,[28] hot tris buffer,[73] perchloric acid,[37] trichoroacetic acid,[76] ethanol,[46, 76] ethanol with Na-EDTA,[77] sulphuric acid for lake sediments,[78] dimethylsulphoxide at acid or neutral pH for spacecraft water supplies,[79, 80] chloroform,[196] n-butanol in tris buffer,[18] N-bromosuccinamide[81] and butanol-octanol buffer[75] for soils. Each extraction procedure has usually been developed specially for the material under investigation. Dilute perchloric acid is a highly efficient extractant for many biological materials and ATP is quite stable in it at 0°C.[37] However, the removal of perchlorate as the insoluble potassium salt should be avoided since ATP co-precipitates and is consequently lost.[82-84] A number of anions and cations are inhibitory, including perchlorate and chloride, and which reduce luminescence.[28, 30, 45, 46, 85] This is remedied by using more luciferase. The importance of using an internal standard in such cases cannot be overemphasised. Phenolic material is frequently a problem since it inhibits luciferase[76] and must therefore be removed from plant extracts before assaying for ATP. Stenlid[86] has reported that iso-flavones also inhibit. The use of internal standards is mandatory for coloured samples, since absorption quenching occurs under these conditions.

The presence of ATPases in tissues, especially plants, makes the estimation of endogenous ATP very difficult and it is usually important to denature them before assay. Thus, in micro-organisms, the injection of the sample into boiling water, boiling buffer or cold acid is often possible, but for multicellular organisms another procedure is required. In our laboratory, freezing small portions of the sample in isopentane (-79°C), then in liquid nitrogen, followed by grinding (-196°C) in a top-drive homogeniser, has proved successful. The powder is then mixed with ice-cold 0.4N perchloric acid and as the powder thaws, perchloric acid rapidly enters the cells, denatures the enzyme and extracts the ATP.[87]

Assay of FMN and NADH

Light emission from the bacterial system appears to come from a protonated flavin, $FMNH^{+}$, since its spectral emission corresponds very closely with the bioluminescence spectrum of the species.[20, 88, 89] Since reduced FMN has a half-life of around 0.1 s in air, it follows that the analytical assay for the reduced form is unlikely to be of value. However, with the aid of NADH-linked dehydrogenase, the system can be coupled to the bacterial luciferase and thus makes possible the assay of both FMN and NADH when these are in turn rate limiting.[34, 38-40, 50] The system may also be used to measure NADPH but the sensitivity is only 5% of that for NADH.[34, 38-40]

The requirement of a long-chain aldehyde is a peculiar feature of the luciferase,[90] and decanal and tetradecanal are frequently employed because they give near optimum light output,[91, 92] although optimal chain length varies somewhat with bacterial type. The reaction is very temperature dependent and has an optimum of about 28°C.[38, 39] The amount of light produced is also dependent on the ionic strength of the buffer as well as on its pH.[39] The reaction is insensitive to cyanide, azide and fluoride at 1 mM and is inhibited by 10^{-5}M Zn^{2+} and Ca^{2+}.[39] The sensitivity of the procedure is in the picomole range,[39] but this depends very much on the luciferase preparation. However, Lee et al.,[20] using a luciferase prepared from Photobacterium fischeri, have obtained a maximum sensitivity of 10^{-16} mole and were limited only by the thermal noise of the photomultiplier. The proportionality between substrate concentration and photon output holds over a wide range, is far superior in sensitivity to fluorescence methods and is not subject to interference.

As mentioned previously, the assay procedure may be used for FMN.[39] However, the number of photons counted is not directly proportional to the amount of FMN present, since on increasing its concentration its yellow colour absorbs increasing amounts of emitted light[39] and there is some evidence that FMN may be a substrate inhibitor for the dehydrogenase.[93] Neither FAD nor riboflavin can replace FMN in the reaction; however, it is claimed that the reduced forms of FAD, riboflavin and other flavins are weakly active in bacterial luminescence but give different emission spectra.[94, 95]

Extraction of NADH[96] and NADPH[97] must be made in alcoholic alkali or with 0.1M NaOH[54] since they are acid labile. NAD[98] and NADP[99] may be extracted using dilute perchloric acid. When NAD/NADH ratios are to be determined, the NADH is measured first and then the NAD is reduced using malic dehydrogenase and a further measurement is made. Mixtures of NADH and NADPH can be measured selectively and specifically by oxidising NADPH with oxidised glutathione and glutathione reductase or specifically by oxidising NADH with acetoacetate and 3-hydroxybutyrate dehydrogenase. The use of internal standards to monitor quenching is of course mandatory.

FMN is usually structurally bound to tissue proteins and can be released by perchloric acid,[100] boiling water or aqueous phenol for yeast.[101] Total phosphoflavin (FMN plus FAD) may be estimated by boiling perchloric acid extracts to convert FAD to FMN. FAD alone can be measured by performing the assay before and after the hydrolysis.[100]

Assay of PAP (adenosine 3'-phosphate 5'-phosphate)

The bioluminescent system of the coelenterate Renilla reniformis has been studied extensively by Cormier and his associates[8, 14-17, 102-106] and it has been shown that light production in cell-free extracts of this organism is absolutely dependent on PAP. The bioluminescent system involves two stages, thus :[102, 103]

$$\text{Luciferyl sulphate} \xrightarrow{\text{Luciferyl sulphokinase}} \text{Luciferin} + \text{Sulphate}$$

$$\text{Luciferin} + O_2 \xrightarrow{\text{Luciferase}} \text{Products} + h\nu(\lambda_{max} = 485\text{ nm})$$

Since PAP is readily formed by the mild hydrolysis of PAPS (adenosine 3'-phosphate 5'-sulphatophosphate), a sensitive method becomes available for these compounds, which are of considerable interest in sulphate metabolism in micro-organisms, plants and animals. Since the K_m for PAP in the reaction is 7.3×10^{-8} M,[102] such a system is likely to be very sensitive. A simple method using the liquid scintillation counter and the Renilla system has now been developed by the author[40] and a fuller account will be published later. The sensitivity attained thus far is a few picomoles of PAP. The procedure depends on adding PAP to the Renilla enzyme complex together with luciferyl sulphate,[107, 108] after which the photon output increases linearly with time and the slope is proportional to the amount of PAP up to about 150 picomoles in the assay procedure.[40] An internal standard is, of course, required and this can be added to the reaction mixture after the unknown sample has been assessed. This method has been of value in showing that PAPS is produced in higher green plants, but depends on the presence of AMP to protect the PAPS from hydrolysis by various nucleotidases.[40]

COUPLED REACTIONS

Strehler and Totter[28] were the first to apply bioluminescence reactions in vitro to determine important intermediates in biological systems. They used firefly extracts to assay not only ATP but also a number of substrates and enzymes by measuring the light produced with a quantum counter cooled in liquid nitrogen.[109, 110] More recently, Strehler has published several reviews and articles based on coupled reactions with the firefly system.[51, 52, 111, 112]

In the coupled system, whether it be for enzyme or substrate, the procedure is dependent on the material to be assayed being present in rate-limiting amounts. Thus to obtain a linear response, the concentration should be less than the K_m value.

The sensitivity achieved is several orders of magnitude greater than for spectrophotometry. Thus NADH measured in a spectrophotometer at 340 nm with an o.d.

of 0.06 in a 1 cm path 2.5 ml cuvette would be about 2.5 x 10^{-8} moles. Since a picomole is readily measured[38, 39] with the Photobacterium system (similar volume), it follows that the bioluminescence system is some 25, 000 times more sensitive.

The applications for coupled reactions fall into three classes :

1. The assay of the substrate (or closely related substrate, such as NADH), e.g. ATP, NADH, NADPH, FMN, PAP.
2. The assay of enzymes. Strehler[51] mentions adenylate kinase and creatine kinase as examples.
3. The assay of related compounds which can be used to mediate a reaction in which ATP or NADH/NADPH is formed or utilised. Strehler[51] lists ADP, phosphocreatine, phosphoenolpyruvate and glucose in this category.

In classes 2 and 3 there have been recent reports concerning the assay of other compounds. Balharry and Nicholas[59] have described the measurement of ATP-sulphurylase and adenosine phosphosulphate (APS); both the enzyme and APS are particularly difficult to assay by other means since they frequently occur at low levels and APS is very labile. Stanley[40, 44] has used the reaction mediated by ATP-sulphurylase to measure pyrophosphate :

$$\text{ATP} + \text{Sulphate} \rightleftharpoons \text{APS} + \text{Pyrophosphate}$$

The enzyme is readily isolated and purified from baker's yeast and the reaction carried out at 20°C. The incubation mixture consists of 1 ml 50 mM Tris-HCl, 1 ml 15 mM sodium arsenate and 1 ml 10 mM $MgCl_2$, all at pH 8.5. This is counted in a vial for 2 min (0.1 or 0.04 min counts); ATP-sulphurylase (amount depends on its activity) and 20 μlitres firefly extract[37] are then added together with 2 nmoles APS[113] and the mixture counted again for 2 min. Addition of 1-100 picomoles of pyrophosphate cause an immediate rise in the count rate as monitored by the liquid scintillation spectrometer, or by a ratemeter or multichannel analyser operated in multiscale. The initial slope of this third segment of the reaction is proportional to the amount of pyrophosphate. A calibration curve and internal standards are of course essential. The use of phosphate buffers should be avoided since pyrophosphate is a frequent contaminant.

Malate and oxaloacetate can be measured in picomole amounts by coupling to the luminous system of Photobacterium.[44] The reaction is mediated by malic dehydrogenase, thus :

$$\text{Oxaloacetate} + \text{NADH} \rightleftharpoons \text{Malate} + \text{NAD}$$

The equilibrium constant at neutral pH is such that malate formation is heavily favoured but conditions for the reverse reaction, to measure malate, can be made more favourable by increasing the pH to 9.5 and using a large excess of NAD or acetyl pyridine NAD. The initial rate at which NADH is utilised or produced is a direct measure of oxaloacetate or malate respectively.

The reaction mixtures used are as follows :

Malate assay	Oxaloacetate assay
2 mlitre 0.05M Tris-HCl, pH 9.5	2 mlitre 0.05M phosphate, pH 7.0
2 μlitres β-mercaptoethanol	2 μlitres β-mercaptoethanol
20 μlitres tetradecanal in ethanol (saturated)	20 μlitres tetradecanal in ethanol (saturated)
2 nmoles FMN	2 nmoles FMN

20 μlitres 1 mg/ml dehydrogenase/ luciferase from Photobacterium	20 μlitres 1 mg/ml dehydrogenase/ luciferase from Photobacterium
5 μlitres 0.5 mg/ml malate dehydrogenase (pig heart; sp. act. 1100 units/mg)	5 μlitres 0.5 mg/ml malate dehydrogenase (pig heart; sp. act. 1100 units/mg)
500 nmoles NAD	10-1000 picomoles NADH (depends on level of oxaloacetate to be assayed)
10-200 picomoles malate	10-1000 picomoles oxaloacetate

Preparation of the tetradecanal solution requires special attention.[39,40] Pig-heart malate dehydrogenase was obtained from Boehringer (Mannheim, Germany), Photobacterium luciferase from Sigma (St. Louis, Missouri, U.S.A.) and tetradecanal from Fluka (Switzerland).

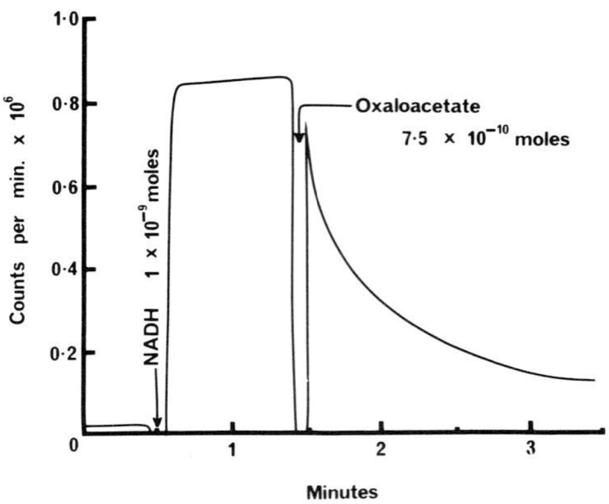

Fig. 2. Reaction sequence for the assay of oxaloacetate (see text for details).

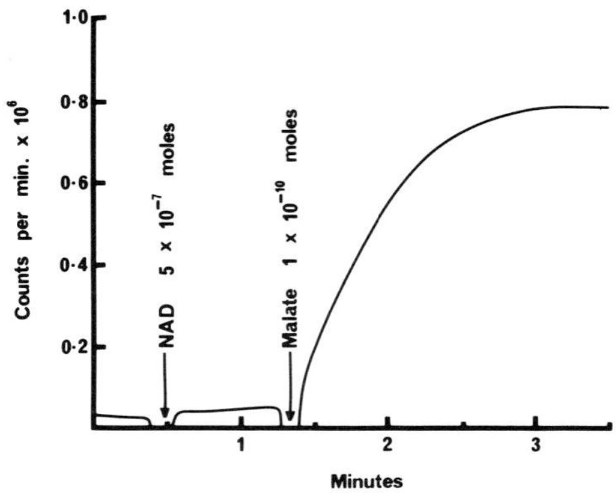

Fig. 3. Reaction sequence for the assay of malate (see text for details).

Figures 2 and 3 show the reaction scheme for oxaloacetate and malate, and calibration curves for these substrates are presented in Figs. 4 and 5. It should be noted that the apparent equilibrium level of NADH after the reaction is less than expected due to the presence of inhibitory or coloured material in the oxaloacetate sample. This illustrates well the necessity for always including a second assay with an internal standard. The non-linearity of the calibration curve for oxaloacetate is due to the level of NADH employed, only 120 picomoles. Thus at higher concentrations of oxaloacetate (>50 picomoles), a significant proportion of the NADH is utilised very quickly and despite the very favourable equilibrium constant the reaction rate slows.

Another technique utilising the Photobacterium complex is that described by Nicholas and Clarke[60] for measuring ammonia. The coupled reaction involved the oxidation of NADH in the presence of glutamic acid dehydrogenase and α-oxoglutaric acid. The sensitivity is around 10^{-8} moles.

Another enzyme which has been measured by a bioluminescence procedure is phosphodiesterase.[114] The assay involves making the enzyme rate-limiting in a reaction sequence in which adenosine 3', 5'-monophosphate (cyclic-AMP) reacts via adenosine 5'-phosphate to give ATP. This enzyme can be readily detected in as little as 0.1 μg of brain tissue.[114]

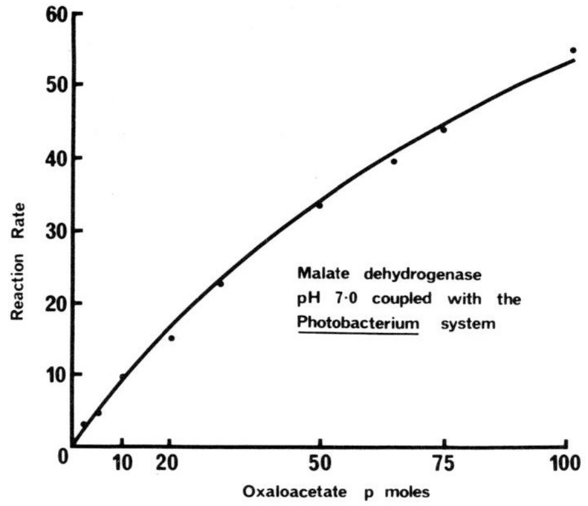

Fig.4. Calibration curve for oxaloacetate.

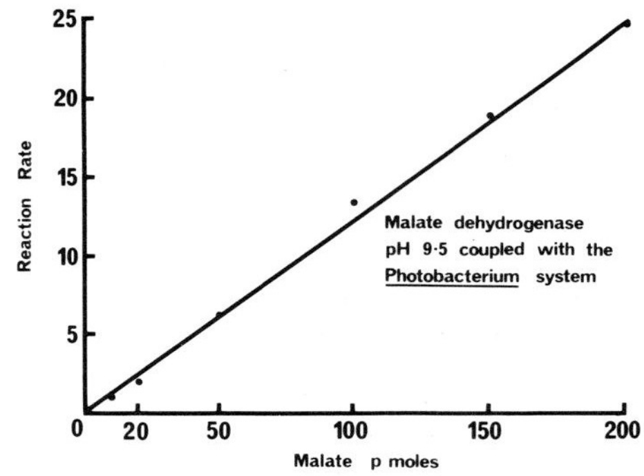

Fig.5. Calibration curve for malate.

Johnson et al.[115] have used phosphodiesterase to assay adenosine 3', 5'-monophosphate (cyclic-AMP) using the following reaction sequence and then measuring the ATP produced:

$$\text{Cyclic-AMP} \xrightarrow{\text{Phosphodiesterase}} \text{5'-AMP}$$

$$\text{5'-AMP} + \text{ATP} \underset{}{\overset{\text{Adenylate kinase}}{\rightleftharpoons}} \text{2 ADP}$$

$$\text{2 ADP} + \text{2 PEP} \underset{}{\overset{\text{Pyruvate kinase}}{\rightleftharpoons}} \text{2 ATP} + \text{2 Pyruvate}$$

PEP = Phosphoenolpyruvate

The sensitivity is around 1 picomole cyclic-AMP. Johnson[116] has published further on the measurement of cyclic-AMP by this procedure and Ebadi[117] has reviewed the method for measuring cyclic-AMP in urine and various tissues.

The balance of AMP, ADP and ATP is especially important in the control of cellular processes and a number of workers have used the firefly luciferase technique to measure mixtures of these compounds. Parkinson and Medley[118] measured these compounds in human plasma using the following enzyme transformations:

$$\text{AMP} + \text{ATP} \overset{\text{Adenylate kinase}}{\rightleftharpoons} \text{2 ADP}$$

$$\text{ADP} + \text{PEP} \xrightarrow{\text{Pyruvate kinase}} \text{ATP}$$

Thus a number of measurements of ATP are required and the amounts of nucleotides are obtained by difference. Pradet[119] has used a similar technique for estimating these compounds in plant tissues. Brooks[120-122] has measured ATP and ADP in spermatazoa and blastocysts and Holmsen et al. their levels in plasma[123] and blood platelets,[77] again employing the standard pyruvate kinase transformation.

The use of a standard light source

In the preceding sections mention has been made of the importance of calibration curves from which results can be obtained. However, to check an instrument from day to day or to work on an absolute basis, a standard light source is required. Hastings and Weber[124] have described a source prepared from hexadecane-1- ^{14}C in a scintillator of PPO, POPOP in either toluene or monoisopropyl biphenyl.

This of course does not give rise to single discrete photons but to scintillations and is thus suitable for photometers rather than for quantum counters. Lee et al.[125] recommend a luminol standard which can be the source of controlled chemiluminescence (single photons). However, a more convenient single-photon source may be a light-emitting diode (LED) of good long-term stability, preferably doped with gallium arsenide and coated with a phosphor to give an emission between 500 and 600 nm. A small mercury battery to drive the LED and the LED itself could fit readily into a scintillation vial.

GENERAL APPLICATIONS

During the last few years bioluminescence has been used to determine cell numbers and biomass in water and soil. An interesting example is a method used in the NASA programme to detect extraterrestrial life by the presence of either ATP[18, 81] or FMN,[50] which are products of living organisms. The measurement of low levels of bacterial contamination of potable water in spacecraft[79, 80] is also possible by this procedure. Chappelle and Levin,[18] in a survey of nineteen species of bacteria, found a mean of 2×10^{-16} g ATP/bacterium with a coefficient of variation of 100% due to abnormally high levels in three of the nineteen species, while Hamilton and Holm-Hansen[126] report a mean of 2.3×10^{-15} g ATP/bacterium (with a coefficient of variation of 100%) for seven species grown in a chemostat. This ten-fold difference may well be associated with the physiological state of the bacteria,[126] and other work supports this suggestion.[127] However, D'Eustachio et al.[128, 129] find that the ATP level is relatively constant at different growth stages and have also reported levels of ATP for a number of bacterial species.

An important aspect of detecting low concentrations of micro-organisms by these methods is that it provides a very sensitive technique for checking the sterility of solutions. This is especially important in the pharmaceutical and beverage industries since the assay is quick in operation, e.g. a result can be obtained within 10 min. Ames[130] has used this technique to test urine samples for bacteruria and to check the effectiveness of various antibiotic treatments. NASA has also used ATP as a measure of urinary tract infection in humans[131] and units have now been specially designed to automatically determine ATP in such samples.[132, 133]

A particularly important aspect of measuring micro-organisms is in the assessment of biomass (bacteria, fungi, algae, protazoa) in fresh and marine waters, and more recently in soils. Holm-Hansen and his co-workers have pioneered this type of work in the aquatic environment[73, 74, 126, 134-140] and it has now been established that ATP is present only in living cells. The nucleotide diffuses rapidly out of dead cells and probably does not adsorb on non-living matter since it is rapidly broken down. A number of workers have established that the ratio of cellular organic carbon to ATP is around 286[3] for marine and freshwater algae,[74, 134, 135, 141-143] bacteria,[72, 126, 127, 144, 196] fungi[145] and tumour cells.[146] However, the ratio of carbon to chlorophyll *a* is around 100[135, 147] and so the ATP assay provides extra evidence for the validity of biomass estimates. Lee et al.[78, 148] have used the ATP procedure for measuring microbial biomass in lake sediments, and Patterson et al.[149] used it for activated sludge. A number of workers have used the method for measuring biomass in soils[56, 75, 81, 150, 151] but absorption of ATP to clay particles appears to be a problem, as well as poor recoveries with certain soil types.

Numerous applications of the technique for measuring ATP in various tissues have been reported including blood,[151] crayfish stretch receptor cells,[41] human, bovine and octopus spermatozoa,[152-154] houseflies,[155] adrenal chromaffin cells,[156] preimplantation mouse embryos,[157] cerebral mitochondria,[158] *Chlorella*,[46] erythrocytes[58] and plants.[76, 87]

Most bioluminescence systems require oxygen and their sensitivity to it varies widely. Very low levels of oxygen are sufficient to cause light emission in the luminous bacteria and thus they make convenient detectors of this gas. Harvey[4, 159] provides evidence that as little as 0.007 mm Hg can be sufficient for bioluminescence, whereas Meyer[160] states that 1 part O_2 in 10^{10} parts N_2 at amospheric pressure is enough. The partial pressure needed for maximum light emission is also small, being only 1-2 mm Hg.[161, 162] More recently Schindler[163-165] has measured O_2 down to 0.0004 µM and states that luminescence is linearly related to O_2 concentration below 0.1 µM.

The luminous bacteria have been used to monitor the effects of ionising radiation,[166, 167] anaesthetics,[168-171] air pollutants[172-176] and rocket fuels in space capsules.[177] The effects of temperature, pressure and narcotics on bioluminescence in intact cells have been thoroughly investigated.[178] The bacteria have also been employed to screen antibiotics since it has been shown that there is a correlation between the therapeutic effectiveness of the drugs and the brightness of luminescence.[179-183] Harvey[5] has listed a number of papers since 1889 on the effects on the luminous bacteria of such diverse materials as tobacco smoke, alcohols, metal salts, ethers and alkaloids. The bacteria may also be useful in the future to screen algacides for their effectiveness in inhibiting photosynthesis (and hence O_2 evolution) in those algae which cause pollution in reservoirs.

Bioluminescence in the hydromedusid *Aequorea* does not involve the usual luciferin-luciferase system. Instead a photoprotein, aequorin, luminesces in the presence of calcium ions.[184, 185] The process does not have an oxygen requirement. Shimomura *et al.*[186] proposed its use as a microtest for Ca^{2+} and it has been used to examine intracellular Ca^{2+} ion transients in electrically stimulated single muscle fibres of the barnacle *Balanus nubilis.*[187-190] They conclude that there is no relationship between Ca^{2+} concentration and tension. This procedure has also been applied to mitochondria[191] and single nerve fibres in *Loligo forbesi.*[192]

Shimomura et al. have suggested the use of *Odontosyllis* extracts as a microtest for cyanide[193, 194] and *Chaetopterus* extracts to assay ferrous ions.[195]

CONCLUSION

The light emission from bioluminescence reactions has been shown to have a wide range of analytical applications. While some of those discussed have to date been measured only in a photometer or even with the naked eye, the liquid scintillation spectrometer could, in most cases, have been used and with increased sensitivity. The areas in which advances can be expected are those for the assay of specific enzymes and substrates and their uses will no doubt expand rapidly when an instrument becomes available commercially which can automatically process samples. This will then make procedures useful not only in the research field but also in the clinical field of microbiochemistry, e.g. in testing for genetic enzyme deletions in amniocentesis samples, where currently routine tests are not available. Thus far more emphasis has been placed on ATP-based assays, but because of the extremely wide range of dehydrogenase enzymes it is likely that the NADH assay system will become more important in the assay of a wide range of compounds with a high degree of specificity and sensitivity, e.g. prostaglandins.

ACKNOWLEDGEMENTS

The author wishes to gratefully acknowledge the financial assistance of the Australian Commonwealth Wheat Industry Research Council and Professor D. J. D. Nicholas for his encouragement throughout this work.

REFERENCES

1 W.D. McElroy, H.H. Seliger and E.H. White, Photochem. Photobiol. **10**, 153 (1969).

2 H.H. Seliger and W.D. McElroy, Light, Physical and Biological Action, Academic Press, New York and London, 1965.

3 H.H. Seliger, in Chemiluminescence and Bioluminescence (Eds. M.J. Cormier, D.M. Hercules and J. Lee), Plenum Press, New York and London, 1973, p.461.

4 E.N. Harvey, Living Light, Princeton University Press, 1940.

5 E.N. Harvey, in Bioluminescence, Academic Press, New York and London, 1952.

6 W.D. McElroy and H.B. Glass (Eds.), Symposium on Light and Life, Johns Hopkins Univ. Press, 1961.

7 W.D. McElroy and H.H. Seliger, Adv. in Enzymology **25**, 119 (1963).

8 M.J. Cormier and J.R. Totter, Ann. Rev. Biochem. **33**, 431 (1964).

9 A.M. Chase, in Photophysiology (Ed. A.C. Giese), Academic Press, New York and London, 1964, Vol.2, p.389.

10 J.W. Hastings, in Current Topics in Biochemistry (Ed. D.R. Sanadi), Academic Press, New York and London, 1966, Vol.1, p.113.

11 F.H. Johnson and Y. Haneda (Eds.) Bioluminescence in Progress, Princeton University Press, 1966.

12 J.W. Hastings, Ann. Rev. Biochem. **37**, 597 (1968).

13 H.H. Seliger and R.A. Morton, in Photophysiology (Ed. A.C. Giese), Academic Press, New York and London, 1968, Vol.4, p.253.

14 M.J. Cormier and J.R. Totter, in Photophysiology (Ed. A.C. Giese). Academic Press, New York and London, 1968, Vol.4, p.315.

15 F.H. Johnson and O. Shimomura, in Photophysiology (Ed. A.C. Giese), Academic Press, New York and London, 1972, Vol.7, p.275.

16 M.J. Cormier, J. Lee and D.M. Hercules (Eds.), Chemiluminescence and Bioluminescence, Plenum Press, New York and London, 1973.

17 F. McCapra, Endeavour **12**, 139 (1973).

18 E.W. Chappelle and G.V. Levin, Biochem. Med. **2**, 41 (1968).

19 G.E. Lyman and J.P. DeVincenzo, Anal. Biochem. **21**, 435 (1967).

20 J. Lee, C.L. Murphy, G.J. Faini and T.L. Baucom, in Liquid Scintillation Counting : Recent Developments (Eds. Philip E. Stanley and Bruce A. Scoggins), Academic Press, New York and London (in press).

21 G.A. Morton, Appl. Optics **7**, 1 (1968).

22 M. Duquesne, P. Vigny and N. Gabillat, Photochem. Photobiol. **11**, 519 (1970).

23 M. Duquesne, I. Tatischeff and N. Recouvreur, Nuc. Instr. and Meth. **41**, 13 (1966).

24 M. Duquesne, and I. Tatischeff, Nuc. Instr. and Meth. **35**, 181 (1965).

25 J.F. James, Monthly Notices of the Royal Astronomical Society **137**, 15 (1967).

26 H.A.W, Tothill, Document Reference R/P029Z70, E.M.I., Ruislip, Middlesex, England.

27 F. Robben, Appl. Optics **10**, 776 (1971).

28 B.L. Strehler and J.R. Totter, Arch. Biochem. Biophys. **40**, 28 (1952).

29 E. Tal, S. Dickstein and F.G. Sulman, Experientia **20**, 652 (1964).

30 L.S. Aledort, R.I. Weed and S.B. Troup, Anal. Biochem. **17**, 268 (1966).

31 S. Addanki, J.F. Sotos and P.D. Rearick, Anal. Biochem. **14**, 261 (1966).

32 E. Schram, Arch. Int. Physiol. Biochim. **75**, 894 (1967).

33 E. Schram, in The Current Status of Liquid Scintillation Counting (Ed. E.D. Bransome, Jr.), Grune & Stratton, New York and London, 1970, p.129.
34 E. Schram, R. Cortenbosch, E. Gerlo and H. Roosens, in Organic Scintillators and Liquid Scintillation Counting (Eds. D.L. Horrocks and C.-T. Peng), Academic Press, New York and London, 1971, p.125.
35 E. Schram and H. Roosens, in Liquid Scintillation Counting (Eds. M.A. Crook, P. Johnson and B. Scales), Heyden & Son, London and New York, 1972, Vol.2, p.115.
36 E. Schram, in Liquid Scintillation Counting : Recent Developments (Eds. Philip E. Stanley and Bruce A. Scoggins), Academic Press, New York and London (in press).
37 P.E. Stanley and S.G. Williams, Anal. Biochem. 29, 381 (1969).
38 P.E. Stanley, in Organic Scintillators and Liquid Scintillation Counting (Eds. D.L. Horrocks and C.-T. Peng), Academic Press, New York and London, 1971, p.607.
39 P.E. Stanley, Anal. Biochem. **39**, 441 (1971).
40 P.E. Stanley, in Liquid Scintillation Counting : Recent Developments (Eds. Philip E. Stanley and Bruce A. Scoggins), Academic Press, New York and London (in press).
41 S.P. Lin and H.P. Cohen, Anal. Biochem. **24**, 531 (1968).
42 E.M.I., Ruislip, Middlesex, England, (a) Photomultiplier Tubes, 1970, p.8; (b) Photomultiplier Tubes, Supplement, 1973.
43 J.L. Denburg and W.D. McElroy, Arch. Biochem. Biophys. **141**, 668 (1970).
44 P.E. Stanley, in Chemiluminescence and Bioluminescence (Eds. M.J. Cormier, D.M. Hercules and J. Lee), Plenum Press, New York and London, 1973, p.494.
45 R. Nielsen and H. Rasmussen, Acta Chem. Scand. **22**, 1757 (1968).
46 J.B. St. John, Anal. Biochem. **37**, 409 (1970).
47 A.M.Chase, in Methods of Biochemical Analysis (Ed. D. Glick), Interscience, New York and London, 1960, Vol.8, p.61.
48 G.W. Mitchell and J.W. Hastings, Anal. Biochem. **39**, 243 (1971).
49 H. Rasmussen and R. Nielsen, Acta Chem. Scand. **22**, 1745 (1968).
50 E.W. Chappelle, G.L. Picciolo and R.H. Altland, Biochem. Med. **1**, 252 (1967).
51 B.L. Strehler, in Methods of Biochemical Analysis (Ed. D. Glick), Interscience, New York and London, 1968, Vol.16, p.99.
52 B.L. Strehler and J.R. Totter, in Methods of Biochemical Analysis (Ed. D. Glick), Interscience, New York and London, 1954, Vol.1, p.431.
53 S.E. Brolin, E. Borglund, L. Tegner and G. Wettermark, Anal. Biochem. **42**, 124 (1971).
54 S.E. Brolin, C. Berne and U. Isacsson, Anal. Biochem. **50**, 50 (1972).
55 H. Prydz and L.O. Froholm, Acta Chem. Scand. **18**, 554 (1964).
56 B.S. Ausmus, Bull. Ecol. Res. Comm. (Stockholm) 17, 223 (1973).
57 R.H. Hammerstedt, Anal. Biochem. **52**, 449 (1973).
58 L. Dufresne and H.J. Gitelman, Anal. Biochem. **37**, 402 (1970).
59 G.J.E. Balharry and D.J.D. Nicholas, Anal. Biochem. **40**, 1 (1971).
60 D.J.D. Nicholas and G.R. Clarke, Anal. Biochem. **42**, 560 (1971).
61 B. Scales, in Liquid Scintillation Counting (Eds. M.A. Crook, P. Johnson and B. Scales), Heyden & Son, London and New York, 1972, Vol.2. p.101.
62 K. Painter, in Liquid Scintillation Counting : Recent Developments (Eds. Philip E. Stanley and Bruce A. Scoggins), Academic Press, New York and London (in press).

63 M.W. Beijerinck, Proc. Acad. Sci. Amst. **4**, 45 (1902).
64 W.D. McElroy, Proc. Natl. Acad. Sci. U.S. **33**, 342 (1947).
65 S. Seto, K. Ogura and Y. Nishiyama, Bull. Chem. Soc. Japan **36**, 331 (1963).
66 E.H. White, F. McCapra and G.F. Field, J. Amer. Chem. Soc. **85**, 337 (1963).
67 N.J. Leonard and R.A. Laursen, Biochemistry **4**, 354 (1965).
68 N.J. Leonard and R.A. Laursen, Biochemistry **4**, 365 (1965).
69 W.D. McElroy and M. Deluca, in Chemiluminescence and Bioluminescence, (Eds. M.J. Cormier, D.M. Hercules and J. Lee), Plenum Press, New York and London, 1973, p.285.
70 J.A. Secrist, J.R. Barrio, N.J. Leonard, C. Villar-Palase and A.G. Gilman, Science **177**, 279 (1972).
71 A.A. Green and W.D. McElroy, Biochim. Biophys. Acta **20**, 170 (1965).
72 H.A. Cole, J.W.T. Wimpenny and D.E. Hughes, Biochim. Biophys. Acta **143**, 445 (1967).
73 O. Holm-Hansen and Ch. R. Booth, Limnol. Oceanogr. **11**, 510 (1966).
74 O. Holm-Hansen, Plant and Cell Physiol. **11**, 689 (1970).
75 J.R. Anderson and P.I. Davies, Bull. Ecol. Res. Comm. (Stockholm) **17**, 271 (1973).
76 G. Guinn and M.P. Eidenbock, Anal. Biochem. **50**, 89 (1972).
77 H. Holmsen, E. Storm and H.J. Day, Anal, Biochem. **46**, 489 (1972).
78 C.C. Lee, R.F. Harris, J.D.H. Williams, D.E. Armstrong and J.K. Syers, Soil Science Am. Proc. **35**, 82 (1971).
79 G.V. Levin, E. Usdin and A.R. Slonim, Aerospace Medicine **39**, 14 (1968).
80 G.V. Levin, C.-S. Chen and G. Davis, Aerospace Medical Research Laboratory: TR–67–71, Aerospace Medical Research Laboratories, WPAFB, Ohio, May, 1971.
81 N.H. MacLeod, E.W. Chappelle and A.M. Crawford, Nature **223**, 267 (1969).
82 P.E. Stanley, unpublished results.
83 J. McD. Stewart, Ph.D. Thesis, Oklahoma State University, University Microfilms, Ann Arbor, Michigan (Diss. Abst. 69–14, 342) (1968).
84 R.L. Bieleski and R.E. Young, Anal. Biochem. **6**, 54 (1963).
85 K. Van Dyke, R. Stitzel, T. McClellen and C. Szustkiewicz, Clin. Chem. **15**, 3 (1969).
86 G. Stenlid, Phytochem. **9**. 2251 (1970).
87 S.G. Williams, Plant Physiol. **45**, 376 (1970).
88 M. Eley, J. Lee, J.-M. Lhoste, C.Y. Lee, M.J. Cormier and P. Hemmerich, Biochemistry **9**, 2902 (1970).
89 C.L. Murphy, M.S. Thesis, University of Georgia (1973).
90 M.J. Cormier and B.L. Strehler, J. Cell. Comp. Physiol. **44**. 277 (1954).
91 J.W. Hastings, J.A. Spudich and G. Malnic, J. Biol. Chem. 238, 3100 (1963).
92 J. Lee and C.L. Murphy, in Chemiluminescence and Bioluminescence (Eds. M.J. Cormier, D.M. Hercules and J. Lee), Plenum Press, New York and London, 1973, p.381.
93 P.E. Stanley, unpublished results.
94 G. Mitchell and J.W. Hastings, J. Biol. Chem. **244**, 2572 (1969).

95 J.W. Hastings, A. Eberhard, T.O. Baldwin, M.Z. Nicoli, T.W. Cline and K.H. Nealson, in Chemiluminescence and Bioluminescence (Eds. M.J. Cormier, D.M. Hercules and J. Lee), Plenum Press, New York and London, 1973, p.369.
96 M. Klingenberg, in Methods of Enzymatic Analysis (Ed. H.-V. Bergmeyer), Academic Press, New York and London, 1963, p.531.
97 M. Klingenberg, in Methods of Enzymatic Analysis (Ed. H.-V. Bergmeyer), Academic Press, New York and London, 1963, p.537.
98 M. Klingenberg, in Methods of Enzymatic Analysis (Ed. H.-V. Bergmeyer). Academic Press, New York and London, 1963, p.529.
99 M. Klingenberg, in Methods of Enzymatic Analysis (Ed. H.-V. Bergmeyer), Academic Press, New York and London, 1963, p.535.
100 E.W. Chappelle and G.L. Picciolo, in Methods in Enzymology (Eds. D.B. McCormick and L.D. Wright), Academic Press, New York and London, 1971, Vol.18B, p.385.
101 A.G. Fazekas and K. Kokai, in Methods of Enzymology (Eds. D.B. McCormick and L.D. Wright), Academic Press, New York and London, 1971, Vol.18B, p.385.
102 M.J. Cormier, J. Biol. Chem. **237**, 2032 (1962).
103 M.J. Cormier, K. Hori and Y.D. Karkanis, Biochemistry **9**, 1184 (1970).
104 K. Hori and M.J. Cormier, Biochim. Biophys. Acta **102**, 386 (1965).
105 K. Hori and M.J. Cormier, Biochim. Biophys. Acta **130**, 420 (1966).
106 J.M. Anderson and M.J. Cormier, in Chemiluminescence and Bioluminescence (Eds. M.J. Cormier, D.M. Hercules and J. Lee), Plenum Press, New York and London, 1973, p.387.
107 K. Hori and M.J. Cormier, Proc. Natl. Acad. Sci. U.S. **70**, 120 (1973).
108 K. Hori and M.J. Cormier, in Chemiluminescence and Bioluminescence (Eds. M.J. Cormier, D.M. Hercules and J. Lee), Plenum Press, New York and London, 1973, p.361.
109 B.L. Strehler, Arch. Biochem. Biophys. **34**, 239 (1951).
110 W.E. Arthur and B.L. Strehler, Arch. Biochem. Biophys. **70**, 507 (1957).
111 B.L. Strehler and W.D. McElroy, in Methods in Enzymology (Eds. S.P. Colowick and N.O. Kaplan), Academic Press, New York and London, 1957, Vol.3, p.871.
112 B.L. Strehler, in Methods of Enzymatic Analysis (Ed. H.-V. Bergmeyer), Academic Press, New York and London, 1963, p.559.
113 C.A. Adams, G.M. Warnes and D.J.D. Nicholas, Anal. Biochem. **42**, 207 (1971).
114 B. Weiss, R. Lehne and S. Strada, Anal. Biochem. **45**, 222 (1972).
115 R.A. Johnson, J.G. Hardman, A.E. Broadus and E.W. Sutherland, Anal. Biochem. **35**, 91 (1970).
116 R.A. Johnson, Adv. Cyclic Nucleotide Res. 2, 81 (1972).
117 M.S. Ebadi, Adv. Cyclic Nucleotide Res. 2, 89 (1972).
118 P.I. Parkinson and E.Medley, in Liquid Scintillation Counting (Eds. M.A. Crook, P. Johnson and B. Scales), Heyden & Son, London and New York, 1972, p.109.
119 A. Pradet, Physiol. Veg. **5**, 209 (1967).
120 D.E. Brooks, J. Reprod. Fert. **23**, 525 (1970).
121 D.E. Brooks and C. Lutwak-Mann, Nature **229**, 202 (1971).
122 D.E. Brooks, J. Reprod. Fert. **25**, 302 (1971).
123 H. Holmsen, I. Holmsen and A. Bernhardsen, Anal. Biochem. **17**, 456 (1966).
124 J.W. Hastings and G. Weber, J. Opt. Soc. Amer. **53**, 1410 (1963).

125 J. Lee, A.A. Wesley, J.F. Ferguson and H.H. Seliger, in Bioluminescence in Progress (Eds. F.H. Johnson and Y. Haneda), Princeton University Press, 1966, p.35.

126 R.D. Hamilton and O. Holm-Hansen, Limnol. Oceanogr. **12**, 319 (1967).

127 W.W. Forrest, J. Bact. **90**, 1013 (1965).

128 A.J. D'Eustachio, D.R. Johnson and G.V. Levin, Bact. Proc. **68**, 131 (1968).

129 A.J. D'Eustachio and D.R. Johnson, Fed. Proc. **27**, 761 (1968).

130 J.S. Ames, M.Sc. Thesis, University of Wisconsin (1970).

131 NASA Technical Support Package 71–10051.

132 NASA Technical Brief 71–10055.

133 G.L. Picciolo, B.N. Kelbaugh, E.W. Chappelle and A.J. Fleig, Goddard Space Flight Center, Greenbelt, Maryland, Publication X–641–71–163 (1971).

134 J. Coombs, P.J. Halicki, O. Holm-Hansen and B.E. Volcani, Exper. Cell Res. **47**, 315 (1967).

135 R.D. Hamilton, O. Holm-Hansen and J.D.H. Strickland, Deep-Sea Research **15**, 651 (1968).

136 O. Holm-Hansen, Limnol. Oceanogr. **14**, 740 (1969).

137 O. Holm-Hansen, in Fertility of Sea (Ed. J.D. Costlow), Gordon & Breach, London, 1971, Vol.1, p.197.

138 O. Holm-Hansen, Univ. Calif. San Diego Report 10P20–124 (1972).

139 O. Holm-Hansen, Bull. Ecol. Res. Comm. (Stockholm) **17**, 215 (1973).

140 J.E. Hobbie, O. Holm-Hansen, T.T. Packard, L.R. Pomeroy, R.W. Sheldon, J.P. Thomas and W.J. Wiebe, Limnol. Oceanogr. **17**, 544 (1972).

141 P.J. Syrett, Arch. Biochem. Biophys. **75**, 117 (1958).

142 R. Daumas and M. Fiala, Marine Biology **3**, 243 (1969).

143 B.R. Berland, D.J. Bonin, P.L. Laborde and S.Y. Maestrini, Marine Biology **13**, 338 (1972).

144 D.P. Kelley and P.J. Syrett, J. Gen. Microbiol. **43**, 109 (1966).

145 E.A. Dawes and P.J. Large, J. Gen. Microbiol. **60**, 31 (1970).

146 P.C.T. Jones, J. Cell. Physiol. **73**, 37 (1969).

147 T.R. Parsons, K. Stephens and J.D.H. Strickland, J. Fish. Res. Bd. Canada **18**, 1001 (1961).

148 C.C. Lee, R.F. Harris, J.D.H. Williams, J.K. Syers and D.E. Armstrong, Soil. Sci. Soc. Am. Proc. **35**, 86 (1971).

149 J.W. Patterson, P.L. Brezonik and H.D. Putman, Environ. Sci. Technol. **4**, 569 (1970).

150 B.S. Ausmus, U.S. IBP Eastern Deciduous Forest Biome, Memo Report No. 71:72, Oak Ridge National Laboratory, 1971.

151 K.G. Doxtader, Seminar at the 69th Ann. Meeting of Amer. Soc. Microbiol., Miami, Florida (1969).

152 E. Beutler and M.C. Baluda, Blood **23**, 688 (1964).

153 D.D. Hoskins, J. Biol. Chem. **248**, 1135 (1973).

154 D.E. Brooks, C. Lutwak-Mann, T. Mann and A.W. Martin, Jr., Proc. Royal Soc. B.**178**, 151 (1971).

155 J.P. Heslop, Biochem. J. **91**, 183 (1964).

156 W.W. Douglas and A.M. Poisner, J. Physiol. **183**, 236 (1966).

157 P. Quinn and R.G. Wales, J. Reprod. Fert. **25**, 133 (1971).

158 H.P. Cohen and J.R. Hyde, Neurology **13**, 355 (1963).

159 E.N. Harvey and T.F. Morrison, J. Gen. Physiol. **6**, 13 (1923).

160 K.P. Meyer, Helv. Phys. Acta **15**, 2 (1942).

161 C.S. Shoup, J. Gen. Physiol. **13**, 27 (1929).
162 H. Shapiro, J. Cell. Comp. Physiol. **4**, 313 (1934).
163 F.J. Schindler, Fed. Proc. **23**, 322 (1964).
164 F.J. Schindler, Ph.D. Dissertation, University of Pennsylvania (1964).
165 F.J. Schindler, in Methods in Enzymology (Eds. R.W. Estabrook and M.E. Pullman), Academic Press, New York and London, 1967 Vol.10, p.629.
166 A.P. Jacobson and K.A. McDermott, J. Lumin. **3**, 419 (1971).
167 I.A. Lerch, Rad. Res. **45**, 63 (1971).
168 D.C. White and C.R. Dundas, Nature 226, 456 (1970).
169 D.C. White and C.R. Dundas, in Cellular Biology and Toxicity of Anaesthetics (Ed. B.R. Fink), Williams & Wilkins Co., 1972, p.168.
170 D.C. White, B. Wardley-Smith and G. Adey, Life Sciences **12**, 453 (1973).
171 M.J. Halsey and E.B. Smith, Nature **227**, 1363 (1970).
172 W.F. Serat, F.E. Budinger, Jr. and P.K. Mueller, J. Bact. **90**, 832 (1965).
173 W.F. Serat, F.E. Budinger, Jr. and P.K. Mueller, Atmos. Environ. **1**, 21 (1967).
174 W.F. Serat, J. Kyono and P.K. Mueller, Atmos. Environ. **3**, 303 (1969).
175 W.F. Serat, in Introduction to Experimental Aerobiology (Ed. R.L. Dimmick), Wiley, New York and London, 1965, p.113.
176 W.F. Serat, F.E. Budinger, Jr. and P.K. Mueller, AIHL Report No. 14, State of California Dept. of Public Health (1965).
177 E. H.-C. Sie, A. Thanos and A. Jordan, in Bioluminescence in Progress (Eds. F.H. Johnson and Y. Haneda), Princeton University Press, 1966, p.195.
178 F.H. Johnson, H. Eyring and M.J. Polissar, The Kinetic Basis of Molecular Biology, Wiley, New York and London, 1954.
179 G. Rake, C.M. McKee and H. Jones, Proc. Soc. Exp. Biol. N.Y. **51**, 273 (1942).
180 G. Rake, H. Jones and C.M. McKee, Proc. Soc. Exp. Biol. N.Y. **52**, 136 (1943).
181 H. Jones, G. Rake and D.M. Hambre, J. Bact. **45**, 461 (1943).
182 N. Atkinson, R.A.W. Sheppard, N.F. Stanley and P. Melvin, Aust. J. Exptl. Biol. & Med. Sci. 22, 223 (1944).
183 F. Kavanaugh, Bull. Torrey Bot. Club **74**, 414 (1947).
184 O. Shimomura, F.H. Johnson and Y. Saiga, J. Cell. Comp. Physiol. **59**, 223 (1962).
185 O. Shimomura, F.H. Johnson and Y. Saiga, Science **140**, 1139 (1963).
186 O. Shimomura, F.H. Johnson and Y. Saiga, Science **140**, 1339 (1963).
187 E.B. Ridgway and C.C. Ashley, Biochem. Biophys. Res. Commun. **29**, 229 (1967).
188 C.C. Ashley and E.B. Ridgway, Nature, **219**, 1168 (1968).
189 C.C. Ashley and E.B. Ridgway, J. Physiol. **209**, 105 (1970).
190 C.C. Ashley, Endeavour **30**, 18 (1971).
191 B. Chance, A. Azzi, I.Y. Lee, C.P. Lee and L. Mela, FEBS Symp. **17**, 233 (1969).
192 P.F. Baker, A.L. Hodgkin and E.B. Ridgway, J. Physiol. **214**, 33P (1971).
193 O. Shimomura, F.H. Johnson and Y. Saiga, J. Cell. Comp. Physiol. **62**, 9 (1963).
194 O. Shimomura, J.R. Beers and F.H. Johnson, J. Cell. Comp. Physiol. 58, 113 (1964).
195 O. Shimomura and F.H. Johnson, In Bioluminescence in Progress (Eds. F.H. Johnson and Y. Haneda), Princeton University Press, 1966, p.495.
196 A.M. Dhople and J.H. Hanks, Appl. Microbiol. **26**, 399 (1973).

DISCUSSION

B.E. Gordon: Has this technique been in fact used for O_2 determination in the low ppm (even 0.01 ppm) region? Also for a flowing system?

P.E. Stanley: My references will quote Meyer[160] who states that 1 part of O_2 in 10^{10} parts of N_2 at atmospheric pressure can be detected! Schindler[163-165] has shown that luminescence is linearly related to O_2 down to 0.0004 μM. These are of course for the bioluminescent bacteria. Higher concentrations above around 1–2 mm Hg need a different organism since this level of O_2 is saturating for the bacteria. I have no knowledge of such a system being used for flow measurements. Perhaps the bacteria could be metered in with the gas and the light measured continuously as the mixture passes a photomultiplier placed close to the mixing point.

J. de Bersaques: What are your experiences in the determination of enzyme activities?

P.E. Stanley: The system is very good for dehydrogenases and ATP-sulphurylase. The appearance of NADH or ATP can be followed continuously. We have used it for measuring malate dehydrogenase and adenylate kinase and as long as the correct controls are used (plus internal standards) these assays are perfectly satisfactory and at least a thousand times more sensitive than conventional spectrophotometric techniques.

Chapter 22

The Continuous Measurement of Carbon-14 Activity in Circulating Blood by Liquid Scintillation Counting

B.F.J. Page

Department of Metabolism and Pharmokinetics,
Huntingdon Research Centre, Huntingdon, England

During the course of work at the Huntingdon Research Centre on the pharmacological aspects of cigarette smoking, a comparison was required between blood levels of nicotine attained by the intravenous route with those achieved via the mouth or the lungs, as it is obtained by smokers. It was felt that this comparison could best be achieved by the administration of Carbon-14 labelled nicotine and subsequent measurement of Carbon-14 levels in blood by liquid scintillation counting. Initial work showed that it would be desirable to be able to monitor the Carbon-14 levels in blood continuously in order to follow rapid fluctuations in blood level of Carbon-14 nicotine during the initial stages of absorption. It was appreciated that this technique would not differentiate between nicotine and other Carbon-14 labelled species resulting from metabolism, but for a comparative study of nicotine absorption this should not matter.

Flow cell devices for the continuous determination of low-energy β-emitting radionuclides in chromatographic column effluents[1, 2] and a flow cell device for determination of high-energy β- and γ-emitting radionuclides in flowing blood,[3] using flow cells fitted to liquid scintillation counters, have been reported. None of these systems was found to be suitable for the determination of low-energy β-emitting radionuclides in flowing blood, and consequently a flow cell for this purpose was constructed from KL211 polystyrene plastic scintillator (Koch-Light Ltd., Colnbrook, England). The U-shaped flow channel was formed in the plastic scintillator so that as large a contact area as possible existed between the blood and the scintillator for events to occur while still presenting the smallest possible area of blood in the field of view of the photomultiplier tubes. As can be seen in Fig. 1, this was achieved by making the flow channel of flat rectangular cross-section (1 in x 0.025 in). The cell (volume 1 ml) was mounted in an I.C.N.-Tracerlab 'Coruflow' SCE542 manual liquid scintillation counter, fitted with dual photomultiplier tubes linked by anti-coincidence circuitry to an I.C.N.-Tracerlab C.M.F.101 dual ratemeter/integrator. Two 'Coruflow' instruments were normally used, one to monitor venous blood and the other to monitor arterial blood, and each instrument

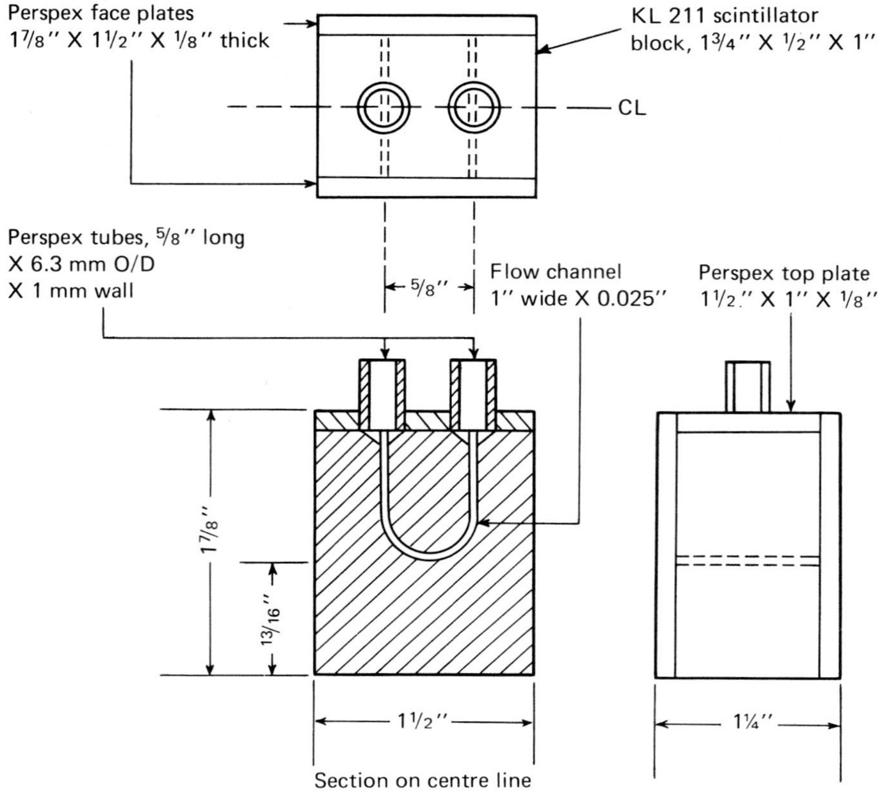

Fig.1. Plastic scintillator flow cell, manufactured by the Scintillator Division of Koch-Light Ltd., Colnbrook, Buckinghamshire, England.

was connected to one channel of the ratemeter/integrator. The signals from the dual ratemeter/integrator were displayed on a Honeywell 'Electronic 194' dual-pen recorder.

The apparatus was calibrated using both water and whole blood and counting efficiency was found to lie between 1.5 and 3%, depending on the individual cell used. There was some variability between cells, which tended to increase with use. The background count-rate of the cells also increased slowly with use and when this was unacceptable the cells were discarded. Alteration of blood flow-rate through individual cells produced no change in the measured Carbon-14 over the range 0–100 ml min^{-1}, as long as there was a constant level of Carbon-14 in the blood passing through the cell. However, when the level of Carbon-14 in the blood passing through the cell was fluctuating, changes in flow produced large differences in the measured Carbon-14 until the Carbon-14 reached a constant level. For this reason, the blood flow through each cell was routinely monitored using a Nycotron blood flowmeter equipped with extracorporeal flow probes. The signals from the blood flowmeters were recorded on a Devices Polygraph recorder.

The baboon was chosen as the experimental animal because it is large enough for the necessary surgical operations and it is a primate. Baboons were anaesthetised with nembutal and given 5000 units of heparin intravenously every 2 h during each experiment, to prevent the blood clotting in the flow cells. The blood was either obtained from the common carotid arteries and internal jugular veins or from the femoral artery and vein. In each instance the total blood supply was routed through the flow cell system and returned to the blood vessel from which it was taken. It was necessary to pump the venous blood through the cell system, using a peristaltic pump. (The pressure drop through the flow cell system was normally 15–25 mm Hg.) The total blood volume external to the animal through each cell system was of the order of 5–10 ml, depending on the vessels cannulated. The blood was routed to and from the cells through black silicone rubber tubing to prevent the passage of radiation into the cell housing which could give an unacceptable increase in background count-rate. We have used this system to study the absorption of nicotine by the baboon from various sites. Figure 2 shows a typical trace obtained following the

Fig. 2. Arterial (A) and venous (V) concentrations of Carbon-14 (relative units) after repeated i.v. administration of ^{14}C-labelled nicotine into the radial vein of a baboon. Black triangles represent doses of nicotine. Each dose was 0.1 ml containing 22.7 μg (5.1 μCi) nicotine.

intravenous administration of ^{14}C-nicotine. N-Methyl- ^{14}C-nicotine was obtained from the Radiochemical Centre, Amersham, at a specific activity of 223 μCi mg^{-1} (free base) and used either as the bitartrate in 0.9% saline solution or buffered to pH 8.5 in sodium bicarbonate buffer. Each dose consisted of 1 ml of solution, administered as 0.1 ml by syringe at one-minute intervals (black triangles on the figure represent each 0.1 ml dose). The trace marked A represents arterial blood Carbon-14 activity and the trace marked V venous blood Carbon-14 activity. In all cases, the ratemeter was used with a time constant of 30 s, in order to give a smoother trace, and blood flows varied between 5 and 40 ml min^{-1} depending on the individual animal.

In the traces shown, the flow did not vary during the experimental period indicated. Figure 2 clearly shows an individual peak as each dose is injected, whereas these have been effectively damped out in the venous side. The blood was obtained for this experiment from the common carotid arteries and internal jugular veins, so the difference between the venous and arterial Carbon-14 levels could represent to some extent uptake by the brain.

Figure 3 shows the trace obtained following administration of ^{14}C-nicotine into the lungs via the trachea. As can be seen, this is essentially similar to that obtained by the i.v. route, though the arterial peaks are almost absent, and up-

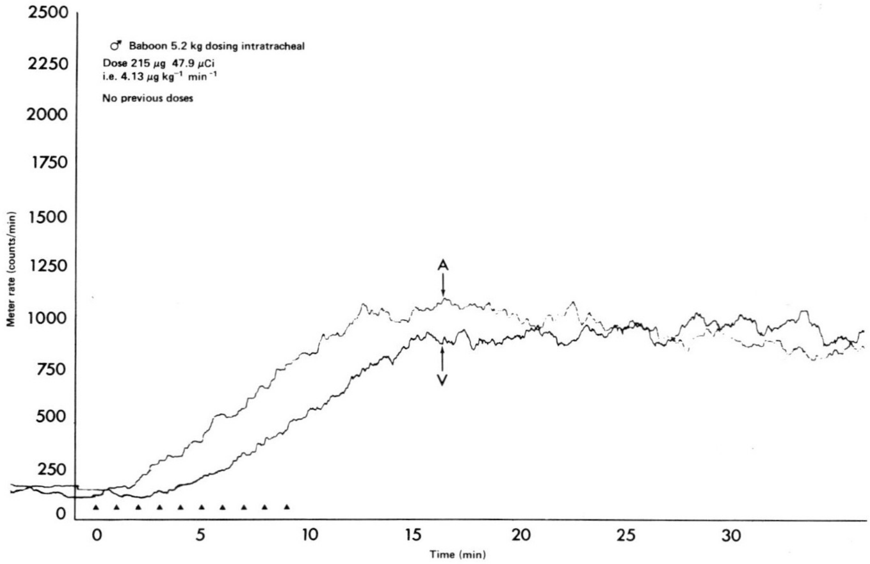

Fig.3. Arterial (A) and venous (V) concentrations of ^{14}C (relative units) after repeated administration of ^{14}C-nicotine into the trachea of a baboon. Black triangles represent doses of nicotine. Each dose was 0.1 ml containing 21.5 μg (4.8 μCi) nicotine.

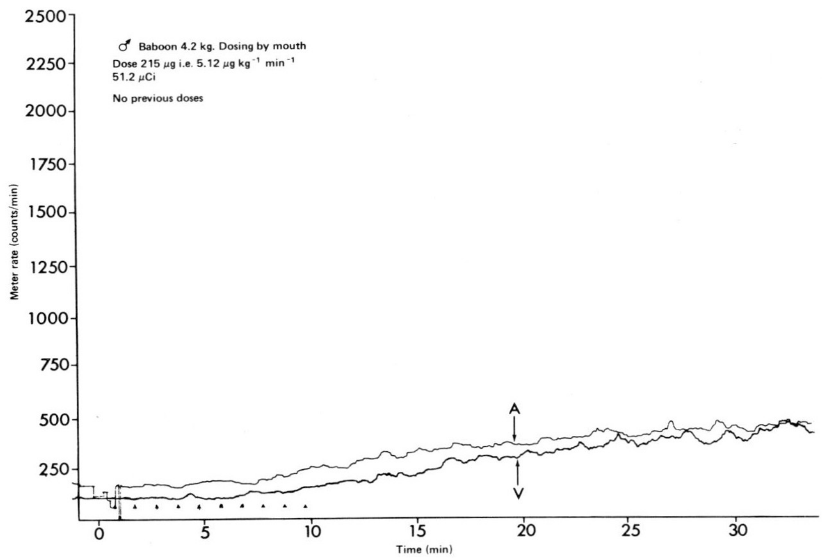

Fig.4. Arterial (A) and venous (V) concentrations of ^{14}C (relative units) after repeated adminstration of ^{14}C-nicotine into the mouth of a baboon. Black triangles represent doses of nicotine. Each dose was 0.1 ml containing 21.5 μg (4.8 μCi) nicotine.

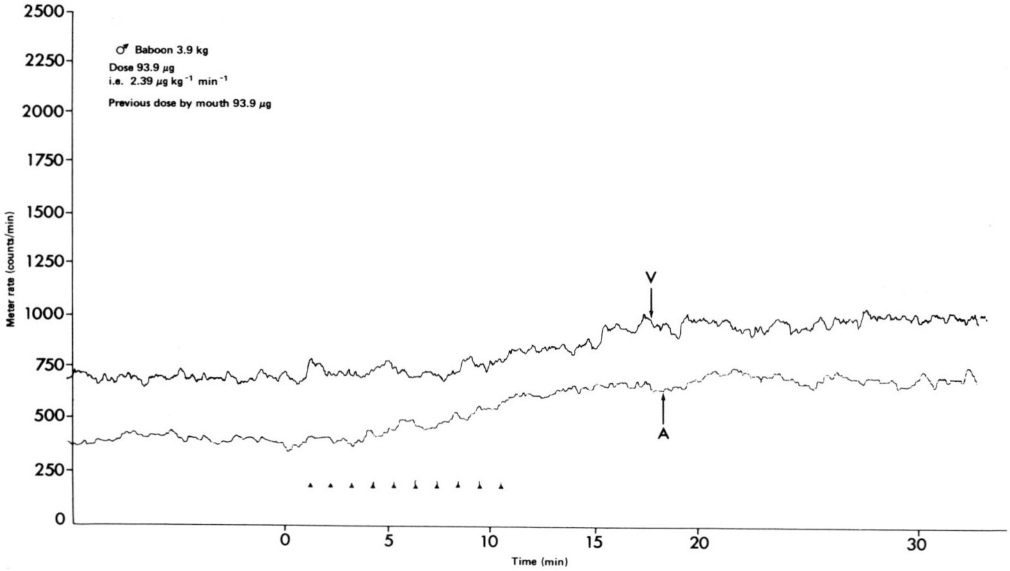

Fig. 5. Arterial (A) and venous (V) concentrations of ^{14}C (relative units) after repeated administration of ^{14}C-nicotine into the nose of a baboon. Black triangles represent doses of nicotine. Each dose was 0.1 ml containing 9.4 µg (2.1 µCi) nicotine.

take is delayed. Figure 4 shows the trace obtained following the sublingual administration of ^{14}C nicotine in sodium bicarbonate buffer. Here the trachea and oesophagus were tied off and blocked with vaseline and bone wax to prevent absorption taking place anywhere but in the mouth. The cells were set up to monitor blood from the femoral artery and femoral vein. Figure 4 clearly shows the absorption is much delayed and very much slower by this route. Figure 5 shows the trace obtained following the administration of the dose into the nose. As can be seen, this route demonstrates an absorption pattern somewhere between that obtained by the i.v./intratracheal routes and the mouth. These experiments are illustrative of the uses to which the technique described can be put. We are at present developing the technique to study not only absorption patterns of drugs, but also uptake of drugs by various organs, and we feel that it should be applicable to a variety of problems.

REFERENCES

1 L. Schutte, J. Chromatogs. **72**, 303 (1972).
2 E. Loos, Anal. Biochem. **47**, 90 (1972).
3 J. Roba, R. Roncucci, and G. Lambelin, Analyt. Lett. **2**, 325 (1969).

DISCUSSION

N.G.L. Harding: I could not read the time scales of your traces, so I wonder if you could comment upon the features of flow cell systems, namely blood coagulation, haemolysis and infection, which could be important in long-term recovery experiments or work with human patients.

B.F.J. Page: Our work to date has only involved terminal experiments with baboons. Anaesthetised animals have been successfully connected to the flow cell system for up to 8 h in the experiments carried out so far. Peristaltic pumps are used to pump the blood through the flow cell systems and to maintain a constant flow rate, so some haemolysis could be occurring. Blood coagulation has not been found to be a problem as long as the animal is kept adequately heparinised.

Chapter 23

The Liquid Scintillation Counting of Iron-55 Using 'Aquasol'

G.A. Sutton and B.R. Harvey

Ministry of Agriculture, Fisheries and Food, Fisheries Radiobiological Laboratory, Lowestoft, Suffolk, England

INTRODUCTION

The work of monitoring and control of radioactive waste disposal to the marine environment, which is the primary responsibility of the Fisheries Radiobiological Laboratory (FRL), involves the detection and assay of numerous artificially produced radionuclides, often at very low levels. Determination is required in a wide variety of environmental materials including water, sediments and biota.

Although Iron-55 appears in the marine environment as a result of discharges of low-level radioactive waste from nuclear power stations, its greatest impact to date has been due to fallout from nuclear weapons testing; for instance, a study of the distribution of Iron-55 in commercial fish species of the North Atlantic by Preston[1] showed levels of $10-10^2$ pCi Iron-55/mg Iron.

In the past, two methods have been used at FRL to count these low-level samples for Iron-55; in both cases preliminary sample preparation involved wet ashing, followed by solvent extraction or ion-exchange chromatography to isolate iron. In one method the iron was electroplated on to a 2.5 cm copper disc and counted using a boron trifluoride proportional counter. Although this method provides an extra stage of decontamination from other radionuclides, the counting efficiency is rather low, even for a thin source (5 mg), and falls rapidly as the source thickness is increased. The other method was that of Eakins and Brown;[2] iron was precipitated as ferric hydroxide, dissolved in a minimum quantity of phosphoric acid and subsequently re-precipitated as the white ammonium ferri-phosphate which, when dispersed in gel scintillant, may be counted in a liquid scintillation spectrometer.

In the present method ferric hydroxide is dissolved in phosphoric acid, which is then incorporated directly into the liquid scintillation cocktail 'Aquasol'; considerable quantities of stable iron can be accommodated with little loss of counting efficiency, thus allowing samples of lower specific activity to be measured than would be possible with other methods. This is of considerable importance when analysing environmental samples in which the specific activity is often low.

EXPERIMENTAL

Preliminary sample treatment

Three essential factors must be considered when preparing samples for liquid scintillation assay of low specific activity Iron-55 :

1. Contamination from other radionuclides must be reduced to a very low level.
2. Stable iron contamination in processing chemicals must be reduced to a minimum and determined by running blanks.
3. Quenching, which may be caused by the constituents of the final iron solution.

When analysing animal tissue these factors, and particularly the second, are most readily controlled by using blood as the source of iron. Fish blood is drained into polyethylene bottles immediately after the fish have been caught; it is then stored at -20°C until required for processing. After thawing, the blood is homogenised in a liquidiser and sub-sampled as required for replicate determinations. 100 ml of blood containing up to 10 mg of iron (more if necessary) is wet-ashed with 50 ml of high purity (Aristar) nitric acid and then 100 ml of an equal mixture of nitric and perchloric acids. The dry residue from this ashing is dissolved in 50 ml of 0.5M HCl buffered to pH 6 with sodium acetate/ammonia solution and the heavy metal complex of iron with diethyl ammonium-diethyl-dithiocarbamate (DDDC) extracted into chloroform. Extraction using 10% DDDC in chloroform is repeated until the organic phase no longer becomes coloured on shaking with the sample solution. The organic layer is evaporated to dryness under an infra-red lamp and the organic residue destroyed by evaporating with concentrated nitric acid and 30% hydrogen peroxide. The final residue — which is white — is evaporated with concentrated hydrochloric acid to remove nitric acid residues, taken up in 5 ml of 2.5M HCl and applied to the top of a 10 ml column of Bio-Rad AG1 x 10 anion-exchange resin (30 cm long) previously conditioned with 2.5M HCl. The column is then washed with 2 bed volumes of 2.5M HCl and the iron fraction is subsequently eluted with 3.5 bed volumes of 0.5M HCl. The stable iron content of this fraction is determined spectrophotometrically. The stable iron blank must be determined repeating the whole process with the reagents only.

A suitable aliquot (e.g. 20 ml) containing a known amount of stable iron is then transferred to a counting vial in which ferric hydroxide is precipitated by the addition of concentrated ammonia (e.g. 3 ml). After centrifuging, the supernate is discarded, the precipitate being taken up in the given quantity of phosphoric acid needed to dissolve it completely.

Selection of optimum sample composition

Samples are prepared by the addition of suitable quantities of Aquasol and water to the solution of iron in phosphoric acid (see Appendix).

In order to achieve optimum counting efficiencies over a wide range of values of stable iron it was found that the relative proportions of water and phosphoric acid had to be varied.

Optimum water content for different amounts of stable iron. Two sets of samples each containing 5 mg of stable iron and the same amount of Iron-55 were prepared (see Appendix), containing 1.6 and 1.4 g of phosphoric acid respectively. 15 g of Aquasol were then added to all samples and varying amounts of water were added to each so that two series of samples were produced, each having a slightly different physical appearance. The counting efficiencies were determined and the results are shown in Fig. 1 (a) and (b).

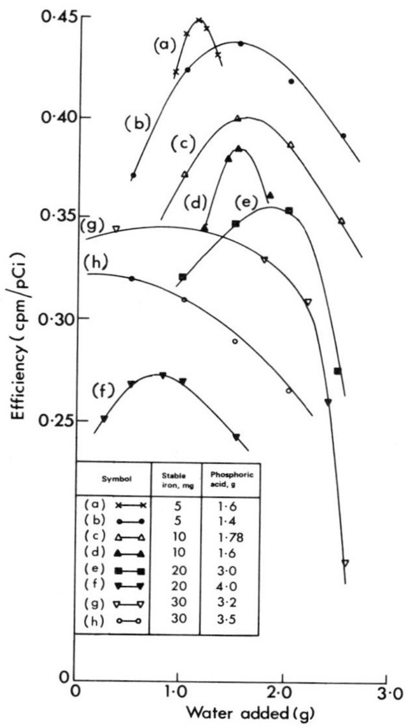

Symbol	Stable iron, mg	Phosphoric acid, g
(a) ×—×	5	1·6
(b) •—•	5	1·4
(c) △—△	10	1·78
(d) ▲—▲	10	1·6
(e) ■—■	20	3·0
(f) ▼—▼	20	4·0
(g) ▽—▽	30	3·2
(h) ○—○	30	3·5

Fig. 1. Variation of counting efficiencies with the water added for various stable iron contents dissolved in a constant amount of phosphoric acid

Further sets of samples were then prepared in the same manner, to determine counting efficiencies for different amounts of stable iron (Fig. 1 (c–h)).

The effect of phosphoric acid. Figure 1 also shows that the phosphoric acid had a marked effect on the counting efficiency. This effect decreases as the amount of stable iron increases, and although the maximum counting efficiency is lower at higher levels of stable iron it is maintained over a much wider range of compositions.

Formation of three-component phase diagrams. The physical characteristics of the counting solution vary with increasing aqueous content, ranging from a clear liquid to a stiff gel. Between these two extremes there is a narrow range in which a two-phase system occurs.

As a starting point for producing the three-component phase diagrams it was assumed that maximum counting efficiency would occur with a clear, colourless sample; this was found to be the case. Figure 2 is a diagrammatic representation of how the counting efficiency varies with phosphoric acid/water content for samples containing 5, 10, 20 and 30 mg of stable iron. The phosphoric acid and water values are plotted along the two axes and the counting efficiencies shown by contours. These contours are formed by drawing a line through samples having the same counting efficiency.

For a sample containing 5 mg of stable iron, the optimum counting efficiency occurs over a very narrow range of phosphoric acid/water values. Slight variations in the amount of either reagent will produce a relatively rapid decrease in the

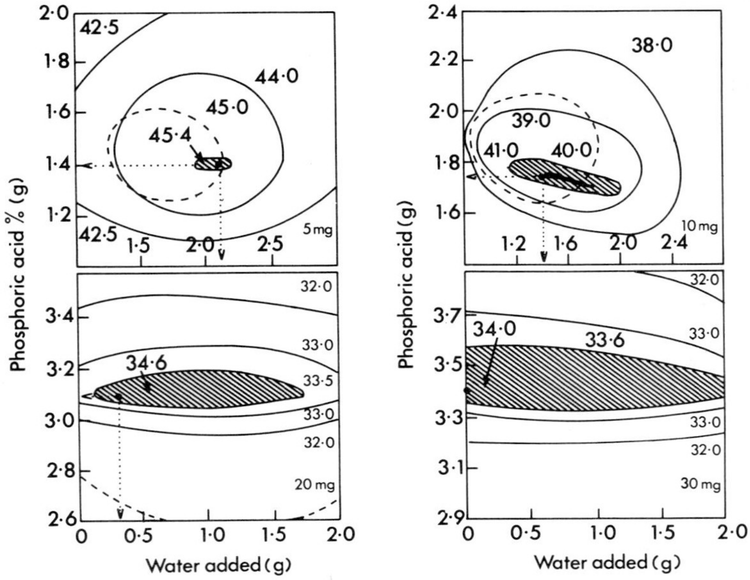

Fig. 2. Diagrammatic representation of the variation of counting efficiency with phosphoric acid and water content. Contours represent efficiency. Broken contours represent single/double phase boundary.

counting efficiency. As the stable iron content of the samples is increased, it is found that the phosphoric acid/water ratio becomes less critical, and variation of either reagent will not produce such a large decrease in the counting efficiency.

Weight for weight, the phosphoric acid has a greater ability to cause gelling of the Aquasol than has water. For low stable iron contents, where the optimum counting efficiency is at a critical position, addition of phosphoric acid to the samples provides a coarse adjustment of the physical characteristics of the sample. Water added drop by drop thus enables the critical position for optimisation to be achieved. As the stable iron content increases, the need for fine adjustment of the physical characteristics of the sample becomes unnecessary because the optimum conditions become less critical, and optimisation can be achieved by using a larger proportion of phosphoric acid.

Characteristics of the final counting solution

Colour and viscosity. By increasing the phosphoric acid content of a sample containing a constant amount of stable iron, the appearance of the final counting solution is changed from a thin cloudy solution, through a clear phase, which gives good counting efficiencies, to a semi-viscous cloudy solution which gives relatively poor counting efficiencies. The viscosity of the solution varies with the stable iron content, i.e. for 5 mg of stable iron the appearance of the solution giving optimum counting efficiency was that of a clear thin solution, whilst for 30 mg of stable iron it was a clear viscous gel. Similar results are obtained by varying the water content, keeping stable iron and phosphoric acid constant.

When in slight excess, phosphoric acid produces gelling and enhances the counting efficiency of the final solution. However, too much phosphoric acid produces a green, apparently insoluble, solid which renders the sample unsuitable for assay.

Formation of two phases. It was found that certain samples suffered from phase separation whilst in the refrigeration compartment of the spectrometer. These samples are indicated by broken contours on Fig. 2. In such samples the problem is resolved by adding a further small quantity of water, which produces a more viscous single phase solution without any serious decrease in the counting efficiency.

Variation of count rate and counting efficiencies with stable iron content

Figure 3 (a) shows the maximum counting efficiencies obtained in these experiments for various amounts of stable iron. The quantities of phosphoric acid and water which produced these maxima are plotted in Fig. 3 (b); for example, for 5 mg of stable iron the sample contained 1.4 g of phosphoric acid and 2.3 g of water, giving an efficiency of 43.4%. From these graphs the operator may select the optimum composition for any sample. At the low stable iron values, maximum counting efficiencies are higher because water has a lower quenching effect than phosphoric acid.

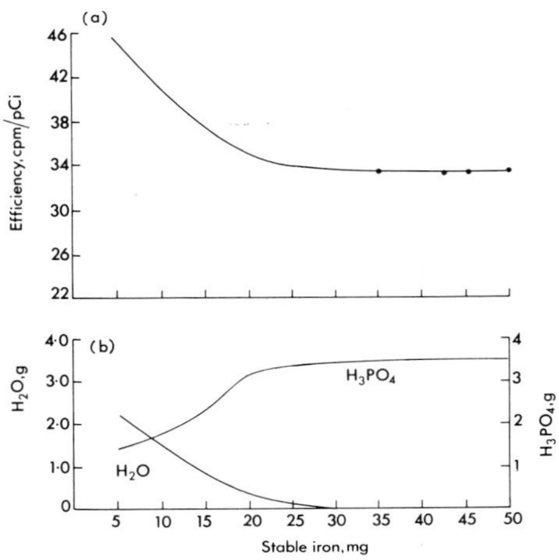

Fig. 3. Optimum sample composition.

COMMENT

The specific activity of environmental samples is often very low and consequently it is desirable to be able to prepare samples for counting which give the maximum counter response. Figure 4 compares the counter response versus stable iron content obtained by the present method and that of Eakins and Brown. It can be seen that for the present method a maximum has not been reached at 50 mg of stable iron; however, this is almost the practical limit for stable iron set by the quantities of reagents needed, relative to the volume of the vials.

Where the amount of stable iron in a sample is, for other reasons, limited to less than 10 mg, the method of gel scintillation (curve b, Fig. 4) offers the advantage of a lower background (about 5 c.p.m.) and slightly higher efficiency. However, where the specific activity is low and sample size is not a limiting factor, then the present method (curve a, Fig. 4) offers considerable advantages even with a higher background (about 10 c.p.m.)

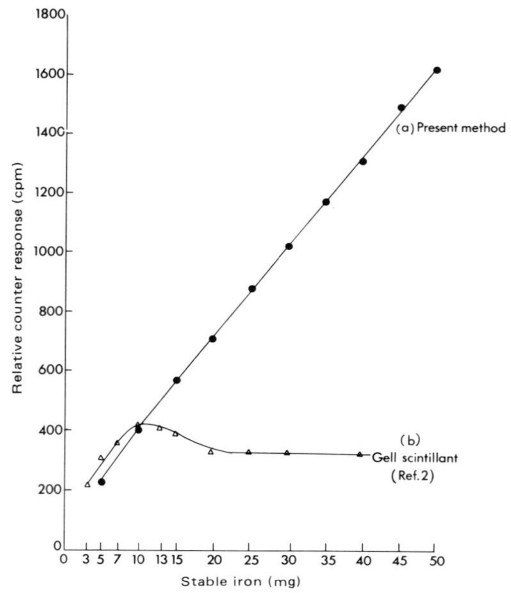

Fig.4. Variation of counter response with weight of stable iron.

REFERENCES

1 A. Preston, Marine Biology **6** (4), 345-9 (1970).
2 J.D. Eakins and D.A. Brown, United Kingdom Atomic Energy Research Establishment Report No. R4946, 1965.
3 R. Loevinger, M. Berman, Nucleonics **9** (1), 26-39 (1951).

APPENDIX: SELECTION OF OPTIMUM COUNTING CONDITIONS (USING A PACKARD TRICARB MODEL 3320)

Gain

With the upper discriminator set at 1000, and the lower at 50 (i.e. full window), a sample was counted at various gain settings (0–100 %). A graph of counts per minute against gain settings was then plotted (Fig. 5).

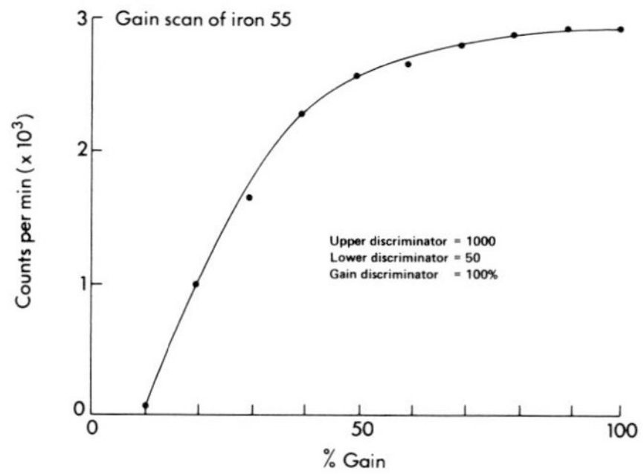

Fig. 5. Selection of optimum gain.

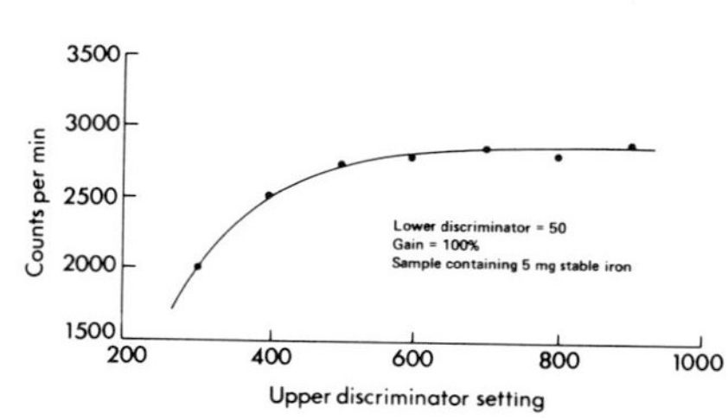

Fig. 6. Selection of optimum upper discriminator setting.

It was found that the maximum count-rate was obtained at a gain setting of 100 % i.e. optimum gain setting for the counting of Iron-55 was 100 %.

Upper discriminator

With the gain set at 100 % and the lower discriminator at 50, a sample was counted for various upper discriminator settings. A graph of count-rate against upper window settings was plotted (Fig. 6). It can be seen from Fig. 6 that the count-rate increases relative to the upper discriminator settings, and reaches a maximum at 1000. However, an upper setting of 1000 is not necessarily the optimum setting, because the count-rate for a sample blank may also increase.

Lower discriminator (Fig. 7)

With the upper discriminator set at 1000 and the gain at 100 %, a sample was counted for various lower discriminator settings (0–400). It was found that, for 100 % gain, the count-rate increases as the lower discriminator setting is decreased. However,

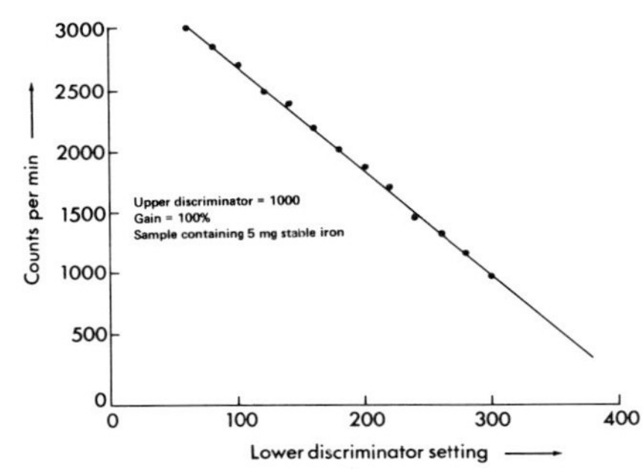

Fig. 7. Selection of optimum discriminator settings for the counting of ^{55}Fe.

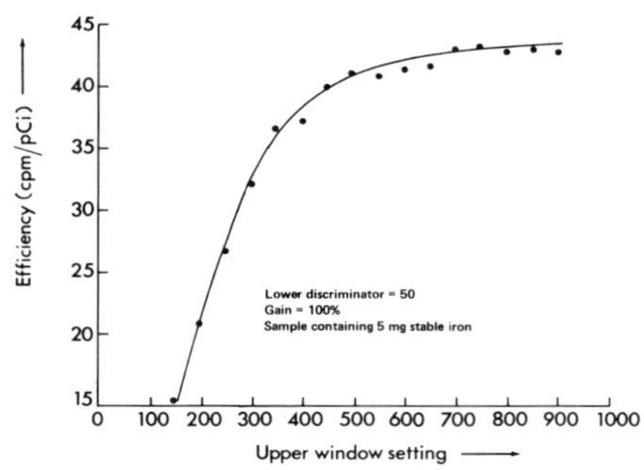

Fig. 8. Variation of efficiency with upper window setting.

the optimum lower setting was not chosen as zero, but 50, in order to maintain a high count-rate while keeping the background interference to a minimum. Thus, for the counting of Iron-55, a lower discriminator setting of 50 was chosen.

Figure of merit (E^2/B)

The ratio E^2/B (where E^2/B=counter sensitivity in counts per unit time per unit of activity) is often taken as a figure of merit criterion for optimization;[3] strictly it applies to conditions when the sample count-rate is small relative to the background. Under more favourable circumstances the sensitivity (efficiency) alone may be the most suitable criterion. Thus both criteria must be considered in the optimisation of a system required to measure a wide range of sample strengths, though for the purposes of this work the system has been optimised for maximum E^2/B.

Both E^2/B and efficiency vary with upper discriminator settings (Figs. 8 and 9). It was found that a maximum occurs for E^2/B between 300 and 350 (upper levels), but there was only a small decrease in counting efficiency if the upper window was lowered from 1000 to 500.

Since the values of the upper discriminator for the optimisation of E^2/B and counting efficiency were not the same, a compromise was made, and an upper discriminator setting of 450 was chosen.

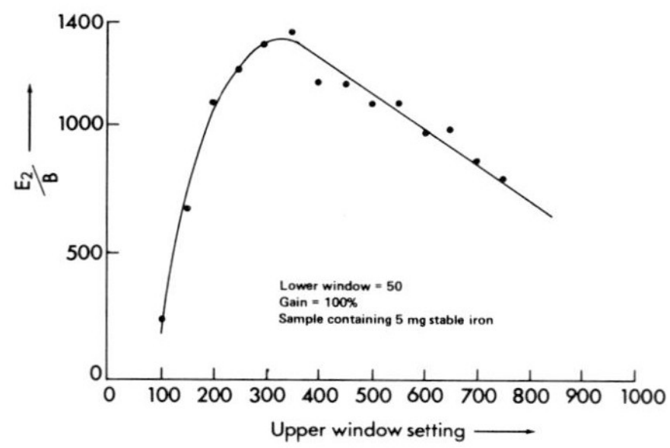

Fig. 9. Variation of E^2/B with upper window setting.

Summary of optimum counting conditions

Gain	=	100 %
Upper discriminator	=	450
Lower discriminator	=	50

DISCUSSION

J.D. Eakins: Was the difficulty you experienced in suspending the ferriphosphate complex only with 20 mg samples or did you have troubles with lesser amounts as well?

G.A. Sutton: The main problem turned out to be due to a poor batch of gel scintillator. With good materials we had little difficulty with samples in the 5–8 mg range, but settling out did occur at levels of 20 mg or more.

Chapter 24

An 'In-Vial' Oxidation Procedure for the Simultaneous Determination of Plutonium alpha-Activity and Plutonium-241 on Smear Samples

J. D. Eakins

Health Physics and Medical Division, Atomic Energy Research Establishment, Harwell, England

INTRODUCTION

The assessment of radioactive surface contamination is one of the control measures used by operational health physicists in order to maintain safety standards in areas where unsealed radioactive sources are present.

The philosophy and practical problems of surface monitoring have been described in detail by Clayton.[1] In general, when the surface is of suitable geometry and the radioactive material emits high-energy β- or γ-radiation, contamination is detected by direct measurement. The presence of α-activity can also be detected on suitable surfaces by direct measurement, provided the surface is clean. However, in the case of low-energy β-emitters such as tritium or Plutonium-241, or for α-activity on surfaces which are dirty, indirect methods are used. Indirect methods may also be used for high-energy β- or γ-emitters where the surface is unsuitable.

The most common method of indirect surface monitoring is by smearing. A defined area is wiped with a suitable material, usually filter paper, and the activity removed by the smear is measured in the appropriate type of counter. An estimate of the original surface activity is then made assuming some fairly arbitrary removal factor, usually between 10 and 20%.

Of particular interest to the health physicist monitoring for plutonium is the ratio of the β-emitting Plutonium-241 to total α-activity. Plutonium α-activity usually consists of Plutonium-238, -239 and -240, the ratios of these depending on the burn-up of the uranium from which it is produced. The amount of Plutonium-241 present also depends on this factor, a high burn-up producing a high β/α ratio. The age of the plutonium also affects the ratio to a certain extent, as Plutonium-241 ($t_{\frac{1}{2}}$ = 13.4y) decays to α-emitting Americium-241 ($t_{\frac{1}{2}}$ = 458y) which remains as an impurity element. The lowest β/α ratio will therefore be found in aged plutonium produced from low burn-up fuel. A rough knowledge of this ratio can thus give useful information to the health physicist on the source of the contamination. In addition, the β/α ratio enables Plutonium-241 contamination to be calculated from the direct measurement of the α-activity alone on suitable surfaces, whereas the direct measurement of this nuclide (E_{max} 21 keV) is extremely difficult.

The usual method of measuring the β/α ratio on a plutonium smear is by a double counting technique. The smear is counted in an internal proportional counter to obtain a measure of the total α- and β-activity, then in an α-scintillation counter to determine the α-activity alone. The β/α ratio can then be calculated if the respective counting efficiencies are known. As surfaces are often dusty or greasy, however, the measurement of α- or low-energy β-activity is complicated by losses due to self absorption.

In selecting a method for the analysis of smear samples for plutonium, two factors should be borne in mind. First, a simple and rapid procedure is necessary as, following a contamination incident, samples may be produced in considerable numbers and the results will be required in a matter of hours. Secondly, samples will generally be taken in areas where separated plutonium is the only radioactive material, so that chemical separation is unnecessary.

Eakins and Lally[2] have described a technique for the simultaneous determination of plutonium α-activity and Plutonium-241 in biological materials by chemical separation and gel scintillation counting. Although this method, which has subsequently been used by other workers,[3] could be applied to smears, a more rapid procedure is required. This paper describes the development of a simple and fairly rapid method for the simultaneous determination of plutonium α- and β-activity in smear samples.

EXPERIMENTAL

Preliminary experiments

A smear sample containing plutonium was analysed according to the method of Eakins and Lally[2] and the gelled ferri-phosphate source counted in a Packard Tricarb Model 3314 liquid scintillation spectrometer. In Fig. 1 a gain scan at 5V and a window setting of 200—400 illustrates the excellent separation of α- and β-emitting plutonium obtainable by this technique. It is not possible to discriminate between the α-particle energies of Plutonium-238, -239 or -240 so only one α-peak is observed. The following experiments were carried out in an attempt to obtain a similar peak separation with a simpler and more rapid procedure.

Artificial smears were prepared from 3 cm Whatman 42 filter papers. If dirty smears were required, these were obtained by smearing the axles of a laboratory truck, which were covered with a thin layer of black dust and grease. The average weight of the deposit on such smears was 15.4 mg. The smears were spiked by the addition of 0.1 ml aliquots of Plutonium-241 and Plutonium-239 and dried under an infra-red lamp. 'Plutonium-239' is used here for convenience to describe the spike material, as it was known to be low in other plutonium isotopes. In the following work, where the composition is unknown the term 'plutonium α-activity' will be used.

A dirty spiked smear was added directly to a glass liquid scintillation counting vial containing NE 220 liquid scintillator. A gain scan of this, at 5 V and a window setting of 200—400, is shown in Fig. 2a. Virtually no separation of the α- and β-activity was obtained. Another dirty smear was placed in a counting vial which was heated in a muffle furnace at 450°C until completely oxidised. This took about 2 h, due to the restricted supply of air in the vial, and a small bead of brownish ash remained. After cooling, liquid scintillator was added and the vial shaken

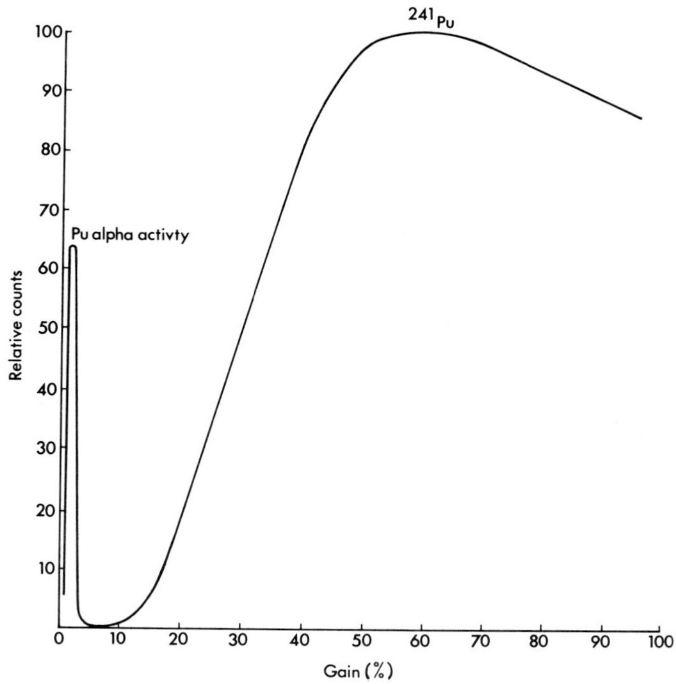

Fig. 1. Gain scan of plutonium from a smear analysed by the ferri-phosphate procedure.

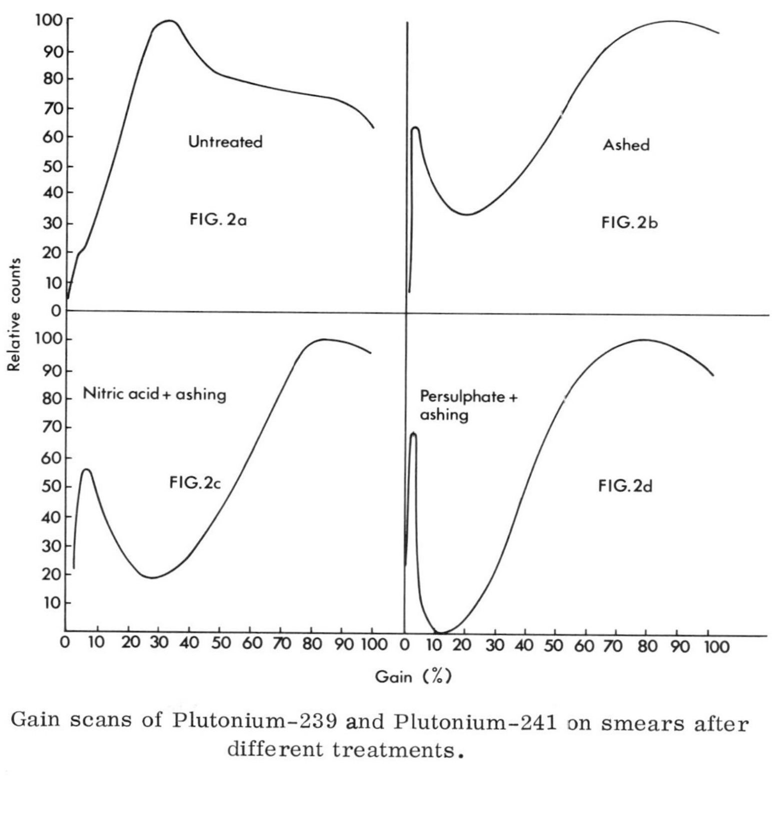

Fig 2. Gain scans of Plutonium-239 and Plutonium-241 on smears after different treatments.

vigorously, but the piece of ash was difficult to disperse. Silica powder was added to the vial to produce a gel and after further shaking the gain scan shown in Fig. 2b was obtained. There is a considerable improvement over that obtained with the untreated smear, but the separation of α- and β-emitting plutonium is still not very good.

A further dirty smear was placed in a counting vial to which 2 ml of nitric acid was added to speed up the oxidation. The vial was heated in the muffle furnace for 30 min, by which time the oxidation was complete, leaving a brownish residue in the bottom. Liquid scintillator was added to the vial, which was shaken up to disperse the residue as much as possible. However, most of it remained stuck to the bottom and sides of the vial. Silica powder was added to gel the source and a gain scan was obtained as before. This is shown in Fig. 2c and indicates a slight improvement on that obtained by ashing without nitric acid.

Ammonium persulphate was then tried as an alternative oxidant. The advantage of this salt is that it has a higher boiling point than nitric acid, completely decomposing into volatile products including oxygen. About 2 g was placed in the bottom of the vial. To prevent the smear ash forming one lump, difficult to disperse, or from adhering to the sides of the vial, silica powder was added to the vial initially. The smear paper was then pushed beneath the powder and a silica wool plug inserted in the neck of the vial to prevent any losses. The vial was placed in the muffle furnace for 30 min, by which time oxidation was complete. Particles of ash, slightly yellow in colour could be observed dispersed in the powder. The silica wool plug was pushed right into the vial and liquid scintillator added. A gel was produced by shaking vigorously and a gain scan carried out which is shown in Fig. 2d. It can be seen that separation of the α- and β-emitting plutonium is very good, similar in fact to that produced by the ferriphosphate technique. A counting vial prepared for oxidation of a smear sample with ammonium persulphate is shown diagrammatically in Fig. 3.

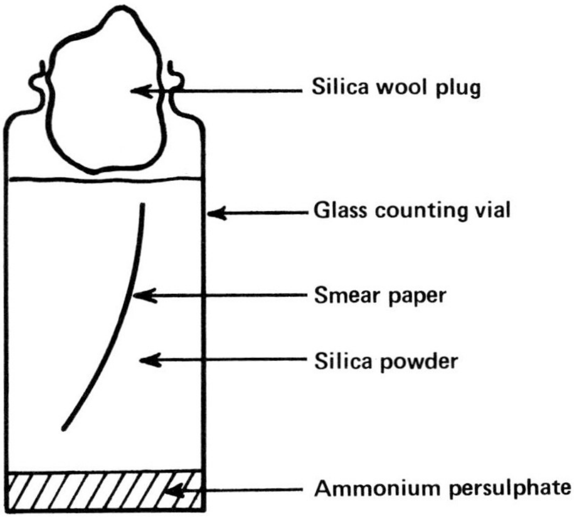

Fig. 3. Counting bottle prepared for 'in-vial' oxidation of smear papers.

Counting efficiencies

Six dirty smears were spiked with Plutonium-239 and a further six with Plutonium-241. All twelve smears were then ashed by the ammonium persulphate procedure, gelled and counted. The Plutonium-239 smears were counted at 3% and the Plutonium-241 at 80% gain. In each case counts were recorded in a restricted window of 200—400 and also in a full window of 50—1000. The results of these counts are given in Tables 1 and 2.

Table 1. Counting efficiency for dirty Plutonium-239 smears.

Sample no.	Counting efficiency (%)	
	200–400	50–1000
1	70.2	102.9
2	66.7	100.2
3	63.9	101.6
4	68.0	103.3
5	68.2	102.0
6	71.0	99.0
Mean and S.D.	68.0 ± 4.6	101.5 ± 3.0

Table 2. Counting efficiency for dirty Plutonium-241 smears.

Sample no.	Counting efficiency (%)	
	200–400	50–1000
1	4.1	11.4
2	3.7	10.0
3	2.9	7.2
4	3.2	8.2
5	4.2	11.7
6	4.2	10.8
Mean and S.D.	3.7 ± 1.0	9.9 ± 3.0

From Table 1 it is clear than Plutonium-239 on smears can be counted at 100% efficiency with a full window. A restricted window was used in the initial work because Eakins and Lally[2] found a better figure of merit when using it with the ferri-phosphate complex. A gain scan of an oxidised dirty Plutonium-241 smear with a full window showed that about 5% of the counts occurred in the Plutonium-239 region, so the restricted window was chosen despite the lower counting efficiency.

The counting efficiency for Plutonium-241 with a full window is higher than that with a restricted window by a factor of almost 3. A gain scan of a Plutonium-239 smear with a full window showed only a 2% contribution to the Plutonium-241 region, so the full window was adopted for Plutonium-241 counting.

In order to assess the average contribution of each nuclide to the other channel, the Plutonium-241 smears were counted at the Plutonium-239 settings and vice versa. The contributions, expressed as a percentage of the counts obtained at the correct gain and window settings, are given in Table 3. These contributions were considered to be negligible, compared with the variations likely to occur in counting efficiencies from one smear to another.

Counting efficiencies for different types of smear

Having established that α- and β-emitting plutonium could be differentiated, even on the worst types of smear, the range of counting efficiencies likely to be

Table 3. Overlap of plutonium α- and β-channels.

Sample no.	Plutonium-241 counts appearing in Plutonium-239 channel (%)	Plutonium-239 counts appearing in Plutonium-241 channel (%)
1	0.11	1.5
2	0.05	1.9
3	0.16	1.8
4	0.11	1.7
5	0.16	2.3
6	0.17	2.4
Mean and S.D.	0.13 ± 0.08	1.9 ± 0.6

encountered with smears from a variety of surfaces was investigated. Six smears of different types were prepared. The first consisted of an unused filter paper. The second was prepared by smearing an apparently clean surface and the remaining four by smearing successively dirtier surfaces. The six samples were spiked with both Plutonium-239 and -241 and the counting efficiencies for each smear calculated. In addition, the ratio of the Plutonium-239 was calculated for each sample. The results of these measurements are shown in Table 4.

Table 4. Counting efficiencies for various smears.

Type of smear	Plutonium-241 (%)	Plutonium-239 (%)	Ratio Plutonium-241/ Plutonium-239
Unused paper	14.3	81.4	0.18 : 1
Clean surface	12.9	76.1	0.17 : 1
Dusty surface	11.9	67.8	0.18 : 1
Very dusty surface	9.9	71.2	0.14 : 1
Greasy surface	11.0	69.2	0.16 : 1
Very greasy surface	8.4	65.5	0.13 : 1
Mean and S.D.	11.8 ± 3.9	71.9 ± 13.6	0.16 ± 0.04 : 1

Both the β- and α-counts are quenched with the increase in solid material on the smear. However, the Plutonium-241 is quenched more than the Plutonium-239, as indicated by the relatively greater standard deviation of the mean. For the accurate determination of plutonium activity on smears it would be necessary to determine the counting efficiencies for each smear. However, it is normally more important to determine the β/α ratio with reasonable accuracy than to determine the absolute activity. A qualitative assessment can be made of the category of smear and the appropriate counting efficiencies used. Alternatively, if the mean counting efficiencies for each nuclide are used, the β/α ratio can be determined to ± 25 %, which is normally adequate for this type of work.

A comparison of Plutonium-241/Plutonium-239 ratios determined by two different methods

Three smears were obtained from plutonium areas and each was divided into two halves. One half was analysed by the ferri-phosphate complex method of Eakins and Lally [2] and the other half by the ammonium persulphate combustion technique. The Plutonium-241/plutonium α-activity ratios obtained by each method are shown in Table 5.

Table 5. Plutonium-241/plutonium α-activity ratios by different methods.

Smear no.	Plutonium-241/plutonium α-activity	
	by ferri-phosphate complex	by 'in-vial' oxidation
1	12 : 1	14 : 1
2	8 : 1	7 : 1
3	23 : 1	20 : 1

The ratios obtained are in reasonable agreement, particularly when it is considered that the combustion method takes less than an hour whereas the other procedure takes about a day.

COMMENT AND CONCLUSIONS

The very good separation of plutonium α- and β-activity obtained by gel scintillation counting, after the 'in-vial' ammonium persulphate oxidation technique, is believed to be due to two factors. The first of these is a reduction in particle size of the smear residue after oxidation. This is because of the support given to the smear paper by the silica powder. When heated in a vial, an unsupported paper curls up and any residue forms a compact lump difficult to disperse. With a supported paper this does not occur and the ash remains distributed in the silica powder. The second factor is related to the composition of the inorganic residue remaining after combustion. This contains a high proportion of iron which normal combustion leaves as a brownish oxide producing considerable colour quenching. Ammonium persulphate converts this iron to a very pale yellow material, presumably ferric sulphate, which produces much less quenching.

No attempt has been made in this work to obtain optimum counting conditions for Plutonium-241 or plutonium α-activity. It seemed pointless to attempt to obtain the maximum figure of merit for a particular smear when individual smears could vary so widely. Instead, counting conditions were used which gave reasonable results for a range of smears.

The procedure described enables Plutonium-241 and plutonium α-activity to be determined on smears in one hour with sufficient accuracy for indirect surface contamination monitoring. The plutonium β/α ratio obtained is good enough for the operational health physicist to estimate Plutonium-241 surface contamination from direct α-counting and also to identify the source of the contamination.

REFERENCES

1 R.F. Clayton, 'Monitoring of Radioactive Contamination on Surfaces', Technical Reports Series No. 120, I.A.E.A., Vienna, 1970.

2 J.D. Eakins and A.E. Lally in Liquid Scintillation Counting, Vol. 2 (Eds. M.A. Crook, P. Johnson and B. Scales), Heyden, London, 1972, pp. 155–65.

3 T.H. Bates in Liquid Scintillation Counting, Vol. 2 (Ed. M.A, Crook, P. Johnson and B. Scales), Heyden, London, 1972, p. 165.

DISCUSSION

G.A. Sutton: Have you considered the use of combustion via the 'Bomb' technique? If not why not?

J.D. Eakins: No we have not. Mainly because the operational health physicist can put his smears directly into the sample vial literally by the dozen — we believe this to be a more speedy method.

Chapter 25

A Bomb Method for Rapid Combustion of Samples

R. Burleigh

Research Laboratory, The British Museum, London, England

INTRODUCTION

This paper describes a highly efficient method for combustion of organic samples up to 20 g or so in weight. The method and apparatus described have been developed primarily for combustion of radiocarbon dating samples but may also have applications to other cases in which relatively large samples must be dealt with.

Before discussing the equipment and its operation and performance it is necessary to consider briefly some practical limitations imposed on samples for radiocarbon dating, the difficulties met in combusting these materials satisfactorily by conventional methods which led to the development of the improved method of combustion and an outline of this development.

For the purpose of obtaining radiocarbon dates, natural Carbon-14 can be measured in a number of ways among which internal gas and scintillation counting are generally preferred. Some more exotic techniques for precise measurement of natural Carbon-14 have also been tried or proposed such as, among others, the use of nuclear track plates,[1] thermoluminescence dosimetry[2] and a bubble chamber method.[3] But the starting point common to the preparative chemistry of all existing methods of measurement is, with one exception, the combustion to carbon dioxide of the sample material for which a date is being sought. The exception is the preparation of acetylene from lithium carbide formed by direct high temperature reaction of wood charcoal with lithium metal.[4] This procedure is not widely used and need not be considered further here.

Because of the low specific activity of natural Carbon-14 — about 14 disintegrations per minute per gramme of carbon — it is not usually practicable to measure less than about 1 g of carbon if the sample concerned has an age of more than a few hundred years. Ideally, in order to arrive at a date with an acceptably small associated error term within a reasonable counting period by gas proportional or scintillation measurement about 5 g of carbon are needed for samples in the age range usually encountered, that is, up to about 50, 000 years old. For most common sample materials this means that a minimum of about 10 g of pretreated sample must be burnt initially. In most cases, though by no means always, it is generally possible to obtain enough archaeological or geological sample material to provide at least this amount for combustion, allowing for some unavoidable wastage during the preliminary treatment for removal of possible age contaminants.

Conventionally the samples are combusted in a scaled-up version of the kind of silica tube furnace used in classical quantitative microchemical analysis for carbon and hydrogen, usually incorporating a wire form copper oxide packing and followed by a variety of alternative wet purification trains depending on the individual laboratory's preferences. However, the combustion of 10 g or more of sample material in such large-scale tube furnaces is a tedious, unproductive process often taking more than 2 h for each sample and, more importantly, requires constant supervision. This type of apparatus is rather troublesome to operate properly and over the long period needed for each combustion there is a strong tendency for bubblers and traps to become blocked and for leaks to develop with the attendant risk of loss or contamination of samples.

For these reasons the idea is attractive of a more or less instantaneous bomb method of combustion, as utilised for example in the estimation of calorific values of fuels but on a larger scale, and with these considerations in mind H. Barker began experimenting with a combustion bomb for radiocarbon samples at the British Museum Research Laboratory more than a decade ago. There were, however, practical difficulties in extracting the mixture of carbon dioxide and excess oxygen at about 10—15 atm pressure from the bomb after a combustion such that there was no appreciable saving in time over the tube furnace method and the idea was not taken any further at the time.

Eventually this difficulty was overcome by means of a semi-automatic system for extraction and separation of the gas mixture from the bomb so that its routine use for sample combustions then became practicable.[5] Meanwhile, V.R. Switsur of the University of Cambridge Radiocarbon Dating Laboratory developed a more sophisticated bomb having a spin ring closure instead of a lid bolted to a flange welded to the body of the vessel as in the case of the original British Museum bomb.[6] The first British Museum bomb was a thick-walled stainless steel vessel converted at AERE, Harwell from its original use as a lithium furnace for large-scale preparation of acetylene[7] and, although properly engineered and tested at the time, was basically of a design which would not now be regarded as fully satisfactory for the purpose to which it had been adapted. This bomb finally failed in use after several hundred successful combustions[8] and was replaced by the purpose-built version incorporating Switsur's spin ring closure principle and this is the equipment which will now be described. Just before leaving the background to the development of these combustion bombs it should be added that Dr. Switsur has now taken this a stage further with a horizontally oriented bomb of advanced design having a hinged door and many other new features, which he has recently described in detail.[9] This equipment[a] could also be used for more fundamental work such as the study of the kinetics of combustion processes.

DESCRIPTION OF THE EQUIPMENT

The combustion bomb, which is commercially available,[b] is shown diagrammatically in Fig. 1 and consists of a stainless steel cylinder some 4.5 in (115 mm) in internal diameter, 15 in (480 mm) in length and of nominal 4 litre volume,

---

[a] High Pressure Combustion Unit manufactured by Phonon Particle Beams Ltd., 356 Wimpole Road, Lords Bridge, Barton, Cambridge CB3 7AE, England.

[b] Manufactured by Interskill Ltd., 121 Cambridge Road, Milton, Cambridge CB4 4AT, England.

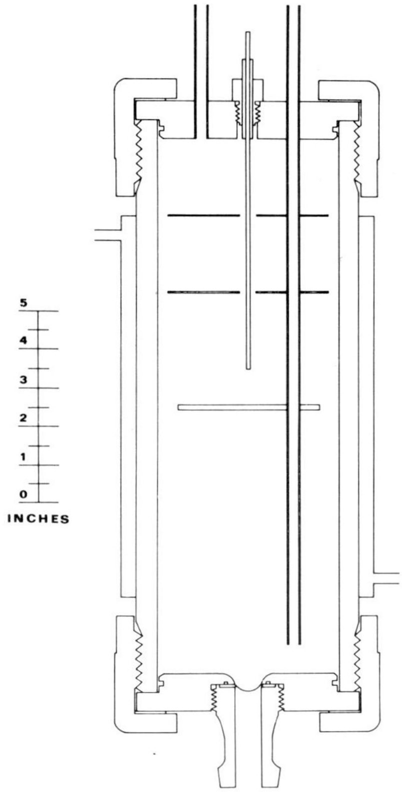

Fig. 1. Diagram of construction of combustion bomb (firing wire and sample crucible not shown).

having 0.5 in (12 mm) walls heavily threaded externally at both ends to take the large aluminium bronze spin rings which secure 1 in (25 mm) thick stainless steel end plates. A cooling water jacket is fitted to the outside of the vessel which is held vertically in a tripod stand bolted in turn to a bench. Gas inlet and outlet tubes with shut-off valves attached and an insulated lead-in for the electrically heated firing wire pass through the top plate. This plate also has an attachment point for a cable and counterweight arrangement so that the whole assembly, including the heavy spin ring, can be easily raised and lowered when placing the sample in position and for servicing and cleaning. In the centre of the bottom plate there is a high pressure relief valve in the form of a stainless steel bursting disc 0.625 in (15 mm) in diameter and two thousandths of an inch (0.025 mm) thick, rated at 1500 lb/in^2. Both the end plates have stepped profiles and are thus retained by the spin rings against the ends of the vessel walls but also form, with an 'O' ring in their inner edges, a piston fit inside the cylinder. The gas inlet tube reaches almost to the bottom of the closed vessel to facilitate purging with oxygen before filling. A ring which supports the sample crucible and above this two heat shield plates are all attached to the inlet tube by bayonet fittings and the tube also provides a convenient earth connection point for the firing wire. All the internal metal components are of stainless steel.

The vessel, without the bursting disc fitted, that is with a blank baseplate, has been subjected to a British Standard Specification static pressure test of 2250 lb/in^2 at the Ministry of Technology, Explosives Research and Development Establishment and carries a test certificate to this effect.

The firing wire is powered by a transformer circuit with an output of about 4 A at 12 V located close to the bomb but operated remotely via switches and relays. The firing wire is a loose coil, about 0.125 in (3 mm) in diameter, of iron or nichrome resistance wire about 0.2 mm thick connected by heavier tinned copper wires between the insulated lead-in and earth. Oxygen from a standard size J cylinder (200 ft^3) is supplied to the bomb through a heavy duty regulator having a maximum outlet pressure of 600 lb/in^2, fitted with a flame arrestor and connected by flexible high pressure oxygen hose to a needle valve and a pressure gauge which allows the filling pressure to be monitored. After combustion the gas mixture is extracted and separated by pumping through a series of refrigerated glass traps, the pressure in this system being controlled by a modified oxygen cylinder regulator and an electromagnetic valve operated from contacts on a mercury manometer. Figure 2 is a diagram of the filling and extraction system.

Obviously it would not be difficult to devise a more elegant arrangement for

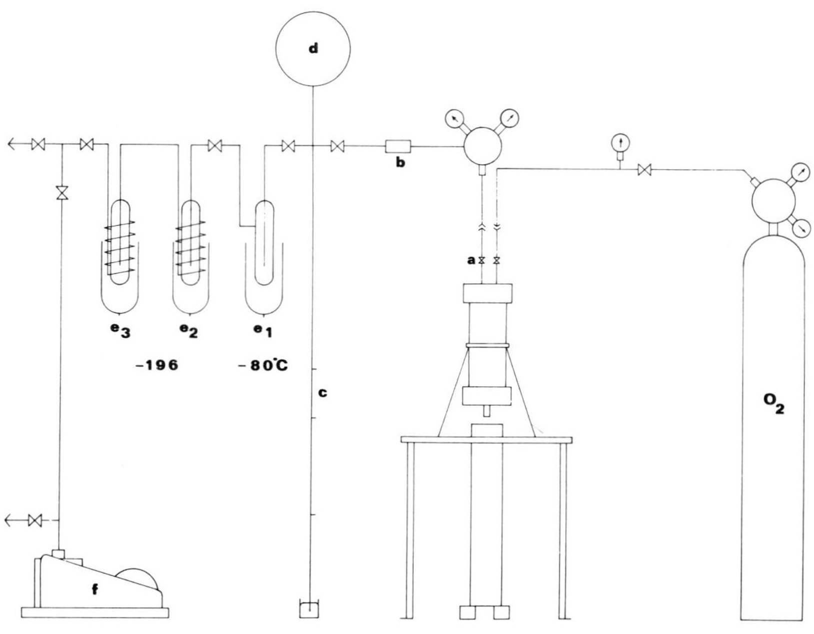

Fig. 2. Schematic layout of combustion bomb filling and extraction system. The procedure for filling the bomb is explained in the text. After a combustion CO_2 is separated from excess O_2 by evacuating the gas mixture through refrigerated traps (e_1, e_2, e_3). By careful control of pumping conditions all CO_2 is retained in traps e_2 and e_3 and O_2 is vented to atmosphere. To initiate this process outlet valve a, an intermediate regulator, and electromagnetic valve b are opened and stopcocks adjusted in the line to pump f (see text for details). Pumping then continues until the bomb is emptied (about 40 min), the gas flow being automatically controlled by valve b which operates, in response to pressure changes, from contacts on manometer c (via relays, not shown). A buffer volume d ensures that pressure changes in the line are not too rapid.

recovery of the carbon dioxide from the bomb, but the system in use is simple and effective and has been trouble-free over a long period.

OPERATION OF THE EQUIPMENT

Samples for radiocarbon dating are, with a few special exceptions, all once-living materials such as wood, charcoal, bone, cloth, plant remains and various other typical substances. These have first to be treated by fairly straightforward chemical methods to remove possible age contaminants. In the case of bone and antler, for reliable results the protein fraction (collagen) must be freed of its inorganic matrix which, since it contains carbonates, is liable to have undergone exchange with (usually) more recent carbon. After treatment the sample materials are generally in the form of coarsely particulate solids. To avoid very rapid combustion, particularly of finely divided materials such as charcoal, it is advisable to moisten and compress the samples into the form of large pellets. We achieve this by means of a press designed for embedding specimens for metallographic examination which happens to be conveniently available in a neighbouring laboratory but presses specifically designed for pelleting combustion samples are obtainable commercially. In our case the pellets are about 1 in (25 mm) in diameter and up to 0.5 in (12 mm) thick and there are one or two such pellets for an optimum sample of about 12 g of typical pretreated material.

The sample pellets are placed in an 80 mm silica crucible supported centrally in the bomb, making sure that there is good thermal contact with the firing wire coil. A small amount of distilled water (nominally 50 ml) is placed in the bomb which is then closed, purged briefly with oxygen and, with the outlet valve shut off, filled to 100—125 lb/in^2 (representing about a five-fold excess of O_2) over a period of about 2 min. With pelleted samples there is a negligible risk of a dust explosion caused by sample material ejected into the bomb atmosphere during filling. The oxygen is not purified before admission to the bomb but is known not to contain a significant quantity of carbon dioxide (stated to be less than 5 v.p.m.). However, to allow for decay of Radon-222 (3.8 day half-life), almost certainly present in small amounts in the oxygen and which might possibly get through physically to the final counting sample, fresh cylinders of oxygen are aged by storage for 4 weeks or so before use.

When filling is complete the inlet valve is closed, the filling line disconnected and after clearing personnel from the area the bomb, with cooling water flowing, is fired remotely from an adjacent room. A pilot light visible from the firing point indicates firing wire continuity. Failure of this to illuminate after operating the firing switch is taken as a simple indication of a successful combustion (in which the wire is always destroyed) and this has proved reliable in practice. As a safety measure the firing switch cannot be operated successfully without first turning a separate key switch. The operator keeps this key in his possession while he is charging the bomb.

After combustion, which probably takes place in about 30 s or less, the mixture of carbon dioxide and excess oxygen is separated by pumping through the cold trap system already mentioned using a large rotary vacuum pump. The extraction line has three fairly large pyrex glass traps and a 5 litre ballast volume to prevent oscillation of the electromagnetic valve which would be caused by rapid pressure changes. The first trap is a simple one, cooled with dry-ice/acetone slush, in which the pressure is kept at about 30 cm Hg to remove water vapour, and the other two are specially designed spiral traps cooled with liquid nitrogen for collection of the carbon dioxide most of which is held back in the first of these (Fig. 2). By careful adjustment of the stopcocks in the extraction line the pressure of the gas mixture pumped through the liquid nitrogen traps can be maintained at about 7.5 mm Hg. This ensures that the excess oxygen does not condense as a liquid in the cold traps but is vented to atmosphere while at the same time allowing virtually complete stripping and retention of carbon dioxide. When these initial settings have been made the system can be left to operate unattended until all the gas has been pumped out of the bomb which takes up to about 40 min. The carbon dioxide is converted via acetylene to benzene for radiocarbon age measurement by liquid scintillation counting.[10]

The safety aspects of the use of the bomb have been considered in some detail. Apart from the need common to all high pressure systems to maintain adequate standards of care and cleanliness in operation, the combustion bomb is regarded as inherently very safe. So far as it can be ascertained the pressure rise on combustion is small. The bursting disc provides a safety factor which is probably at least an order of magnitude greater than the pressure increases which occur in normal operation. The bomb is mounted over an asbestos tube about 5 in (125 mm) in diameter into which the orifice of the bursting disc assembly is directed. If the disc did burst the escaping hot gas mixture and any minor debris would be ducted harmlessly away, probably the most serious consequence being the loss of a sample.

The bursting disc is exposed to quite strongly acidic conditions each time a sample is combusted. Since it is a component too expensive to replace routinely, the disc is microscopically examined at regular intervals for signs of stress corrosion. This could weaken the disc and if neglected might cause its failure at normal operating pressures resulting in the otherwise avoidable loss of a sample. So far no such corrosion has been observed and indeed one disc deliberately burst after 35 combustions still failed very close to the rated pressure of 1500 lb/in^2. Damage caused by normal handling of the relatively fragile disc is most likely to make its replacement necessary long before it has been significantly weakened by corrosion.

PERFORMANCE OF THE EQUIPMENT

The performance of the combustion bomb is very satisfactory. The equipment described here has been in routine use in our laboratory for some 18 months and about 150 combustions have been carried out with it in this time. Compared with the alternative tube furnace method the bomb method is easy to operate and requires only about 20 min of full-time attention by an operator for each combustion against 2 h or more of constant supervision for an equivalent tube furnace combustion. There is no doubt that the kinds of sample material commonly met in radiocarbon dating are completely burnt in the bomb. Generally little or no ash remains after the combustion[a] and almost all (probably better than 99%) of the carbon dioxide can certainly be recovered. The purity of the gas is high as traces of oxides of nitrogen and sulphur formed during the combustion are mostly dissolved in the small amount of distilled water deliberately placed in the bomb for this purpose

---

[a] It is of some interest to note that in certain cases, usually when charcoal is burnt, numerous glassy microspheres, generally greenish or brownish in colour and of sub-millimetre dimensions, are afterwards found in the sample crucible. Superficially, these closely resembly some forms of lunar spheres in appearance and it is known that similar spherules can be produced in a number of other ways, for example in the manufacture of mineral wool[11] and occasionally by steam locomotives.[12] In this particular case the spherules must be derived from fusion of small amounts of silica and metallic oxides sometimes present in the sample materials.

before firing. These impurities are usually much more difficult to remove satisfactorily from carbon dioxide obtained by tube furnace combustions. After final drying carbon dioxide from the bomb combustion can be reacted directly, without the need for further purification, to form acetylene via lithium carbide (and subsequently to benzene for liquid scintillation counting). Carbon dioxide obtained by bomb combustion would of course still require further purification if the intention was to measure the Carbon-14 activity directly by internal gas counting.

In our laboratory the optimum sample size for our purposes is 12 g and this amount yields about 9 litres of carbon dioxide from most common sample materials. Samples of comparable carbon content (30—50%) up to at least 20 g in weight could easily be combusted and the filling pressure of oxygen could be increased up to 250 lb/in^2 if necessary. With some difficulty larger samples containing as little as 5% or less of carbon could most probably be successfully combusted. Any samples much smaller than about 1 g in weight would probably still be more conveniently combusted in a tube furnace.

Finally, the equipment can be easily disassembled for cleaning after each combustion so there is ordinarily no risk of memory effects, at least when dealing with radiocarbon dating samples. Provided that total activities higher than a few tens of disintegrations per minute are not introduced to the bomb there is no reason to suppose that there will be memory problems in other applications.

CONCLUSIONS

The bomb combustion method described has, at least in its application to radiocarbon dating samples, distinct advantages over the numerous variants of the more conventionally used large-scale tube furnace combustion method in terms of time saved and improved yield and purity of the resultant carbon dioxide. These advantages soon repay the fairly high initial cost of the equipment (about £1000). It seems possible, therefore, that this type of equipment may also have applications in other fields, particularly where the problem may be one of dealing with large samples of intrinsically low radioactivity. The productivity of the equipment does not of course compare with that of commercial automatic combustion units intended for smaller and normally higher activity samples and a maximum of about five or six combustions per normal working day seems a probable figure. However, this will often be more than adequate in low level applications in which there may follow several more stages of sample preparation chemistry and where finally, to obtain the required precision, it may take a week or so to evaluate a small group of samples by scintillation counting.

ACKNOWLEDGEMENTS

For the benefits of their most helpful advice and discussions I am greatly indebted to my colleagues at the British Museum Research Laboratory, Mr. H. Barker, who initiated this work, Mr. N.D. Meeks who constructed much of the apparatus and has carried out most of the combustions and to whose expertise the success of the work is largely due, and Mrs. J. Lang who has provided metallographic information and facilities; to Dr. V.R. Switsur of the Sub-Department of Quaternary Research, University of Cambridge and to Mr. J.M. Cleland of Interskill Limited, Milton, Cambridge.

REFERENCES

1 D. Jeffreys, D. Larson and J.D. French, Am. J. Phys. 40 1400 (1972).
2 J. Winter, Archaeometry 14, 281 (1972).
3 M.J. Aitken, Physics and Archaeology, Interscience, New York, 1961, p.115.
4 E.R. Swart, Experientia 20, 47 (1963).
5 H. Barker, R. Burleigh and N. Meeks, Nature 221, 49 (1969).
6 V.R. Switsur, M.A. Hall and R.G. West, Radiocarbon 12, 590 (1970).
7 H. Barker, Nature 172, 631 (1953).
8 R. Burleigh, in Proc. 8th Intern. Conf. on Radiocarbon Dating, Lower Hutt, Wellington, New Zealand, 1972, Vol. 1, p.110.
9 V.R. Switsur, in Proc. 8th Intern. Conf. on Radiocarbon Dating, Lower Hutt, Wellington, New Zealand, 1972, Vol.1, p.120.
10 R. Burleigh, in Liquid Scintillation Counting, Vol. 2, Heyden, London, 1972, pp.142–4.
11 C.A. Cross, Nature 233, 185 (1971).
12 S.A. Durrani, Department of Physics, University of Birmingham, personal communication.

DISCUSSION

C.R. Jones: What was the cost of manufacture of the bomb?

R. Burleigh: £800–£1,000.

B. Scales: Do you have any problems in maintaining a steady low pressure which is sufficient to prevent condensation, but gives good recovery of carbon dioxide in the liquid nitrogen-cooled traps?

R. Burleigh: No. Liquid oxygen is a hazard which we certainly take note of, but at the low pressures maintained by our large rotary pump, the recovery of carbon dioxide in our carefully designed traps is better than 99%.

SHORT COMMUNICATION

A Micromethod for Urease

J. De Bersaques

Department of Dermatology, University of Ghent, Belgium

The determination of urease activity by measurement of the $^{14}CO_2$ set free from Carbon-14-labelled urea appears to be a simple, sensitive and specific technique. The $^{14}CO_2$ has been measured after precipitation as Barium-Carbon-14-carbonate. Recently,[2] a method was described in which the reaction was stopped by the addition of sulphuric acid and the escaping $^{14}CO_2$ was trapped in Hyamine 10-X and counted by liquid scintillation counting. The whole reaction took place in a 10 or 25 ml Erlenmeyer flask fitted with a well for the trapping agent.

By using the microtechniques developed by Linderstrøm-Lang and Holter,[3] viz. by putting a liquid alkaline 'seal' in the neck of the reaction tube to trap the $^{14}CO_2$, the reaction volumes can be reduced considerably and the sensitivity is very much increased. Such techniques were used for the study of enzymatic reactions in which $^{14}CO_2$ is liberated, viz. for glutamic decarboxylase[4] and glycolysis.[5]

In our experiments, small throw-away tubes were made from 5 mm i.d. borosilicate glass tubing and siliconised (Silikon Lösung Serva, Heidelberg). The reaction mixture, 10 µl with 20 mM urea, 40 mM pH 7.0 Tris-HCl buffer, 70, 000 c.p.m. Carbon-14-urea (55 mCi/mM, CEA France) and urease (crystalline, type C-3, Sigma Chem. Co., St. Louis), was present at the bottom of the tube. A 5 µl drop of 1 N sulphuric acid was placed on the side wall. The 30 µl liquid seal was put in a separate piece of glass tubing, linked to the reaction tube by a short length of latex tubing. This two-piece construction was adopted in order to avoid contamination of the seal by the radioactive solution. A glass stopper was placed at the other end of the second tube (Fig. 1). The tubes could be incubated horizontally on the bottom of a water bath. After the chosen incubation period, the acid was tipped into the reaction mixture to destroy the enzyme and liberate $^{14}CO_2$. After a suitable volatilisation period, the latex tubes were removed and the glass tube with the alkaline seal was dropped into a counting vial with scintillator and counted (Tri-Carb liquid scintillation spectrometer, model 2111, Packard Instrument Co.) at room temperature.

Numerous agents and methods have been advocated for the trapping and liquid scintillation counting of $^{14}CO_2$: KOH, NaOH, hydroxide of Hyamine 10-X, ethanolamine, Primene, phenylethylamine, etc. in various homogeneous and inhomogeneous scintillating mixtures. Some of these alkaline solutions were thought to be less suitable, either because of their viscosity, making it difficult to use micropipettes, because of streaking due to low surface tension, or because

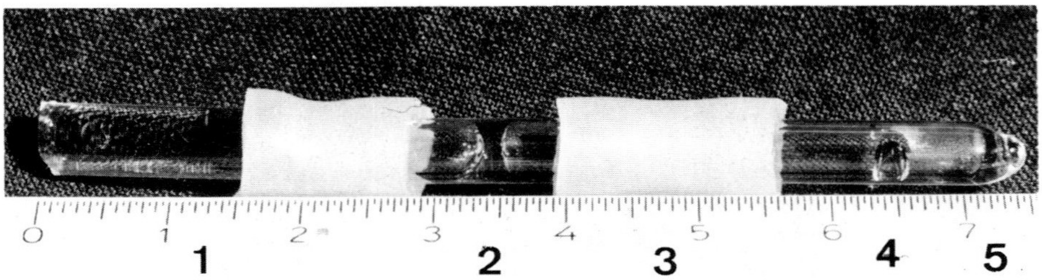

Fig. 1. Assay device (scale in cm). 1. Glass stopper. 2. Tube with alkaline seal. 3. Latex tubing. 4. Drop of 1 N sulphuric acid. 5. Reaction mixture at the bottom of the reaction tube.

of their content in organic solvents, which might inhibit the enzyme; for these reasons, Hyamine 10-X in methanol was not used.

Most of the papers published on $^{14}CO_2$ absorption are concerned with the maximal capacity of the trapping solution, while in our experiments the total quantity of CO_2 was very small: 0.2 μmole at 100% hydrolysis. It appeared that with some combinations of alkali and scintillator severe losses of CO_2 and radioactivity occurred, although the amount of alkali was theoretically more than sufficient:

—with phenylethylamine and a toluene scintillator: 50% loss after 40 min;
—with phenylethylamine and a dioxane scintillator: 15% loss after 40 min;
—with 0.3 N KOH in methanol and a dioxane scintillator: 20% loss after 20 h.

Such a decrease in counts has been mentioned only once,[6] when $^{14}CO_2$ trapped in Primene was diluted with toluene scintillator and kept at room temperature.

Loss of counts was only minimal with 1 N KOH or NaOH in water as the trapping agent and 5 ml of a scintillating solution made up from 4 volumes toluene, 2 volumes Triton X-100 and 1 volume 0.1 N NaOH, with 6 g/litre PPO (a formula derived from Ref. 7). The efficiency, measured with Carbon-14-benzoic acid, was 75%. The blank was about 15 c.p.m. above background after 1 h incubation.

The time needed to capture quantitatively the $^{14}CO_2$ released by the acid was less than 10 min. The radioactivity measured was proportional with time and amount of enzyme, up to about 20% hydrolysis of the substrate. The sensitivity of the method was such that 2×10^{-6} Sumner units gave 1000 c.p.m. after 1 h incubation at 37°C.

Recently, Wilson and Kronick[8] described a device made from scintillator plastic, enabling continuous measurement by scintillation counting of the $^{14}CO_2$ released from an enzymatic reaction. This means that CO_2 already escapes from the reaction mixture kept at pH 7 and that the addition of sulphuric acid is not essential. However, when our system was used without acid the radioactivity measured in the NaOH seal was always 10 to 30% lower than when sulphuric acid had been added.

A direct method of continuous measurement was also tried: the reaction tube was put into the scintillation vial, slightly inclined downwards to avoid penetration of scintillator into the tube (Fig. 2), and covered with 5 ml scintillator. The vial was then counted continuously at room temperature, or at chosen intervals, in which case the incubation can be done at 37°C. In the latter case, there is a very real danger of splashing some of the radioactive mixture into the scintillator during the manipulations to and from the counter or because of the movements in the counter itself.

A lag period of about 15 min when incubating at 37°C and of 30 min when

Fig. 2. Scintillation vial with reaction tube (without scintillator) for direct measurements.

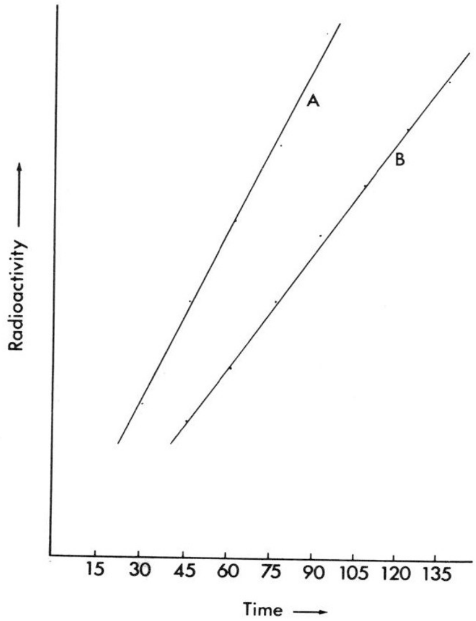

Fig. 3. Increase of radioactivity in function of time (in min), measured with a device as shown in Fig. 2 (with scintillator). A Incubation at 37°C
B Incubation at 26°C (room temperature).

incubating at room temperature was apparent before the increase in radioactivity proceeded in a linear fashion and at maximum speed (Fig. 3). This maximum speed was equal to that measured with the first, indirect method. The delay in equilibration in our experiments was much longer than that of 2 min mentioned by Wilson and Kronick.[8]

REFERENCES

1 R.J. Roon and B. Levenberg, J. Biol. Chem. **243**, 5213 (1968).
2 J.A. McDonald, K.V. Speeg, J.W. Campbell, Enzymologia **42**, 1 (1972).
3 K. Linderstrøm-Lang and H.C.R. Holter, Trav. Lab. Carlsberg **19**, 14, 1 (1933).
4 R.W. Albers and R.O. Brady, J. Biol. Chem. **234**, 926 (1959).
5 K. Adachi and H. Uno, Adv. Biol. Skin IX, 511 (1969).
6 R.A. Opperman, R.F. Nystrom, W.O. Nelson and R.E. Brown, Intern. J. Applied Radiation and Isotopes **7**, 38 (1959).
7 J. Murray, Intern. J. Applied Radiation and Isotopes 22, 209 (1971).
8 S.H. Wilson and M.N. Kronick, Anal. Biochem. **43**, 460 (1971).

Subject Index

Index to Contributors

Kaartinen, N., 174